Studies in Systems, Decision and Control

Volume 648

The series "Studies in Systems, Decision and Control" (SSDC) covers both new developments and advances, as well as the state of the art, in the various areas of broadly perceived systems, decision making and control–quickly, up to date and with a high quality. The intent is to cover the theory, applications, and perspectives on the state of the art and future developments relevant to systems, decision making, control, complex processes and related areas, as embedded in the fields of engineering, computer science, physics, economics, social and life sciences, as well as the paradigms and methodologies behind them. The series contains monographs, textbooks, lecture notes and edited volumes in systems, decision making and control spanning the areas of Cyber-Physical Systems, Autonomous Systems, Sensor Networks, Control Systems, Energy Systems, Automotive Systems, Biological Systems, Vehicular Networking and Connected Vehicles, Aerospace Systems, Automation, Manufacturing, Smart Grids, Nonlinear Systems, Power Systems, Robotics, Social Systems, Economic Systems and other. Of particular value to both the contributors and the readership are the short publication timeframe and the world-wide distribution and exposure which enable both a wide and rapid dissemination of research output.

Indexed by SCOPUS, DBLP, WTI Frankfurt eG, zbMATH, SCImago.

All books published in the series are submitted for consideration in Web of Science.

Hieu Trinh · Van Thanh Huynh · Samson Yu ·
Tyrone Fernando

Unknown Inputs Estimation in Linear Time-Delay Systems Using Generalized Functional Observers

Springer

Hieu Trinh
School of Engineering
Deakin University
Geelong, VIC, Australia

Van Thanh Huynh
School of Engineering
Deakin University
Geelong, VIC, Australia

Samson Yu
School of Engineering
Deakin University
Geelong, VIC, Australia

Tyrone Fernando
Department of Electrical
Electronic and Computer Engineering
University of Western Australia
Crawley, WA, Australia

ISSN 2198-4182 ISSN 2198-4190 (electronic)
Studies in Systems, Decision and Control
ISBN 978-3-032-11869-1 ISBN 978-3-032-11870-7 (eBook)
https://doi.org/10.1007/978-3-032-11870-7

This Springer imprint is published by the registered company Springer Nature Switzerland AG
The registered company address is: Gewerbestrasse 11, 6330 Cham, Switzerland

The authors express their deep gratitude to their families for their love and encouragement throughout the preparation of this book.

Preface

Modern engineering systems are becoming increasingly complex, interconnected, and subject to uncertainties. From renewable energy grids and autonomous vehicles to aerospace systems and advanced manufacturing, engineers must ensure that these systems operate safely, reliably, and efficiently despite disturbances, incomplete measurements, and unknown influences. In such contexts, the ability to accurately reconstruct the hidden internal dynamics of a system—together with the inputs that drive it—is not just a theoretical challenge but a practical necessity. State and input estimation therefore provides the foundation for advanced monitoring, fault detection, and control strategies that are indispensable in today's technological landscape.

State estimation plays a central role in control systems because it provides the information required for effective decision-making when not all system variables or driving inputs can be directly measured. In most practical applications, measuring every state and input is either infeasible, prohibitively expensive, or intrusive to the system' operation. By reconstructing both unmeasured states and unknown inputs from available measurements, enables the implementation of advanced feedback control strategies, improves reliability, and enhances overall system performance. This capability is particularly vital in safety-critical domains such as aerospace, power systems, robotics, and process control, where hidden disturbances or unmeasured inputs can significantly affect system behavior. Without accurate knowledge of the states and the inputs driving them, many modern control techniques—including optimal, robust, and fault-tolerant control—would be impractical or impossible to implement in real-world settings.

Research on estimating the states of dynamical systems has been active for more than half a century, beginning with the pioneering work of Kalman in the early 1960s. Much of the classical literature has focused on linear time-invariant systems under the assumption that both inputs and outputs are measurable. While this theory underpins a wide range of practical applications, real-world systems often introduce further challenges: some inputs may be unmeasurable, certain states may be inaccessible or costly to measure, and the dynamics may depend not only on the current state but also on its past values. Theoretical developments to analyze such systems—driven by unknown inputs and limited measurements—remain relatively under-explored.

This book, written in an exploratory style, examines these scenarios and provides insights aimed at equipping the reader with a foundation for extending the theoretical framework to greater depths.

In the early 1990s, Hou and Müller laid the groundwork for estimating unknown inputs in linear time-invariant systems. The theory developed in this book builds on their pioneering contributions, extending the design methodology by exploiting additional degrees of freedom available in observer design. Furthermore, the analysis is advanced by introducing the concept of generalized functionals, which offers a powerful framework for addressing time-delay systems where delays may appear in the states, inputs, and outputs.

The theory of unknown input and state estimation, which is the subject of this book, is increasingly attracting attention due to the pressing need for advanced theoretical tools capable of handling complex dynamical systems. These tools form the basis for developing new control methodologies suitable for modern engineering challenges. This book presents recent advances alongside fundamental theory, offering researchers a valuable resource to understand the essential concepts and to explore state-of-the-art developments in unknown input and state estimation. It will serve as a useful reference for researchers and postgraduate students working in control systems and may also support specialized senior undergraduate courses in control engineering and applied mathematics with a research focus.

The authors gratefully acknowledge the contributions of colleagues and friends whose discussions have enriched this work. In particular, Hieu Trinh thanks Assoc. Profs. Phan Thanh Nam and Dinh Cong Huong for their collaboration and interest in the subject.

Geelong, Australia
September 2025

Hieu Trinh
Van Thanh Huynh
Samson Yu
Tyrone Fernando

Competing Interests The authors have no competing interests to declare that are relevant to the content of this manuscript.

Contents

Chapter 1
Background, Motivations, Preliminaries and Organization of the Book

1.1 Background

We first start with the following simple linear system with an unknown input

$$\dot{x}(t) = Ax(t) + Dd(t), \tag{1.1}$$

$$y(t) = Cx(t), \tag{1.2}$$

where $x(t) \in \mathbb{R}^n$, $y(t) \in \mathbb{R}^p$, $p > 1$, and $d(t) \in \mathbb{R}$ are, respectively, the state vector, output vector and an unknown input. Matrices $A \in \mathbb{R}^{n\times n}$, $C \in \mathbb{R}^{p\times n}$ and $D \in \mathbb{R}^n \neq 0$ is a column matrix.

Our objective is to design an unknown input observer, denoted by $\hat{d}(t)$, to asymptotically estimate the unknown input $d(t)$. For the sake of simplicity, we consider one unknown input in the system, and the known control input vector, namely, $u(t)$, is omitted in (1.1). Subsequent results to be derived in this book will cover a wider class of dynamic systems, which include multiple unknown inputs, time-delay systems, etc.

The problem of estimating unknown inputs is an important and well-known problem in the literature, and numerous results have been reported. It is well-known that to asymptotically estimate the unknown inputs without any assumption being made on their bounds or derivatives, one must use output measurements and also at least the first-derivatives of the output measurements (see, for example, [1, 2]). In this regard, the research done by Hou and Müller [1] gives a straightforward method for designing an unknown input observer $\hat{d}(t)$ of the following structure

$$\hat{d}(t) = Q_1\dot{y}(t) + Q_2y(t) + Q_3w(t), \tag{1.3}$$

where $Q_1, Q_2 \in \mathbb{R}^{1\times p}$, $Q_3 \in \mathbb{R}^{1\times(n-p)}$ and $w(t) \in \mathbb{R}^{(n-p)}$ is a state vector of an asymptotic state observer estimating the state vector of a reduced-order disturbance-free system.

H. Trinh et al., *Unknown Inputs Estimation in Linear Time-Delay Systems Using Generalized Functional Observers*, Studies in Systems, Decision and Control 648,
https://doi.org/10.1007/978-3-032-11870-7_1

It should be noted that many existing unknown input observers require the estimated unknown input, $\hat{d}(t)$, to be dynamically coupled to the state observers, $\hat{x}(t)$ (see, for example, [3]). In contrast, the unknown input observer (1.3) has a desirable feature that allows the estimation of the unknown inputs to be independently carried out, by simply designing a reduced-order state observer of a disturbance-free reduced-order system.

On the other hand, observer (1.3) may contain first-derivatives of all the output measurements. This worst-case scenario occurs when Q_1 contains all non-zeros elements which implies the requirement of the first-derivatives of all available outputs. Note that to construct the unknown input as reported in [1], a reduced-order state observer of order $(n - p)$ is required.

1.2 Motivations

In this book, one of our aims is to reduce the complexity of the unknown input observer (1.3) by proposing the following simpler and more practical unknown input observer

$$\hat{d}(t) = \beta_i \dot{y}_i(t) + \gamma_i y_i(t) + \hat{z}(t), \tag{1.4}$$

where $\beta_i \neq 0$ and γ_i are scalars, $\hat{z}(t) \in \mathbb{R}^m$, $1 \leq m \leq (n - p)$, is a scalar asymptotic reduced-order functional observer of a disturbance-free system, and $y_i(t)$ is a single output measurement to be selected out of all p output measurements.

Clearly, observer (1.4) is simpler and more practical than observer (1.3). Here, instead of possibly using first-derivatives of all the output measurements, we only need to take the first-derivative of one output measurement, $\dot{y}_i(t)$. Furthermore, for high-order complex systems with multiple outputs, $p > 1$, a reduced-order scalar functional observer can often be less complex than a reduced-order state observer (see, for example, [4–9]). Thus, the practical significance of observer (1.4) over (1.3) is very clear based on the above two salient features. To our knowledge, such an attempt to reduce the complexity of unknown input observers has not yet been reported in the literature.

In this book, firstly, based on a simple but elegant matrix decomposition technique, we provide a straightforward procedure for designing observer (1.4). We then consider the problem of estimating an unknown input in the presence of multiple unknown inputs. Novel design methods will be developed to design unknown input observers. We then take on an exploratory journey and derive new results for estimating unknown inputs using novel generalized functional observers for a wide range of time-delay systems, including non-commensurate time delays in both the state and output vectors.

1.3 Preliminaries

To obtain the unknown input observer (1.3), the design method [1] involved a two-stage matrix decomposition as briefly described below. Firstly, a state transformation $x(t) = T\bar{x}(t)$, $T \in \mathbb{R}^{n\times n}$ is invertible, where

$$T = \begin{bmatrix} E & D \end{bmatrix}, \tag{1.5}$$

and $E \in \mathbb{R}^{n\times(n-1)}$ is an arbitrary matrix of full-column rank is applied to (1.1)–(1.2). It is easy to see that $T^{-1}D = \begin{bmatrix} 0_{n-1} \\ 1 \end{bmatrix}$.

Applying the above state transformation and by partitioning $\bar{x}(t) = \begin{bmatrix} \bar{x}_1(t) \\ \bar{x}_2(t) \end{bmatrix}$, $\bar{x}_1(t) \in \mathbb{R}^{(n-1)}$, $\bar{x}_2(t) \in \mathbb{R}$ and defining $\bar{A} = T^{-1}AT = \begin{bmatrix} \bar{A}_{11} & \bar{A}_{12} \\ \bar{A}_{21} & \bar{A}_{22} \end{bmatrix}$, where $\bar{A}_{11}$, $\bar{A}_{12}$, $\bar{A}_{21}$ and $\bar{A}_{22}$ are of appropriate dimensions, the following equations are obtained

$$\dot{\bar{x}}_1(t) = \bar{A}_{11}\bar{x}_1(t) + \bar{A}_{12}\bar{x}_2(t), \tag{1.6}$$
$$\dot{\bar{x}}_2(t) = \bar{A}_{21}\bar{x}_1(t) + \bar{A}_{22}\bar{x}_2(t) + d(t), \tag{1.7}$$
$$y(t) = CE\bar{x}_1(t) + CD\bar{x}_2(t). \tag{1.8}$$

Provided that $CD \neq 0$ (i.e., the observer matching condition is satisfied [1]), consider an invertible matrix $V \in \mathbb{R}^{p\times p}$, where

$$V = \begin{bmatrix} CD & P \end{bmatrix}, V^{-1} = \begin{bmatrix} V_1 \\ V_2 \end{bmatrix}. \tag{1.9}$$

Here, $P \in \mathbb{R}^{p\times(p-1)}$ is an arbitrary matrix of full-column rank, and $V_1 \in \mathbb{R}^{1\times p}$ and $V_2 \in \mathbb{R}^{(p-1)\times p}$.

Secondly, by applying a second matrix decomposition to the output equation (1.8) (i.e., pre-multiplying both sides of (1.8) by V^{-1}) and after rearranging, we have

$$\bar{x}_2(t) = V_1 y(t) - V_1 CE\bar{x}_1(t), \tag{1.10}$$
$$V_2 y(t) = V_2 CE\bar{x}_1(t). \tag{1.11}$$

Substituting (1.10) into (1.6), and together with (1.11), a reduced-order *disturbance-free* system is obtained as follows

$$\dot{\bar{x}}_1(t) = \tilde{A}_1\bar{x}_1(t) + \bar{A}_{12}V_1 y(t), \tag{1.12}$$
$$\bar{y}(t) = \tilde{C}_1\bar{x}_1(t), \tag{1.13}$$

where $\bar{y}(t) = V_2 y(t)$, $\tilde{A}_1 = \bar{A}_{11} - \bar{A}_{12}V_1 CE$, and $\tilde{C}_1 = V_2 CE$.

Finally, by substituting (1.10) and (1.12) into (1.7), the unknown input, $d(t)$, is obtained as follows

$$d(t) = V_1\dot{y}(t) - (\bar{A}_{22}V_1 + V_1CE\bar{A}_{12}V_1)y(t) + (\bar{A}_{22}V_1CE - V_1CE\tilde{A}_1 - \bar{A}_{21})\bar{x}_1(t). \tag{1.14}$$

If the matrix pair $(\tilde{A}_1, \tilde{C}_1)$ is observable, then according to Hou and Müller [1], by designing a minimum-order asymptotic state observer to estimate the state vector $\bar{x}_1(t)$ of disturbance-free system (1.12)–(1.13), an unknown input observer of the form (1.3) is obtained.

As elaborated in the Motivations section, our aim is to reduce the complexity of observer (1.3) by replacing it with (1.4).

1.4 Organization of the Book

This book considers the design of unknown input observers to asymptotically estimate unknown inputs of a wide range of dynamical systems, including systems with multiple time delays in both the state and output vectors.

As will be detailed later in the book, decomposing the unknown input vector for time-delay systems results in three components: (1) a linear combination of outputs, (2) a linear combination of the first-derivative of the outputs, and (3) a generalized functional which contains multiple state-delayed terms, for example, $z(t) = Fx(t) + F_dx(t-\tau) + F_{dd}x(t-2\tau) + \cdots$, or $z(t) = Fx(t) + F_dx(t-\tau) + F_hx(t-h)$, where τ and h are constant non-commensurate time delays. Given that the outputs and the first-derivative of the outputs are known or can be determined, an effective way to estimate the unknown input is to estimate the generalized functional directly. As currently is the case in the literature, state/functional observer design methods for time-delay systems (see, for example, [10–13]) mainly deal with the problem of estimating the state vector, $x(t)$, or a linear combination of the state vector, i.e., $z(t) = Fx(t)$. This is the first time that the problem of estimating a generalized functional is reported. Indeed, the need to estimate a generalized functional arose from the unknown input asymptotic estimation problem. Since it was not being given any attention before, its significance was not fully realized and reported. Through our research reported in this book, we conduct detailed investigations into this problem and show that generalized functional observers can bring significant extra degrees of freedom and relax their observer existence conditions. An interesting, but not surprising, finding is that we can design observers for time-delay systems with an order higher than n (where n is the number of state variables). This is achieved through extra degrees of freedom the state-delayed terms brought into $z(t)$. By using generalized functional observers, the unknown inputs of time-delay systems can be directly estimated.

In the first part of the book, for dynamical systems with unknown inputs, we consider the problem of estimating an unknown input in the presence of multiple unknown inputs in the system. The objective is designing an unknown input observer to asymptotically estimate a targeted unknown input, while it is insensitive to the rest of the unknown inputs. In this part, the key contribution is on a novel matrix decomposition technique to analytically express or decompose the targeted unknown input into the most simplified form, which contains the least number of output differentiators and a linear function of the state vector. As a result, the targeted unknown input can be estimated with the least number of first-derivatives of outputs and a reduced-order functional observer.

The second part of the book considers the design of unknown input observers to asymptotically estimate unknown inputs of a wide range of time-delay systems, including systems with non-commensurate time delays in both the state and output vectors. By utilizing a novel matrix decomposition technique as presented in the first part of the book, we can derive reduced-order (sometimes, disturbance-free) time-delay systems with multiple time delays in both the state and output vectors, and can derive analytical forms for the targeted unknown inputs. In its own form, it contains the least number of output differentiators and a generalized functional containing multiple state-delayed terms. Then to estimate the unknown input, an effective way is to directly estimate the generalized functional, using generalized functional observers. This leads to the third part of the book.

The third part of the book takes us on an exploratory journey to investigate the design of generalized functional observers to asymptotically estimate generalized functionals of a wide class of time-delay systems, including multiple time delays and/or non-commensurate time delays in both the state and output vectors. As we have mentioned earlier, the generalized functional observers design problem allows us to design observers with more general structures with less conservative existence conditions. As a first application, we use generalized observers to solve the problem of designing dynamic output feedback controllers for time-delay systems using a less restrictive state feedback control law with a state-delayed term, i.e., $u(t) = Fx(t) + F_d x(t - \tau)$, where a generalized functional observer is used to estimate the control law $u(t)$.

This book has a distinctive expository style of writing. In a sense, it is more like a journal of our research journey. Our intention is to provide simple explanations of how a problem arises and is solved via simple mathematical derivations. The derived results are illustrated via extensive numerical examples with simulations to verify their correctness. More often than not, they are used to explain the design procedure and provide more insights into the considered problems. We hope to take the reader on an exploratory journey with us via our simple approach to solving those problems in the book. We take the reader through various cases, scenarios and illustrative examples to show the significance of the new generalized functional estimation problem.

The first and second parts of the book have 8 chapters dedicated to the design of unknown input observers for linear systems and linear time-delay systems. The third part of the book has 10 miscellaneous chapters devoted to the design of generalized functional observers to asymptotically estimate generalized functionals of a wide class of time-delay systems. The first of those miscellaneous chapters provides a detailed account of existing results on the design of state/functional observers for time-delay systems from our own perspective. The results compiled in this chapter are useful for the design of unknown input observers for time-delay systems as presented in Chap. 4. But as we consider a wider class of time-delay systems, we need to design generalized functional observers. Thus, the next 9 miscellaneous chapters are a collection of discussions and results on the significance of the new estimation problem and design of generalized functional observers for time-delay systems. These chapters are more like a journal of our research. We also aim to make the book as self-contained as possible and thus we include a short chapter with some results reported in the literature by Mondié and Kharitonov [14] on the stability and stabilization of time-delay systems using Linear Matrix Inequalities (LMIs). This is useful since to design observers for time-delay systems, we need to ensure asymptotic/exponential stability of estimation error, which is also a time-delay system.

Finally, the rest of the chapters in this book are as described below.

Chapter 2 provides our first result on the asymptotic estimation of an unknown input for a class of linear systems using a simplified observer structure. By a novel simple matrix decomposition technique, we show how the unknown input can be expressed in its simplest form, which contains only one output measurement, its first-derivative, and a linear function of the state vector. Thus, the unknown input can be estimated by using the first-derivative of only one output measurement and a minimal-order functional observer. Therefore, the result is the simplest designed unknown input observer for dynamical systems.

Chapter 3 considers the problem of estimating an unknown input in the presence of multiple unknown inputs in the system. The designed unknown input observer asymptotically estimates a targeted unknown input, while it is insensitive to the rest of the unknown inputs. We provide a detailed analysis of two design approaches and discuss their advantages and drawbacks. Together, they offer the designer a viable tool for designing an unknown input observer with the least number of output differentiators and a reduced-order functional observer. The results derived in this chapter can be used to design distributed unknown input observers for interconnected dynamical systems. Here, at each local subsystem, an unknown input observer is designed to estimate the local unknown inputs, while the designed observer is insensitive to the external unknown inputs that go into the other subsystems.

Chapter 4 discusses the design of unknown input observers for a class of time-delay systems with a constant time delay. Based on the matrix decomposition technique presented in Chap. 2, we can effectively obtain a reduced-order disturbance-free time-delay system, and then obtain a closed form analytical expression for the unknown input. Here, the unknown input is in the simplest form containing only one output measurement, its delayed output measurement, the first-derivative of one output measurement, and a linear function of the state vector that contains a state-delayed term, i.e., $z(t) = Fx(t) + F_d x(t-\tau)$ (from now on, we shall call any vector $z(t)$ which contains at least one state-delayed term as a generalized functional). True to the expository nature of this book, we look at possible ways to realize the estimation of the generalized functional by examining two design options, Option 1 and Option 2, where Option 2 comprises four scenarios. Scenario 4 of Option 2 points to the need to estimate the generalized functional directly, as this is a more effective and economical way. This makes us consider the design of generalized functional observers for a wide range of time-delay systems, which is the bulk of the third part of the book.

Chapter 5 discusses the design of unknown input observers for a class of time-delay systems, where a constant time delay, τ, now also occurs in the output vector, i.e., $y(t) = Cx(t) + C_d x(t - \tau)$. To solve this problem, we first opt for a matrix decomposition technique presented in Chap. 2, but now we need to impose a constraint on matrix C_d, i.e., $V_1 C_d D = 0$. This constraint implies either $V_1 C_d = 0_{1\times n}$ or $C_{di} D = 0$. Consequently, we discuss the design of unknown input observers with the above two structural constraints. About the first constraint, i.e., $V_1 C_d = 0_{1\times n}$, the decomposition results in a disturbance-free time-delay system with a time delay, τ, in both the state and output vectors. The unknown input is expressed in its entirety that contains a generalized functional. Thus, the unknown input can be estimated by designing a generalized functional observer to estimate this generalized functional. As for the second constraint, i.e., $C_{di} D = 0$, it covers a wider class of time-delay systems than the first constraint. On the other hand, the decomposition results in a more complex disturbance-free time-delay system with multiple time delays in both the state and output vectors. Moreover, the unknown input in its entirety now contains a generalized functional, which has multiple state-delayed terms, hence the need to design generalized functional observers.

Chapter 6 discusses the problem of designing an unknown input observer to asymptotically estimate an unknown input, $d(t)$, for a class of time-delay systems where only delayed output vector is available, i.e., $y(t) = C_d x(t - \tau)$. Our solution approach enables us to estimate the unknown input after a time delay, τ. Thus, we can have $\hat{d}(t - \tau)$ asymptotically converge to $d(t - \tau)$.

Chapter 7 considers another approach to design unknown input observers for a class of time-delay systems as considered in Chap. 5. This approach is named an Expanded Dimensional Approach, where an auxiliary state vector $\zeta(t) \equiv x(t - \tau)$ is augmented with the original state vector $x(t)$ to form an expanded time-delay system. In its expanded dimensional form, just like in Chap. 5, we investigate two design approaches to design an unknown input observer. The objective of Design Approach 1 is to use the output measurements and the first-derivative of only one output measurement. To do this, we require the same structural constraint $V_1 C_d D = 0$ as in Chap. 5. However, compared with both cases discussed in Chap. 5, the Expanded Dimensional Approach results in a high-dimensional time-delay system, and it still contains a delayed unknown input. Moreover, the design of generalized functional observers for time-delay systems with an unknown input is subject to more stringent observer existence conditions. Thus, this finding highlights the attractiveness of the two approaches presented in Chap. 5.

On the other hand, as for Design Approach 2, we show that the Expanded Dimensional Approach can overcome the requirement $V_1 C_d D = 0$ imposed in Chap. 5. To achieve this, it is necessary to use first-derivatives of two output measurements and a larger number of output measurements, i.e., $p > 2$, and hence clearly a trade-off is required. Accordingly, we obtain a disturbance-free time-delay system, and the unknown input can be asymptotically estimated by designing a generalized functional observer, which is the topic discussed extensively in the third part of the book.

Chapter 8 discusses the unknown input estimation problem for a class of time-delay systems with two non-commensurate delays, τ and h, in both the state and output vectors. Due to the presence of non-commensurate delays, the design of an unknown input observer is more complicated than the one considered in Chap. 5. We provide some discussions on their designs as well as some open problems to be considered in the future. To express the unknown input in its entirety, we consider three cases where we impose some constraints on output matrices C_d and C_h. In Case 1, the constraint imposed on C_d and C_h is $V_1 C_d D = 0$, $V_1 C_h D = 0$ and $V_1 C_d E = V_1 C_h E = 0_{1\times(n-1)}$. Subject to this constraint, we obtain a disturbance-free time-delay system with two non-commensurate time delays. The unknown input $d(t)$ also contains a generalized functional of the form $z_{\tau h}(t) = Fx(t) + F_d x(t - \tau) + F_h x(t - h)$. Thus, the estimation of $d(t)$ rests with the successful design of a generalized functional observer to estimate $z_{\tau h}(t)$. On the other hand, with a weaker constraint imposed on C_d or C_h (i.e., Case 2, where either $V_1 C_d E \neq 0_{1\times(n-1)}$ or $V_1 C_d E \neq 0_{1\times(n-1)}$), we obtain a disturbance-free time-delay system but with mixed non-commensurate time delays in both state and output vectors, and likewise for $d(t)$. However, as for both Cases 2 and 3 discussed in this chapter, the design of generalized functional observers for such complicated systems with mixed non-commensurate time delays remains an open research problem for future research.

Chapter 9 considers the problem of designing an unknown input observer to estimate a specific unknown input, $d_i(t)$, in the presence of multiple unknown inputs for a class of time-delay systems with both the state vector and output vector subject to a constant time delay. Thus, this chapter extends the results in Chap. 5 to deal with multiple unknown inputs.

Chapter 10 (the first of the miscellaneous chapters) provides a detailed account of existing results on the design of state/functional observers for time-delay systems from our own perspective. The results compiled in this chapter are useful for the design of unknown input observers for time-delay systems as presented in Chap. 4.

Chapter 11 considers the design of various types of functional observers to estimate a function of the state vector for a class of time-delay systems with a state-delayed term in the output vector. This term can be easily dealt with in our approach, which is an adapted solution first proposed in [12]. We present existence conditions for the proposed observers and outline a procedure for determining their observer gain matrices. Via a numerical example, we show that our approach is less conservative than the Output Injection approach of [11].

Chapter 12 considers the design of various types of functional observers for a class of time-delay systems with multiple delays in both state and output vectors. By a novel idea of estimating directly a new generalized functional that has multiple state-delayed terms, we can relax their observer existence conditions. Through a numerical example, we show the significance of the new estimation problem, where we can expand the dimension of the designed observer beyond the original dimension of the state vector, i.e., n. This new finding offers an effective way to design generalized functional observers for a wide class of time-delay systems. This will be further investigated in the subsequent chapters.

Chapter 13 provides a detailed investigation into the problem of estimating a generalized functional $z(t) = Fx(t) + F_d x(t - \tau)$ for a class of time-delay systems with or without a state-delayed term in the output vector. Thanks to this unified framework, we can design a wider range of observers than currently available in the literature. The results derived in this chapter are used for estimating unknown inputs as discussed in Chaps. 4 and 5. As a byproduct of the derived results in this chapter, we also show that we can design delayed state observers without using any complicated state transformations such as the method reported in [15].

Chapter 14, as a byproduct of the results presented in Chap. 13, we look at the problem of designing dynamic output feedback controllers for time-delay systems via the novel application of using $\hat{z}(t)$ to directly realize a general state feedback control law, $u(t) = Fx(t) + F_d x(t - \tau)$, for time-delay systems. The objective of the control law is to ensure that one of the selected outputs of the system asymptotically tracks any step (or ramp) reference command input, $r(t)$. Via several numerical examples, we show the advantages of our approach via estimating $\hat{z}(t)$ directly. With our approach, we can have a simpler observer-based control scheme with lesser hardware components and better transient responses. This finding, again, highlights the significance of the generalized functional estimation problem reported in this book.

In Chap. 15, we continue our research into the design of generalized functional observers. We look at time-delay systems with multiple delays in both the state and output vectors. For this class of time-delay systems, we design generalized functional observers to directly estimate a generalized functional $z(t) = Fx(t) + F_d x(t - \tau) + F_{dd} x(t - 2\tau) + F_{3d} x(t - 3\tau)$. Via various numerical examples, we show how such generalized functional observers can be designed and their advantages. The solution to this problem is used in the unknown input estimation problem for time-delay systems considered in Chap. 5, where the second constraint, i.e., $C_{di} D = 0$, is satisfied.

In Chap. 16, we further demonstrate the generality and advantages of estimating a more generalized functional, $z_{2\tau}(t) = Fx(t) + F_d x(t - \tau) + F_{dd} x(t - 2\tau)$ for time-delay systems considered in Chapter 13. This class of time-delay systems has a single time delay, τ, in both state and output vectors. Here, compared with $z(t)$ considered in Chap. 13, $z_{2\tau}(t)$ has an extra term, namely, $F_{dd} x(t - 2\tau)$. We provide extensive numerical examples to discuss the design of various types of generalized functional observers for this class of time-delay systems. Through these examples, we show the generality and salient features of our design method and the generality of our results over those reported in [10, 12]. The solution to this problem is used in Chap. 6 to asymptotically estimate delayed unknown input, where only delayed output vector is available.

Chapter 17 considers the problem of designing a generalized functional observer to estimate a generalized functional $z_{\tau h}(t) = Fx(t) + F_d x(t - \tau) + F_h x(t - h)$ of a time-delay with two non-commensurate delays, τ and h, in the state and output vectors. Here, $\hat{z}_{\tau h}(t)$ is used in Chap. 8 to estimate the unknown input of a class of time-delay systems with two non-commensurate delays. Thanks to our unified framework, we can readily obtain observer existence conditions, and via three numerical examples, we demonstrate in detail the design of various types of generalized functional observers.

Chapter 18 extends the design of generalized functional observers for a class of time-delay systems considered in Chap. 15 to include unknown inputs. Thus, the designed generalized functional observers are now required to be insensitive to an external disturbance vector. This problem arises from the unknown input estimation problem of time-delay systems with more than one unknown input as discussed in Chap. 9. We provide generalized observer existence conditions and demonstrate the design of generalized functional observers via some numerical examples. These examples are used in Chap. 9 for estimating the unknown input.

In Chap. 19, we look at the problem of designing generalized functional observers for a class of systems with mixed non-commensurate time delays in both state and output vectors. We propose a general observer structure and derive their existence conditions. However, we stop short of finding numerical examples to illustrate the design procedure. Thus, the problem of designing generalized functional observers to estimate a generalized functional that contains mixed non-commensurate time delays remains an open research problem for future research.

Chapter 20 provides some existing results in the literature for the stability and stabilization of time-delay systems using LMIs. This is useful in observer designs for time-delay systems, as we need to ensure asymptotic/exponential stability of the error time-delay system.

References

1. Hou, M., Müller, P.C.: Design of observers for linear systems with unknown inputs. IEEE Trans. Automat. Control **37**(6), 871–875 (1992)
2. Hou, M., Patton, R.J.: Input observability and input reconstruction. Automatica **34**(6), 789–794 (1998)
3. Ha, Q.P., Trinh, H.: State and input simultaneous estimation for a class of nonlinear systems. Automatica **40**(10), 1779–1785 (2004)
4. Darouach, M., Fernando, T.: Functional detectability and asymptotic functional observer design. IEEE Trans. Automat. Control **68**(2), 975–990 (2023)
5. Fernando, T.L., Trinh, H., Jennings, L.: Functional observability and the design of minimum order linear functional observers. IEEE Trans. Automat. Control **55**(5), 1268–1273 (2010)
6. Jennings, L.S., Fernando, T.L., Trinh, H.M.: Existence conditions for functional observability from an eigenspace perspective. IEEE Trans. Automat. Control **56**(12), 2957–2961 (2011)
7. Rotella, F., Zambettakis, I.: Minimal single linear functional observers for linear systems. Automatica **47**(1), 164–169 (2011)
8. Trinh, H., Fernando, T.: Functional Observers for Dynamical Systems. Springer, Berlin (2012)
9. Trinh, H., Tran, T.D., Nahavandi, S.: Design of scalar functional observers of order less than $(\nu - 1)$. Int. J. Control **79**(12), 1654–1659 (2006)

10. Darouach, M.: Linear functional observers for systems with delays in state variables. IEEE Trans. Automat. Control **46**(3), 491–496 (2001)
11. Hou, M., Zítek, P., Patton, R.J.: An observer design for linear time-delay systems. IEEE Trans. Automat. Control **47**(1), 121–125 (2002)
12. Nam, P.T., Pathirana, P.N., Trinh, H.: ε-bounded state estimation for time-delay systems with bounded disturbances. Int. J. Control **87**(9), 1747–1756 (2014)
13. Trinh, H.: Linear functional state observers for time-delay systems. Int. J. Control **72**(18), 1642–1658 (1999)
14. Mondié, S., Kharitonov, V.L.: Exponential estimates for retarded time-delay systems: an LMI approach. IEEE Trans. Automat. Control **50**(2), 268–273 (2005)
15. Huong, D.C., Trinh, H.: Method for computing state transformations of time-delay systems. IET Control Theory Appl. **9**(16), 2405–2413 (2015)

Chapter 2
Unknown Input Observers Design—Single Unknown Input Case

In this chapter, we look at system (1.1)–(1.2), i.e.,

$$\dot{x}(t) = Ax(t) + Dd(t),$$
$$y(t) = Cx(t),$$

and we provide a procedure for determining $P \in \mathbb{R}^{p\times(p-1)}$ and $E \in \mathbb{R}^{n\times(n-1)}$, so that the unknown input can be estimated by using (1.4).

2.1 Determination of P and E: A Novel Matrix Decomposition Technique

Let us now relook at (1.10), i.e.,

$$\bar{x}_2(t) = V_1 y(t) - V_1 C E \bar{x}_1(t).$$

By letting

$$V_1 C E = 0_{1\times(n-1)}, \tag{2.1}$$

we have $\bar{x}_2(t) = V_1 y(t)$ and thus $\dot{\bar{x}}_2(t) = V_1 \dot{y}(t)$. As a result, the unknown input $d(t)$ as defined by (1.14) is further simplified and given below

$$d(t) = V_1 \dot{y}(t) - \bar{A}_{22} V_1 y(t) - \bar{A}_{21} \bar{x}_1(t). \tag{2.2}$$

Furthermore, instead of designing an asymptotic state observer to estimate $\bar{x}_1(t)$ for system (1.12)–(1.13) as was done by Hou and Müller [1], we should now design a minimal-order scalar functional observer to estimate the following function

H. Trinh et al., *Unknown Inputs Estimation in Linear Time-Delay Systems Using Generalized Functional Observers*, Studies in Systems, Decision and Control 648,
https://doi.org/10.1007/978-3-032-11870-7_2

$$z(t) = -\bar{A}_{21}\bar{x}_1(t). \tag{2.3}$$

Furthermore, by enforcing the following constraint on matrix V_1,

$$V_1 = \frac{1}{\alpha_i}\begin{bmatrix} 1 & 0_{1\times(p-1)} \end{bmatrix}, \tag{2.4}$$

and $\alpha_i \neq 0$, we can see that an unknown input observer in the form (1.4) is obtained.

Thus, the key to designing the simplest unknown input observer (1.4) is to find two matrices $P \in \mathbb{R}^{p\times(p-1)}$ and $E \in \mathbb{R}^{n\times(n-1)}$ to satisfy conditions (2.1) and (2.4). The solution to this problem is rather straightforward and, in the following, we provide a simple matrix decomposition technique to find them.

Under the assumption that $\text{rank}(CD) = \text{rank}(D)$, there always exists at least one output measurement, $y_i(t) = C_i x(t)$, such that $C_i D = \alpha_i \neq 0$, where $C_i \in \mathbb{R}^{1\times n}$ and α_i is a scalar. Thus, let us now pick an output measurement, e.g., $y_i(t)$, such that $C_i D = \alpha_i \neq 0$ and its first-derivatives (i.e., $\dot{y}_i(t)$) will be used for the reconstruction of the unknown input. Then the output vector, $y(t)$, is rearranged as follows

$$y(t) = \begin{bmatrix} y_i(t) \\ y_r(t) \end{bmatrix} = \begin{bmatrix} C_i \\ C_r \end{bmatrix} x(t), \tag{2.5}$$

where $y_r(t) \in \mathbb{R}^{p-1}$ and $C_r \in \mathbb{R}^{(p-1)\times n}$. Subsequently, without any ambiguity, we let $C = \begin{bmatrix} C_i \\ C_r \end{bmatrix}$.

It is clear that with V_1 takes the form (2.4), we only need to use the first-derivative of the output $y_i(t)$ since $V_1\dot{y}(t) = \frac{1}{\alpha_i}\dot{y}_i(t)$. Now, let V be expressed as follows

$$V = \begin{bmatrix} CD & P \end{bmatrix} = \begin{bmatrix} \alpha_i & P_1 \\ C_r D & P_2 \end{bmatrix}.$$

By letting $P_1 = 0_{1\times(p-1)}$ and $P_2 = I_{p-1}$, we then have $V = \begin{bmatrix} \alpha_i & 0_{1\times(p-1)} \\ C_r D & I_{p-1} \end{bmatrix}$ and thus V^{-1} can be readily obtained as follows

$$V^{-1} = \begin{bmatrix} V_1 \\ V_2 \end{bmatrix} = \begin{bmatrix} \frac{1}{\alpha_i} & 0_{1\times(p-1)} \\ -\frac{1}{\alpha_i}C_r D & I_{p-1} \end{bmatrix}. \tag{2.6}$$

Thus, from (2.6), we obtain V_1, which is as given in (2.4), and $V_2 = \begin{bmatrix} -\frac{1}{\alpha_i}C_r D & I_{p-1} \end{bmatrix}$.

Next, we look at how to find $E \in \mathbb{R}^{n\times(n-1)}$ such that $V_1 C E = 0_{1\times(n-1)}$. Note that $V_1 C$ is a row of dimension $1 \times n$, and thus there always exists a full-column rank matrix $E \in \mathbb{R}^{n\times(n-1)}$ such that $V_1 C E = 0_{1\times(n-1)}$. Matrix E is obtained by using either an orthogonal basis or a rational basis for the null-space of $V_1 C$ (i.e., singular value decomposition or the reduced row echelon form). Then matrix $T = \begin{bmatrix} E & D \end{bmatrix}$ is obtained, and provided that it has full rank, the design of an unknown input observer can be systematically carried out. It is clear that a necessary condition for an invertible matrix T is that both E and D have full-column rank. Since E and D are full-column rank matrices, the necessary condition is met. In fact, with the above matrix decomposition for V_1 and E, matrix T always has full rank. The proof is obvious for the case where the output matrix C takes the canonical form $C = \begin{bmatrix} I_p & 0_{p\times(n-p)} \end{bmatrix}$ (note that a state transformation can first be applied to transform system (1.1)–(1.2) such that C is in the required canonical form).

Furthermore, with V_1 in the form (2.4), we now have $-\bar{A}_{22}V_1 y(t) = -\frac{\bar{A}_{22}}{\alpha_i} y_i(t)$ and $\bar{A}_{12}V_1 y(t) = \frac{\bar{A}_{12}}{\alpha_i} y_i(t)$. Hence, from (2.2), the estimation of the unknown input $d(t)$ can now be realized by using the following simplest observer

$$\hat{d}(t) = \frac{1}{\alpha_i}\dot{y}_i(t) - \frac{\bar{A}_{22}}{\alpha_i} y_i(t) + \hat{z}(t), \tag{2.7}$$

where $\hat{z}(t)$ is a minimal-order functional observer that asymptotically estimates the function $z(t) = -\bar{A}_{21}\bar{x}_1(t)$ of the following disturbance-free reduced-order system

$$\dot{\bar{x}}_1(t) = \bar{A}_{11}\bar{x}_1(t) + \frac{\bar{A}_{12}}{\alpha_i} y_i(t), \tag{2.8}$$

$$\bar{y}(t) = \tilde{C}_1\bar{x}_1(t). \tag{2.9}$$

The problem of designing a reduced-order scalar functional observer, $\hat{z}(t)$, to estimate the linear function, $z(t)$, of system (2.8)–(2.9) with or without known input vector has been very well addressed in the literature (see, for example, [2–5]). We can readily design a reduced-order scalar functional observer of the following structure

$$\hat{z}(t) = Kw(t) + M\bar{y}(t), \tag{2.10}$$

$$\dot{w}(t) = Nw(t) + G\bar{y}(t) + J_i y_i(t), \tag{2.11}$$

where $w(t) \in \mathbb{R}^m$, and $N \in \mathbb{R}^{m\times m}$ is a stable matrix. Here, the order of the functional observer is m, $1 \le m \le (\nu - 1)$, where ν is the observability index of the matrix pair $(\bar{A}_{11}, \tilde{C}_1)$ (see, [4]). As such, a reduced-order functional observer can be designed by solving the following constrained generalized Sylvester equation

$$NL + G\tilde{C}_1 - L\bar{A}_{11} = 0, \tag{2.12}$$

$$\bar{A}_{21} + KL + M\tilde{C}_1 = 0, \tag{2.13}$$

for the unknown constant matrices N, L, G, K and M of appropriate dimensions. The detail is omitted here and the reader can refer to [5] (or some recent results in the literature [6, 7]) for a design algorithm to obtain a minimal-order, m, and the observer matrices. Furthermore, J_i is obtained as $J_i = \frac{L\bar{A}_{12}}{\alpha_i}$.

2.2 Algorithm 1: Estimation of One Unknown Input

The following is a step-by-step design algorithm. For ease of presentation and subsequent analysis but without loss of generality, we assume that matrix C takes the canonical form, i.e., $C = \begin{bmatrix} I_p & 0_{p\times(n-p)} \end{bmatrix}$.

Algorithm 1: Estimation of one unknown input

1. Select one output measurement, $y_i(t)$, such that $C_i D = \alpha_i \neq 0$.
2. Rearrange the output vector according to (2.5). Let $C = \begin{bmatrix} C_i \\ C_r \end{bmatrix}$.
3. Obtain $V = [CD\ P]$, $P = \begin{bmatrix} 0_{1\times(p-1)} \\ I_{p-1} \end{bmatrix}$. From (2.6), obtain V_1 and V_2.
4. Obtain matrix E, where E is a rational basis for the null-space of $V_1 C$. Obtain matrix $T = [E\ D]$.
5. Obtain $\bar{A} = T^{-1}AT$ and submatrices $\bar{A}_{11}$, $\bar{A}_{12}$, $\bar{A}_{21}$ and $\bar{A}_{22}$.
6. Obtain the disturbance-free system (2.8)–(2.9) where $\tilde{C}_1 = V_2 C E$.
7. Obtain K, M, J_i, N and G by designing a reduced-order functional observer to estimate the function $z(t)$.
8. Estimation of the unknown input is given by (2.7).

2.3 Numerical Example 1

We consider a numerical example to show the simplicity of our design method and the advantages of using unknown input observer (1.4). Let us consider a 6th-order system with three outputs and one unknown input, where matrices A, C and D are given below

$$A = \begin{bmatrix} -7 & -1 & -2 & 0 & 0 & -1 \\ 6 & -12 & -2 & 0 & 1 & -2 \\ -2 & 1 & -1 & 0 & 0 & 1 \\ -5 & 1 & 0 & -5 & 3 & 4 \\ 8 & -4 & -1 & 1 & -8 & 2 \\ 0 & -2 & -4 & 0 & 2 & -6 \end{bmatrix}, C = \begin{bmatrix} I_3 & 0_{3\times3} \end{bmatrix}, D = \begin{bmatrix} 1 & 2 & 0 & 0 & 1 & 0 \end{bmatrix}^T.$$

In this example, the condition $\text{rank}(CD) = \text{rank}(D)$ is satisfied. We also have $CD = \begin{bmatrix} 1 \\ 2 \\ 0 \end{bmatrix}$, which implies that we can select either the first-derivative of $y_1(t)$ or $y_2(t)$ to be used for the unknown input reconstruction. For demonstrative purpose, let us pick the first output $y_1(t)$, where we have $C_1 D = \alpha_1 = 1$. By following Algorithm 1,we have thus completed Steps 1 and 2.

From Step 3, we obtain $V = \begin{bmatrix} 1 & 0 & 0 \\ 2 & 1 & 0 \\ 0 & 0 & 1 \end{bmatrix}$. Hence $V^{-1} = \begin{bmatrix} V_1 \\ V_2 \end{bmatrix}$ is obtained, where $V_1 = \begin{bmatrix} 1 & 0 & 0 \end{bmatrix}$ and $V_2 = \begin{bmatrix} -2 & 1 & 0 \\ 0 & 0 & 1 \end{bmatrix}$.

From Step 4, we have $V_1 C = \begin{bmatrix} 1 & 0_{1\times5} \end{bmatrix}$ and hence a rational basis for the null-space of $V_1 C$ can be easily obtained. We obtain $E = \begin{bmatrix} 0_{1\times5} \\ I_5 \end{bmatrix}$. Hence matrix $T = \begin{bmatrix} E & D \end{bmatrix}$ is obtained and it is easy to see that it is invertible as expected.

From Step 5, with the state transformation $x(t) = T\bar{x}(t)$, we obtain the following sub-matrices

$$\bar{A}_{11} = \begin{bmatrix} -10 & 2 & 0 & 1 & 0 \\ 1 & -1 & 0 & 0 & 1 \\ 1 & 0 & -5 & 3 & 4 \\ -3 & 1 & 1 & -8 & 3 \\ -2 & -4 & 0 & 2 & -6 \end{bmatrix}, \quad \bar{A}_{12} = \begin{bmatrix} 1 \\ 0 \\ 0 \\ 1 \\ -2 \end{bmatrix}, \quad \bar{A}_{21} = \begin{bmatrix} -1 & -2 & 0 & 0 & -1 \end{bmatrix} \text{ and}$$

$\bar{A}_{22} = -9$.

From Step 6, the disturbance-free reduced-order system (2.8)–(2.9) is obtained where $\tilde{C}_1 = \begin{bmatrix} I_2 & 0_{2\times3} \end{bmatrix}$ and $\frac{\bar{A}_{12}}{\alpha_1} = \bar{A}_{12}$.

From Step 7, we design a reduced-order scalar functional observer for the disturbance-free reduced-order system (2.8)–(2.9) to estimate the following function

$$z(t) = -\bar{A}_{21}\bar{x}_1(t) = \begin{bmatrix} 1 & 2 & 0 & 0 & 1 \end{bmatrix} \bar{x}_1(t).$$

Based on the design method and Example 1 as given in [5], a first-order functional observer can be designed, where

$$L = \begin{bmatrix} -2 & 2 & 0 & 0 & 1 \end{bmatrix}, K = 1, N = -4, M = \begin{bmatrix} 3 & 0 \end{bmatrix}, G = \begin{bmatrix} 12 & -2 \end{bmatrix} \text{ and } J_1 = -4.$$

Finally, from Step 8 and noting that $\bar{y}(t) = V_2 y(t)$, the following unknown input observer is obtained

$$\begin{aligned} \hat{d}(t) &= \dot{y}_1(t) + 9y_1(t) + \hat{z}(t), \\ \hat{z}(t) &= w(t) + \begin{bmatrix} -6 & 3 & 0 \end{bmatrix} y(t), \\ \dot{w}(t) &= -4w(t) + \begin{bmatrix} -28 & 12 & -2 \end{bmatrix} y(t). \end{aligned}$$

The above is the simplest unknown input observer that can be used to asymptotically estimate the unknown input, $d(t)$. In comparison, a third-order state observer would be required with the method presented by Hou and Müller [1]. Furthermore, based on their method with an arbitrary matrix P, one might end up using first-derivatives of all three outputs. The advantages of our observer are thus very clear.

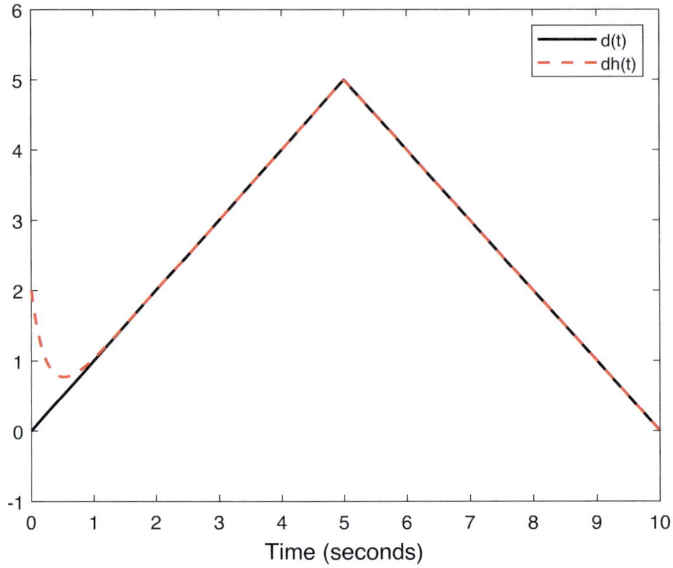

Fig. 2.1 Trajectories of $d(t)$ and $\hat{d}(t)$ (Numerical Example 1, Chap. 2)

Figure 2.1 shows the trajectories of the disturbance $d(t)$ and its estimate $\hat{d}(t)$. It is clear from the figure that asymptotic estimation is achieved.

Remark 1 It is expected that $\hat{d}(t)$ converges to $d(t)$ with the same rate as $\hat{z}(t)$ converges to $z(t)$. Let $e_d(t) = \hat{d}(t) - d(t)$, then it is clear that $e_d(t)$ can be expressed as follows

$$e_d(t) = \hat{d}(t) - d(t) = e(t), \tag{2.14}$$

where $e(t) = \hat{z}(t) - z(t)$. Since $\hat{z}(t)$ is designed to ensure that $\dot{e}(t) = Ne(t)$, where N is a stable matrix. Therefore, the error $e_d(t)$ converges to zero with the same rate as $e(t)$ converges to zero.

Figure 2.2 also shows the trajectories of a fast time-varying disturbance $d(t)$ and its estimate $\hat{d}(t)$. It is clear from the figure that asymptotic estimation is achieved.

2.4 Concluding Remark

In this chapter, we have derived our first result on the asymptotic estimation of an unknown input, $d(t)$, for a class of linear systems using the most simplified observer structure. Without loss of generality, we have let the output matrix C take the

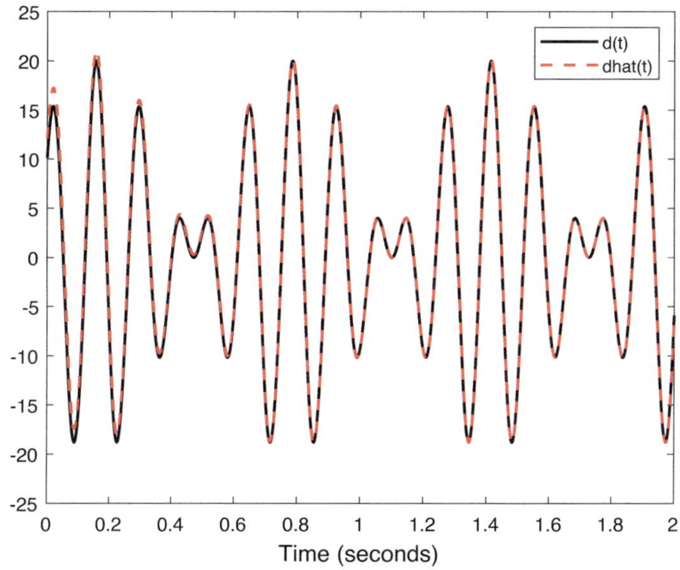

Fig. 2.2 Trajectories of $d(t)$ and $\hat{d}(t)$ (Numerical Example 1, Chap. 2, Remark 1)

canonical form $C = \begin{bmatrix} I_p & 0_{p\times(n-p)} \end{bmatrix}$, and we show that by choosing $P = \begin{bmatrix} 0_{1\times(p-1)} \\ I_{p-1} \end{bmatrix}$ and $E = \begin{bmatrix} 0_{1\times(n-1)} \\ I_{n-1} \end{bmatrix}$, we can express the unknown input in its simplest form, where

$$d(t) = \frac{1}{\alpha_i}\dot{y}_i(t) - \frac{\bar{A}_{22}}{\alpha_i}y_i(t) + z(t).$$

Thus, the unknown input can be estimated by using the first-derivative of one output, $y_i(t)$, and a reduced-order functional observer, $\hat{z}(t)$, which asymptotically estimates the function $z(t) = -\bar{A}_{21}\bar{x}_1(t)$ of the disturbance-free reduced-order system (2.8)–(2.9). Our obtained result will lay a foundation for developing more results into the design of unknown inputs for a wide class of complex systems, which will be presented in the subsequent chapters in this book.

References

1. Hou, M., Müller, P.C.: Design of observers for linear systems with unknown inputs. IEEE Trans. Automat. Control **37**(6), 871–875 (1992)
2. Fernando, T.L., Trinh, H., Jennings, L.: Functional observability and the design of minimum order linear functional observers. IEEE Trans. Automat. Control **55**(5), 1268–1273 (2010)

3. Rotella, F., Zambettakis, I.: Minimal single linear functional observers for linear systems. Automatica **47**(1), 164–169 (2011)
4. Trinh, H., Fernando, T.: Functional Observers for Dynamical Systems. Springer, Berlin (2012)
5. Trinh, H., Tran, T.D., Nahavandi, S.: Design of scalar functional observers of order less than $(\nu - 1)$. Int. J. Control **79**(12), 1654–1659 (2006)
6. Darouach, M., Fernando, T.: Functional detectability and asymptotic functional observer design. IEEE Trans. Automat. Control **68**(2), 975–990 (2023)
7. Darouach, M., Fernando, T.: On functional observability and functional observer design. Automatica **173**, 112,115 (2025)

Chapter 3
Estimation of an Unknown Input in the Presence of Multiple Unknown Inputs

In this chapter, we look at system (1.1)–(1.2) with multiple unknown inputs, and we provide a comprehensive analysis on the design of an unknown input observer in the presence of multiple unknown inputs in the system. Here, the designed unknown input observer asymptotically estimates a targeted unknown input, but it is insensitive to the rest of the unknown inputs.

3.1 Multiple Unknown Inputs

We now consider system (1.1)–(1.2) with multiple unknown inputs, i.e., $d(t) \in \mathbb{R}^q$, $q > 1$, matrix $D \in \mathbb{R}^{n \times q}$ has full column rank, i.e., $\mathrm{rank}(D) = q$, and C takes the canonical form.

In many practical systems, we often face a situation where we need to estimate a particular unknown input in the presence of multiple unknown inputs and/or disturbances in the system. For example, in the load frequency control of multi-area power systems, we need to know (i.e., observe/estimate) the real-time load demand in one local power area which is an unknown input in the presence of multiple load demands which are also unknown inputs and intermittent energy resources in the whole system. In this case, we don't need to estimate all the unknown inputs, but rather a selected unknown input is required to be estimated.

On the other hand, when system (1.1)–(1.2) is subject to structured real-valued parameter variations, i.e., $A + \Delta A(t)$, then $\Delta A(t)x(t)$ can be expressed as (see, for example, [1])

$$\Delta A(t)x(t) = \Omega g(t)\Phi x(t),$$

where $\Omega \in \mathbb{R}^{n \times v}$ is a full-column rank matrix, $g(t) \in \mathbb{R}^{v \times v}$ and $\Phi \in \mathbb{R}^{v \times n}$ is a full-row rank matrix. By letting $d(t) = g(t)\Phi x(t) \in \mathbb{R}^v$ and treating it as an unknown input vector, we have the system (1.1)–(1.2) with multiple unknown inputs.

H. Trinh et al., *Unknown Inputs Estimation in Linear Time-Delay Systems Using Generalized Functional Observers*, Studies in Systems, Decision and Control 648,
https://doi.org/10.1007/978-3-032-11870-7_3

Thus, in this chapter, we solve the problem of designing a practical unknown input observer, $\hat{d}_i(t)$, to estimate an unknown input, $d_i(t)$, in the presence of multiple unknown inputs. Our aim is to design the simplest unknown input observer (with a systematic design procedure) that will use the least number of first-derivatives of the available output measurements and with a minimal-order functional observer. We will extend the result in Chap. 2 to solve this problem. For ease of presentation, let us consider the estimation of the first unknown input, i.e., $d_1(t)$, out of the q unknown inputs, and the first-derivative of $y_1(t)$ is used in the unknown input observer. It is expected that the designed observer retains all the attractive features as per Chap. 2, while being *insensitive* to other unknown inputs, i.e., $d_r(t) \in \mathbb{R}^{(q-1)}$.

3.2 Main Results

Firstly, let us partition D as $D = [D_1 \;\; D_r]$, where $D_1 = \begin{bmatrix} \alpha_1 \\ D_{12} \end{bmatrix}$, $D_r = \begin{bmatrix} D_{r1} \\ D_{r2} \end{bmatrix}$, $\alpha_1 \neq 0$, $D_{12} \in \mathbb{R}^{(n-1)}$, $D_{r1} \in \mathbb{R}^{1\times(q-1)}$, $D_{r2} \in \mathbb{R}^{(n-1)\times(q-1)}$, and $d(t) = \begin{bmatrix} d_1(t) \\ d_r(t) \end{bmatrix}$. We investigate two possible approaches to design an unknown input observer to asymptotically estimate $d_1(t)$ in the presence of $d_r(t)$. The advantages and drawbacks of each approach will be explained in the subsequent sections. Together, they offer the designer a viable tool for designing an effective unknown input observer.

3.2.1 *Design Approach 1*

In this approach, we reuse matrices T and V_1 as derived in Sect. 2.2 in Chap. 2 to achieve the decoupling of $d_1(t)$ from the rest of the unknown inputs, i.e.,

$$d_1(t) = \frac{1}{\alpha_1}\dot{y}_1(t) - \frac{\bar{A}_{22}}{\alpha_1}y_1(t) - \bar{A}_{21}\bar{x}_1(t). \tag{3.1}$$

Now, to arrive at (3.1), let us select T as follows

$$T = \begin{bmatrix} E & D_1 \end{bmatrix} = \begin{bmatrix} 0_{1\times(n-1)} & \alpha_1 \\ I_{(n-1)} & D_{12} \end{bmatrix}.$$

Thus, T^{-1} is obtained as

$$T^{-1} = \begin{bmatrix} -\alpha_1^{-1}D_{12} & I_{(n-1)} \\ \alpha_1^{-1} & 0_{1\times(n-1)} \end{bmatrix}, \tag{3.2}$$

and hence $T^{-1}D$ is obtained as follows

$$T^{-1}D = T^{-1}[D_1 \ D_r] = \begin{bmatrix} 0_{(n-1)} & D_{r2} - \alpha_1^{-1} D_{12} D_{r1} \\ 1 & \alpha_1^{-1} D_{r1} \end{bmatrix}. \tag{3.3}$$

It is clear from (3.3) that to decouple $d_1(t)$ completely from the rest of the unknown inputs (see, (3.1)), it is necessary that $D_{r1} = 0_{1\times(q-1)}$. A simple interpretation of this condition is that the output $y_1(t)$ is not reachable from $d_r(t)$, i.e., $C_1 D_r = 0_{1\times(q-1)}$. If this condition is satisfied (which can be easily verified), then (3.1) is obtained, and we also obtain the following reduced-order system

$$\dot{\bar{x}}_1(t) = \bar{A}_{11}\bar{x}_1(t) + \frac{\bar{A}_{12}}{\alpha_1} y_1(t) + D_{r2} d_r(t), \tag{3.4}$$

$$\bar{y}(t) = \tilde{C}_1 \bar{x}_1(t). \tag{3.5}$$

Note that the system described by (3.4)–(3.5) is not a disturbance-free reduced-order system (cf. (2.8)–(2.9)) as it contains $d_r(t)$. Therefore, we need to design a *disturbance-decoupled* functional observer for system (3.4)–(3.5) to asymptotically estimate the function $z(t) = -\bar{A}_{21}\bar{x}_1(t)$ required in (3.1) to estimate $d_1(t)$ (see, [2, 3] for observer design of (3.4)–(3.5)). Accordingly, a functional observer of the form (2.10)–(2.11) is designed subject to the satisfaction of both Eqs. (2.12)–(2.13) and an extra equation as described below [2]

$$LD_{r2} = 0. \tag{3.6}$$

Note that condition (3.6) is to ensure that the designed functional observer remains *insensitive* to the unknown input vector $d_r(t)$. The reader can refer to [2] for a design algorithm to obtain observer matrices. The design detail of disturbance-decoupled functional observers for system (3.4)–(3.5) is therefore omitted here.

Remark 1 The main advantage of Design Approach 1 over Design Approach 2 (to be detailed later) is that the observer matching condition regarding $\text{rank}(CD) = q$ is not needed. Furthermore, the design method presented in [2] does not require the observer matching condition of reduced-order system (3.4)–(3.5) to be satisfied, i.e., $\text{rank}(\tilde{C}_1 D_{r2}) = (q-1)$. It was also demonstrated in [2] that it is possible to design functional observers for system (3.4)–(3.5) even for the case where there are more unknown inputs than the available outputs. This further highlights the attractiveness of our design approach based on linear functional observers instead of reduced-order state observers.

3.2.2 Algorithm 2: Estimation of One Unknown Input in the Presence of Multiple Unknown Inputs Using Design Approach 1

Following Sect. 3.2.1, if $D_{r1} = 0_{1\times(q-1)}$, we can now design an unknown input observer to estimate $d_1(t)$ in the presence of multiple unknown inputs, $d_r(t)$. The below design algorithm is for $i = 1$, i.e., we select and use the first-derivative of output $y_1(t)$.

Algorithm 2: Estimation of one unknown input in the presence of multiple unknown inputs

1. Perform Steps 1–5 of Design Algorithm 1.
2. Obtain reduced-order system (3.4)–(3.5).
3. Obtain matrices K, M, J_1, N and G by designing a disturbance-free functional observer, $\hat{z}(t)$, to estimate the function $z(t) = -\bar{A}_{21}\bar{x}_1(t)$ for system (3.4)–(3.5).
4. Estimation of the unknown input $d_1(t)$ is given as below

$$\hat{d}_1(t) = \frac{1}{\alpha_1}\dot{y}_1(t) - \frac{\bar{A}_{22}}{\alpha_1}y_1(t) + \hat{z}(t). \tag{3.7}$$

3.2.3 Numerical Examples 1 and 2

Numerical Example 1

To demonstrate the salient features of our design procedure based on Design Approach 1, we reconsider Example 1 (Chap. 2) but now with two unknown inputs, $d_1(t)$ and $d_2(t)$, where the additional input matrix D_r is given as follows

$$D_r = \begin{bmatrix} 0 & 0 & 0 & 1 & 1 & 0 \end{bmatrix}^T.$$

We wish to estimate the unknown input $d_1(t)$ in the presence of $d_2(t)$. Clearly, in this example, $\text{rank}(D) = 2$ (note that $D = [D_1 \;\; D_r]$, where $D_1 = \begin{bmatrix} 1 & 2 & 0 & 0 & 1 & 0 \end{bmatrix}^T$), but the observer matching condition is not satisfied since $\text{rank}(CD) = 1 \neq 2$. As $D_{r1} = 0$, we can therefore proceed with the steps given in Algorithm 2. Since Steps 1–5 of Algorithm 1 have already been performed in Example 1 (Chap. 2), we proceed to Step 2 of Algorithm 2 to obtain the reduced-order system (3.4)–(3.5), where $D_{r2} = \begin{bmatrix} 0 & 0 & 1 & 1 & 0 \end{bmatrix}^T$.

It should be noted that the observer matching condition for reduced-order system (3.4)–(3.5) is also not satisfied, since $\text{rank}(\tilde{C}_1 D_{r2}) = 0$. As a result, we cannot design a reduced-order asymptotic state observer to estimate the state vector, $\bar{x}_1(t)$ (see, for example, [4, 5]). Fortunately, based on the design method of [2] (or [3]), we can still

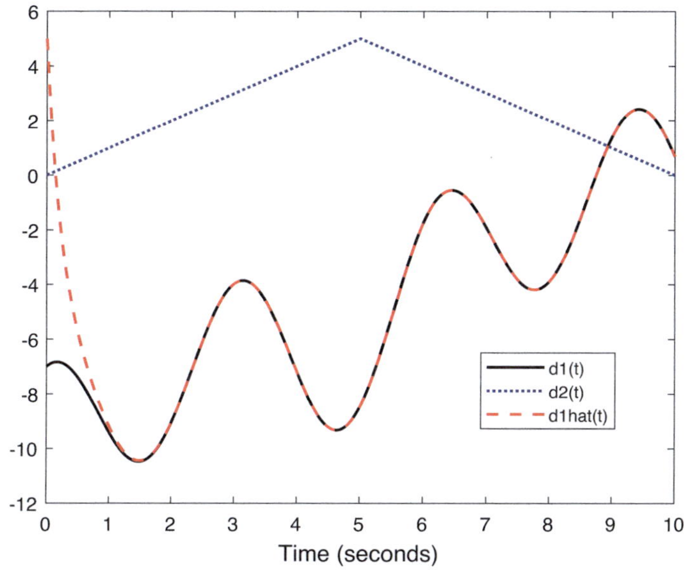

Fig. 3.1 Trajectories of $d_1(t)$, $d_2(t)$ and $\hat{d}_1(t)$ (Numerical Example 1, Chap. 3)

design an asymptotic linear functional observer to estimate the linear function $z(t)$. This fact highlights the advantage of our design approach over previous studies on unknown input observers.

Now, by Step 3 of Algorithm 2, we can use the design method in [2], and we obtained the observer matrices as given in Step 7 of Algorithm 1. Here, it is easy to check that with $L = \begin{bmatrix} -2 & 2 & 0 & 0 & 1 \end{bmatrix}$, condition (3.6) is satisfied, i.e., $LD_{r2} = 0$. Thus, our designed unknown input observer, $\hat{d}_1(t)$, is insensitive to the unknown input $d_2(t)$, but it always provides asymptotic estimation of $d_1(t)$.

By Step 4 of Algorithm 2, the estimation of $d_1(t)$ is as given in Step 8 of Algorithm 1, where $\hat{d}(t)$ is replaced by $\hat{d}_1(t)$.

Figure 3.1 shows the trajectories of the disturbances $d_1(t)$, $d_2(t)$ and the estimate $\hat{d}_1(t)$. It is clear from the figure that the unknown input observer tracks $d_1(t)$, while being insensitive to $d_2(t)$. Observe that asymptotic estimation is achieved even for abrupted disturbance signals.

Remark 2 As a matter of fact, the designed scalar functional observer in this example can handle up to four independent unknown inputs in $d_r(t)$, while still being able to robustly estimate $d_1(t)$. This is clear from (3.6), i.e., $LD_{r2} = 0$. Since L is a row matrix of dimension $1 \times (n-1)$. Therefore, $D_{r2} \in \mathbb{R}^{(n-1)\times(n-2)}$ exists and it is the null-space of L. Using this result, we can also characterize the unknown inputs that enable the designed unknown input observer to be insensitive to them. To illustrate this point, let us choose a matrix D_r as below

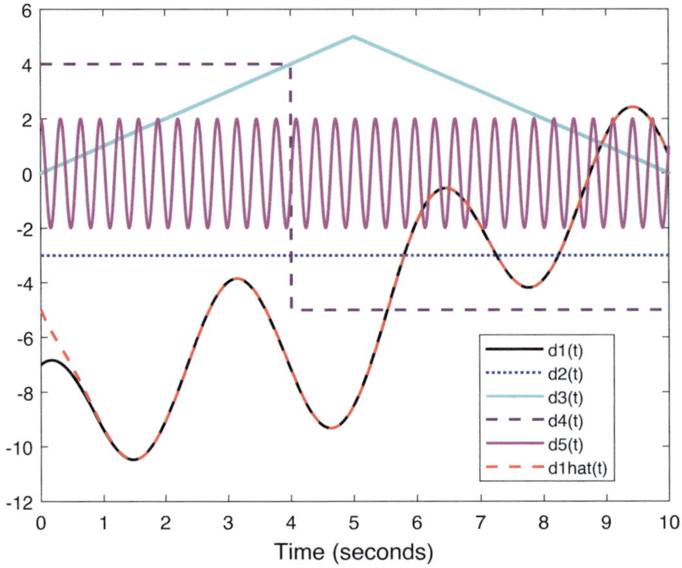

Fig. 3.2 Trajectories of $d(t)$ and $\hat{d}_1(t)$ (Numerical Example 1, Remark 2, Chap. 3)

$$D_r = \begin{bmatrix} 0 & 0 & 0 & 0 \\ 1 & 0 & 0 & 0.5 \\ 1 & 0 & 0 & 0 \\ 0 & 1 & 0 & 0 \\ 0 & 0 & 1 & 0 \\ 0 & 0 & 0 & 1 \end{bmatrix}.$$

Then, we can see that $D_{r1} = \begin{bmatrix} 0 & 0 & 0 & 0 \end{bmatrix}$, and

$$D_{r2} = \begin{bmatrix} 1 & 0 & 0 & 0.5 \\ 1 & 0 & 0 & 0 \\ 0 & 1 & 0 & 0 \\ 0 & 0 & 1 & 0 \\ 0 & 0 & 0 & 1 \end{bmatrix}.$$

It is easy to check that $LD_{r2} = 0_{1\times 4}$. Hence, the designed unknown input observer $\hat{d}_1(t)$ can still robustly estimate the unknown input $d_1(t)$ in the presence of four independent unknown inputs in the system.

Figure 3.2 shows the trajectories of the disturbances $d_i(t)$, $i = 1, 2, 3, 4, 5$ and the estimate $\hat{d}_1(t)$. It is clear from the figure that $\hat{d}_1(t)$ tracks $d_1(t)$, while being insensitive to the rest of the disturbances.

Remark 3 When $D_{r1} \neq 0_{1\times(q-1)}$, Eq. (3.1) contains some unknown inputs, i.e., $\alpha_1^{-1} D_{r1} d_r(t)$, in its right hand side. In this case, the output $y_1(t)$ is reachable from at least one of the remaining unknown inputs $d_r(t)$. This implies that $\dot{y}_1(t)$ (i.e., our desired choice) cannot be used to reconstruct the unknown input $d_1(t)$, even though $C_1 D_1 \neq 0$. To be still able to use Design Approach 1, we will need to examine matrix CD and select an output measurement, $y_i(t)$ (if it exists) such that $C_i D_1 = \alpha_i \neq 0$ and $C_i D_r = 0_{1\times(q-1)}$. Then, by re-ordering the state vector, $x(t)$ and using a simple coordinate change where the positions of the state variables $x_1(t)$ and $x_i(t)$ are swapped, matrix D can be reordered as $D = \begin{bmatrix} \alpha_i & D_{r1} \\ D_{12} & D_{r2} \end{bmatrix}$, where $D_{r1} = 0_{1\times(q-1)}$. Hence, the rest of the design procedure can be followed, where the designed unknown input observer (3.7) now uses $\dot{y}_i(t)$ instead of $\dot{y}_1(t)$. Even though our preferred first-derivative of $y_1(t)$ is not used, we can still retain all the attractive features of Design Approach 1, i.e., only first-derivative of one output measurement is used and the designed functional observer is of a low-order.

Numerical Example 2

To illustrate Remark 3, we consider the following numerical example, where matrices A, D and C are given below

$$A = \begin{bmatrix} 0 & -9 & -21 & 7 \\ 1 & -10 & -1 & -2 \\ 5 & -4 & -9 & 1 \\ 7 & -7 & -8 & -1 \end{bmatrix}, \ D = \begin{bmatrix} 1 & 1 & 2 \\ 1 & 0 & 0 \\ 0 & 1 & 0 \\ 0 & 0 & 1 \end{bmatrix} \text{ and } C = \begin{bmatrix} I_2 & 0_{2\times 2} \end{bmatrix}.$$

In this example, we wish to estimate the unknown input $d_1(t)$ in the presence of both unknown inputs $d_2(t)$ and $d_3(t)$. Note that there are more unknown inputs than the available output measurements. It is clear in this example that $\text{rank}(D) = 3$ but the observer matching condition, i.e., $\text{rank}(CD) = 2 \neq \text{rank}(D)$, hence it is not possible to design an unknown input observer to asymptotically estimate $d_1(t)$ using existing design methods available in the literature.

To proceed, we first consider matrix CD, where $CD = \begin{bmatrix} 1 & 1 & 2 \\ \mathbf{1} & \mathbf{0} & \mathbf{0} \end{bmatrix}$. The second row of CD shows that $y_2(t)$ is reachable from $d_1(t)$ but not reachable from the two unknown inputs $d_2(t)$ and $d_3(t)$. Thus, according to Remark 3, we should select $y_2(t)$ and use its first-derivatives to estimate $d_1(t)$. Accordingly, we rearrange the positions of the state variables, and this is achieved by using a simple coordinate state transformation matrix, $\begin{bmatrix} 0 & 1 & 0 & 0 \\ 1 & 0 & 0 & 0 \\ 0 & 0 & 1 & 0 \\ 0 & 0 & 0 & 1 \end{bmatrix}$. We have

$$\begin{bmatrix} \dot{x}_2(t) \\ \dot{x}_1(t) \\ \dot{x}_3(t) \\ \dot{x}_4(t) \end{bmatrix} = \begin{bmatrix} -10 & 1 & -1 & -2 \\ -9 & 0 & -21 & 7 \\ -4 & 5 & -9 & 1 \\ -7 & 7 & -8 & -1 \end{bmatrix} \begin{bmatrix} x_2(t) \\ x_1(t) \\ x_3(t) \\ x_4(t) \end{bmatrix} + \begin{bmatrix} \mathbf{1} & \mathbf{0} & \mathbf{0} \\ 1 & 1 & 2 \\ 0 & 1 & 0 \\ 0 & 0 & 1 \end{bmatrix} \begin{bmatrix} d_1(t) \\ d_2(t) \\ d_3(t) \end{bmatrix},$$

$$y(t) = \begin{bmatrix} y_2(t) \\ y_1(t) \end{bmatrix} = \begin{bmatrix} I_2 & 0_{2\times 2} \end{bmatrix} \begin{bmatrix} x_2(t) \\ x_1(t) \\ x_3(t) \\ x_4(t) \end{bmatrix}.$$

Thus, by selecting $y_2(t)$ and its first-derivatives to be used to reconstruct $d_1(t)$, we can now follow the design Algorithm 2. For completeness, we show the rest of the design procedure. We obtain the following matrices

$$E = \begin{bmatrix} 0 & 0 & 0 \\ 1 & 0 & 0 \\ 0 & 1 & 0 \\ 0 & 0 & 1 \end{bmatrix}, T = \begin{bmatrix} E & D_1 \end{bmatrix}, D_1 = \begin{bmatrix} 1 \\ 1 \\ 0 \\ 0 \end{bmatrix}, V = \begin{bmatrix} 1 & 0 \\ 1 & 1 \end{bmatrix}, V_1 = \begin{bmatrix} 1 & 0 \end{bmatrix}, V_2 = \begin{bmatrix} -1 & 1 \end{bmatrix},$$

$$\bar{A}_{11} = \begin{bmatrix} -1 & -20 & 9 \\ 5 & -9 & 1 \\ 7 & -8 & -1 \end{bmatrix}, \bar{A}_{12} = \begin{bmatrix} 0 \\ 1 \\ 0 \end{bmatrix}, \bar{A}_{21} = \begin{bmatrix} 1 & -1 & -2 \end{bmatrix}, \text{ and } \bar{A}_{22} = -9.$$

From Step 2 of Algorithm 2, we obtain a reduced-order system (3.4)–(3.5) where

$$D_{r2} = \begin{bmatrix} 1 & 2 \\ 1 & 0 \\ 0 & 1 \end{bmatrix}, \bar{y}(t) = V_2 y(t) = y_1(t) - y_2(t), \tilde{C}_1 = \begin{bmatrix} 1 & 0 & 0 \end{bmatrix}.$$

From Step 3 of Algorithm 2, we can now design a first-order functional observer to estimate the function $z(t) = -\bar{A}_{21}\bar{x}_1(t)$ where $K = 1$, $N = -5$, $L = \begin{bmatrix} -1 & 1 & 2 \end{bmatrix}$, $G = 15$, $J_2 = 1$, $M = 0$. It is easy to check that with the above derived observer matrices, conditions (2.12)–(2.13) and (3.6) are satisfied.

Finally, from Step 4 of Algorithm 2 and Remark 3, the following unknown input observer, $\hat{d}_1(t)$, is obtained

$$\begin{aligned} \hat{d}_1(t) &= \dot{y}_2(t) + 9y_2(t) + \hat{z}(t), \\ \hat{z}(t) &= w(t), \\ \dot{w}(t) &= -5w(t) + 15y_1(t) - 14y_2(t). \end{aligned}$$

Finally, Fig. 3.3 shows the trajectories of the disturbances $d_i(t)$, $i = 1, 2, 3$ and the estimate $\hat{d}_1(t)$.

3.2.4 Design Approach 2

In this approach, the composition of T and V may need to include some or all columns of D_r (cf. just D_1 as in Design Approach 1). This is necessary since it covers various scenarios where we need to use differentiations (i.e., first-derivatives) of more than

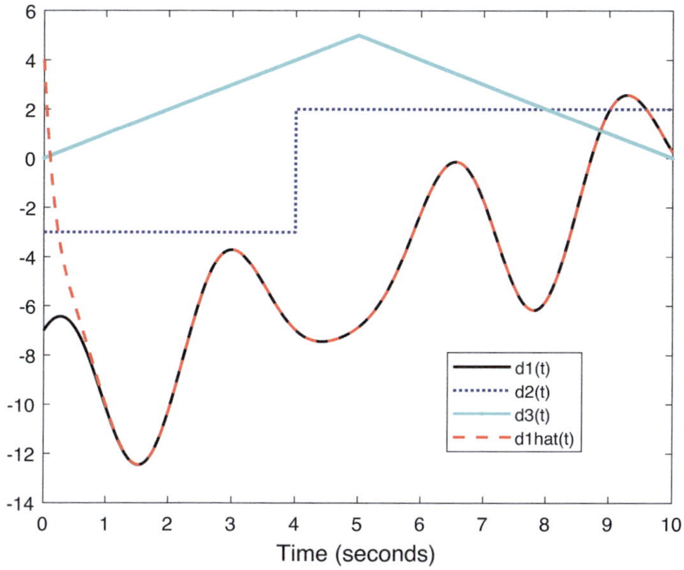

Fig. 3.3 Trajectories of $d(t)$ and $\hat{d}_1(t)$ (Numerical Example 2, Chap. 3)

one output measurement to estimate the unknown input $d_1(t)$. From a practical point of view, the worst-case scenario is where differentiations of q output measurements are needed. Before we discuss a method for the systematic design of an unknown input observer, $\hat{d}_1(t)$, that uses the least number of differentiators, let us first consider the following scenario as discussed in Remark 4 below.

Remark 4 We discuss the case where T and V contain all columns of D_r. In this instance, it is clear that in order to decompose system (1.1)–(1.2) into the form (1.12)–(1.14), the observer matching condition is needed, i.e., $\text{rank}(CD) = q$ and $p > q$. Subject to this condition, there always exist at least q output measurements, denoted by $y_q(t) = C_q x(t)$ where $C_q \in \mathbb{R}^{q \times n}$, such that $\text{rank}(C_q D) = q$. Accordingly, we rearrange the output vector, $y(t)$, as follows

$$y(t) = \begin{bmatrix} y_q(t) \\ y_r(t) \end{bmatrix} = \begin{bmatrix} C_q \\ C_r \end{bmatrix} x(t) \triangleq Cx(t), \tag{3.8}$$

where $y_r(t) \in \mathbb{R}^{p-q}$ and $C_r \in \mathbb{R}^{(p-q)\times n}$. Now, by letting $P = \begin{bmatrix} 0_{q\times(p-q)} \\ I_{p-q} \end{bmatrix}$, hence $V = \begin{bmatrix} C_q D & 0_{q\times(p-q)} \\ C_r D & I_{p-q} \end{bmatrix}$ and V^{-1} is obtained as follows

$$V^{-1} = \begin{bmatrix} V_1 \\ V_2 \end{bmatrix} = \begin{bmatrix} (C_q D)^{-1} & 0_{q\times(p-q)} \\ -C_r D (C_q D)^{-1} & I_{p-q} \end{bmatrix}. \tag{3.9}$$

From (3.9), $V_1 = \left[(C_q D)^{-1} \; 0_{q\times(p-q)} \right] \in \mathbb{R}^{q\times p}$ and $V_2 = \left[-C_r D(C_q D)^{-1} \; I_{p-q} \right] \in \mathbb{R}^{(p-q)\times p}$. Finally, by selecting $E \in \mathbb{R}^{n\times(n-q)}$ to be a rational basis of the null-space of $V_1 C$, we have $V_1 C E = 0_{q\times(n-q)}$ and $T = \left[E \; D \right]$. As a result, we obtain the following equations

$$\dot{\bar{x}}_1(t) = \bar{A}_{11}\bar{x}_1(t) + \bar{A}_{12} V_1 y(t), \tag{3.10}$$

$$\bar{y}(t) = V_2 C E \bar{x}_1(t), \tag{3.11}$$

$$d(t) = V_1 \dot{y}(t) - \bar{A}_{22} V_1 y(t) - \bar{A}_{21}\bar{x}_1(t), \tag{3.12}$$

where $\bar{x}_1(t) \in \mathbb{R}^{n-q}$ and $\bar{y}(t) = V_2 y(t) \in \mathbb{R}^{p-q}$. Denoting $(X)^1$ as the first row of matrix X, from (3.12), $d_1(t)$ is obtained as

$$d_1(t) = (V_1)^1 \dot{y}(t) - (\bar{A}_{22} V_1)^1 y(t) - (\bar{A}_{21})^1 \bar{x}_1(t).$$

Therefore, to design an unknown input observer to estimate $d_1(t)$, we can design a minimal-order scalar functional observer to estimate the function $z(t) = -(\bar{A}_{21})^1 \bar{x}_1(t)$ of the disturbance-free system (3.10)–(3.11). Hence, $\hat{d}_1(t)$ is given below

$$\hat{d}_1(t) = (V_1)^1 \dot{y}(t) - (\bar{A}_{22} V_1)^1 y(t) + \hat{z}(t). \tag{3.13}$$

While the above approach results in a minimal-order functional observer (cf. a reduced-order state observer), it is not the simplest observer since $(V_1)^1$ may still contain up to q non-zero elements, i.e.,

$$(V_1)^1 = \left[v_1 \; v_2 \; \cdots \; v_q \; 0_{1\times(p-q)} \right], \; v_i \neq 0.$$

Therefore, the worst-case scenario is where first-derivatives of q output measurements are needed to estimate $d_1(t)$. On the other hand, by choosing P as $P = \begin{bmatrix} 0_{q\times(p-q)} \\ I_{p-q} \end{bmatrix}$, we have reduced the number of differentiators from p down to q. This is an improvement over the method proposed in [5], where first-derivatives of all the output measurements might be needed. Also, if needed, we can estimate all the unknown inputs, i.e., $d(t)$. This can be realized by designing a functional observer to estimate the vector $z(t) \in \mathbb{R}^q$, where $z(t) = -A_{21}\bar{x}_1(t)$. Lastly, we note that to realize the disturbance-free system (3.10)–(3.11), we need the observer matching condition to be satisfied. This concludes Remark 4.

In Remark 4, we showed that by including all of the columns of D_r in T and V, we can obtain a reduced-order disturbance-free system (3.10)–(3.11). However, the estimation of $d_1(t)$ may require up to q differentiators. From the technical and economical perspective, we prefer to use the least number of differentiators for estimating $d_1(t)$. Thus, when it is not possible to use only one differentiator as in Design Approach 1, we then seek the least number of differentiators, starting with two differentiators. This is done by a process, where some coordinate changes are performed so that the rows and columns of matrix D are rearranged, resulting in the least number of differentiators.

To facilitate a systematic development of the design of an unknown input observer for $d_1(t)$ that uses the least number of differentiators, let us first discuss the case where we seek two differentiators. Denoting $\mathrm{Col}(X)_i \leftrightarrow \mathrm{Col}(X)_j$ $(\mathrm{Row}(X)_i \leftrightarrow \mathrm{Row}(X)_j)$ as the operations, where the two columns (two rows) of matrix X in positions i and j swap their positions. Here, we introduce two transformations. The first transformation, $d^N(t) = Qd(t)$, $Q \in \mathbb{R}^{q\times q}$, is used to rearrange the last $(q-1)$ columns of matrix D (note that the first column of D remains unchanged as we wish to estimate $d_1(t)$). The second transformation, $x^N(t) = Mx(t)$, $M \in \mathbb{R}^{n\times n}$, is used to switch the positions of the state variables and hence the rows of matrix D. As a result of these transformations, and whenever it is possible to have only two differentiators, the new matrix $\bar{D} = MDQ^{-1}$ takes the following form

$$\bar{D} = MDQ^{-1} = \begin{bmatrix} \bar{D}_{11} & \bar{D}_{r1} \\ \bar{D}_{12} & \bar{D}_{r2} \end{bmatrix}, \tag{3.14}$$

where $\bar{D}_{11} \in \mathbb{R}^{2\times 2}$ is an invertible matrix, $\bar{D}_{r1} = 0_{2\times(q-2)}$, $\bar{D}_{12} \in \mathbb{R}^{(n-2)\times 2}$ and $\bar{D}_{r2} \in \mathbb{R}^{(n-2)\times(q-2)}$.

Thanks to the above transformations, it is now possible to use only two differentiators to estimate the unknown input $d_1(t)$. Note that in this case, the composition of T and V contains the first two columns of $\bar{D}$, i.e., $\bar{D}_1 \triangleq \begin{bmatrix} \bar{D}_{11} \\ \bar{D}_{12} \end{bmatrix}$. Then based on similar lines as in Design Approach 1, we can decompose the system into the form (3.4)–(3.5), in which a reduced-order disturbance-free functional observer can be designed, and the estimation of $d_1(t)$ can be realized by using two differentiators. The detail is omitted here.

For the case where partition (3.14) cannot be realized, it implies that it is not possible to use two differentiators to estimate $d_1(t)$. In such cases, we sequentially increase the number of differentiators and repeat the row and column operations until the following partition is obtained

$$\bar{D} = MDQ^{-1} = \begin{bmatrix} \bar{D}_{11} & \bar{D}_{r1} \\ \bar{D}_{12} & \bar{D}_{r2} \end{bmatrix}, \tag{3.15}$$

where $\bar{D}_{11} \in \mathbb{R}^{h\times h}$ is an invertible matrix, $\bar{D}_{r1} = 0_{h\times(q-h)}$, $\bar{D}_{12} \in \mathbb{R}^{(n-h)\times h}$ and $\bar{D}_{r2} \in \mathbb{R}^{(n-h)\times(q-h)}$. Here, h is the smallest integer in the range $2 \le h \le q$. By this approach, we can ensure that the least number of differentiators, h, are used for the estimation of $d_1(t)$.

3.2.5 Numerical Example 3

To illustrate Remark 4 and the above development, let us just pick two matrices C and D as

$$C = \begin{bmatrix} I_5 & 0_{5\times 2} \end{bmatrix} \text{ and } D = \begin{bmatrix} 1 & 2 & 3 & 4 \\ -1 & 0 & -2 & 3 \\ 1 & 0 & 0 & 1 \\ 0 & 2 & 1 & 1 \\ 0 & 0 & 0 & 1 \\ 0 & 1 & 0 & 1 \\ 0 & -1 & 1 & 0 \end{bmatrix}.$$

Since $\text{rank}(CD) = \text{rank}(D) = 4$, we can choose V as $V = \begin{bmatrix} CD & P \end{bmatrix}$, where $P = \begin{bmatrix} 0_{4\times 1} \\ 1 \end{bmatrix}$, and the decomposition is complete with $E = \begin{bmatrix} 0_{4\times 3} \\ I_3 \end{bmatrix}$ and $T = \begin{bmatrix} E & D \end{bmatrix}$. However, as discussed in Remark 4, this approach results in four non-zero elements in $(V_1)^1$, where

$$(V_1)^1 = \begin{bmatrix} -\frac{1}{6} & -\frac{1}{6} & 1 & \frac{1}{6} & 0 \end{bmatrix}.$$

Thus, to be able to estimate the unknown input $d_1(t)$, this approach requires first-derivatives of the four outputs $y_1(t)$, $y_2(t)$, $y_3(t)$ and $y_4(t)$.

Let us now seek to reduce the number of required differentiators by using some co-ordinate transformations as discussed in Design Approach 2. First, let us consider the operation $\text{Col}(D)_2 \leftrightarrow \text{Col}(D)_4$, i.e., the second and 4th columns of D are interchanged. This results in a transformation matrix Q, where $Q(1, 1) = 1$, $Q(2, 4) = 1$, $Q(3, 3) = 1$ and $Q(4, 2) = 1$, and the rest of its elements are zero. With this operation, we obtained

$$DQ^{-1} = \begin{bmatrix} 1 & 4 & 3 & 2 \\ -1 & 3 & -2 & 0 \\ 1 & 1 & \mathbf{0} & \mathbf{0} \\ 0 & 1 & 1 & 2 \\ 0 & 1 & \mathbf{0} & \mathbf{0} \\ 0 & 1 & 0 & 1 \\ 0 & 0 & 1 & -1 \end{bmatrix}.$$

Next, we consider the rows of DQ^{-1} and seek to perform some row operations so that the new matrix $\bar{D}$ can be expressed in the form (3.14). As highlighted in bold, it is indeed possible to obtain an invertible matrix $\bar{D}_{11}$ and $\bar{D}_{r1} = 0_{2\times 2}$. There are a number of possible ways to achieve this. For instance, the following row operations:

$\text{Row}(DQ^{-1})_3 \leftrightarrow \text{Row}(DQ^{-1})_1$,
$\text{Row}(DQ^{-1})_5 \leftrightarrow \text{Row}(DQ^{-1})_2$,
$\text{Row}(DQ^{-1})_2 \leftrightarrow \text{Row}(DQ^{-1})_3$,
$\text{Row}(DQ^{-1})_1 \leftrightarrow \text{Row}(DQ^{-1})_5$,

or equivalently, the following transformation matrix M, where $M(1,3)=1$, $M(2,5)=1$, $M(3,2)=1$, $M(4,4)=1$, $M(5,1)=1$, $M(6,6)=1$) and $M(7,7)=1$. As a result, we can obtain

$$\bar{D}=MDQ^{-1}=\begin{bmatrix} \mathbf{1} & \mathbf{1} & \mathbf{0} & \mathbf{0} \\ \mathbf{0} & \mathbf{1} & \mathbf{0} & \mathbf{0} \\ -1 & 3 & -2 & 0 \\ 0 & 1 & 1 & 2 \\ 1 & 4 & 3 & 2 \\ 0 & 1 & 0 & 1 \\ 0 & 0 & 1 & -1 \end{bmatrix},$$

which is in the form (3.14), where $\bar{D}_{11}=\begin{bmatrix} 1 & 1 \\ 0 & 1 \end{bmatrix}$ is an invertible matrix and $\bar{D}_{r1}=0_{2\times 2}$.

Also note that with the above row operations, the new output vector, $y^N(t)$, can now be expressed as

$$y^N(t)=\begin{bmatrix} y_3(t) \\ y_5(t) \\ y_2(t) \\ y_4(t) \\ y_1(t) \end{bmatrix}=Cx^N(t),$$

which means that only first-derivatives of $y_3(t)$ and $y_5(t)$ are used to estimate $d_1(t)$. Now, as in Design Approach 1, we only need to use the first two columns of $\bar{D}$, i.e., $\bar{D}_1$. By letting V as $V=\begin{bmatrix} C\bar{D}_1 & P \end{bmatrix}$, where $P=\begin{bmatrix} 0_{2\times 3} \\ I_3 \end{bmatrix}$, and hence the decomposition is complete with $E=\begin{bmatrix} 0_{2\times 5} \\ I_5 \end{bmatrix}$ and $T=\begin{bmatrix} E & \bar{D}_1 \end{bmatrix}$. Note that with the decomposition matrices T and V, we obtained a fifth-order system of the form (3.4)–(3.5) with three outputs, $\bar{y}=V_2y^N(t)$, and two unknown inputs $d_2(t)$ and $d_3(t)$, in which we can proceed to design a disturbance-free scalar functional observer. Finally, note that $(V_1)^1$ now only contains two non-zero elements, and

$$(V_1)^1=\begin{bmatrix} 1 & -1 & 0 & 0 & 0 \end{bmatrix}.$$

Thus, instead of using four differentiators as in the case of Remark 4, we now use two differentiators, which take the first-derivatives of outputs $y_3(t)$ and $y_5(t)$. From the practical perspective, this is a useful result. Finally, as in Design Approach 1, we note that the design of the disturbance-free functional observer does not necessarily require that the observer matching condition related to the two unknown inputs $d_2(t)$ and $d_3(t)$ be satisfied.

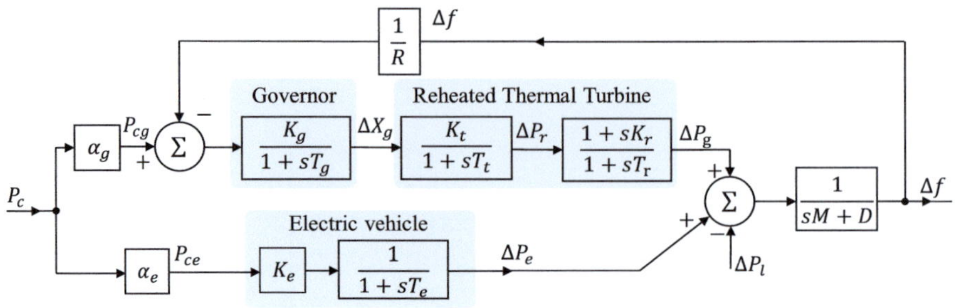

Fig. 3.4 Transfer function model of a thermal-based power system with EVs [6]

3.2.6 A Practical Example

We consider a practical example to illustrate the design of an unknown observer to estimate the load demand in a power system with a reheated thermal turbine and a fleet of electric vehicles (EVs). In [6], a novel load frequency control (LFC) scheme using EVs to assist thermal turbine units to stabilize frequency and interchange power of a multi-area power system was presented. Figure 3.4 is taken from [6], which shows a general transfer function model of a thermal power plant and a first-order dynamics model for the discharging of EVs.

A state-space model can be derived, where

$$\dot{x}(t) = Ax(t) + Bu(t) + Dd(t), \tag{3.16}$$

$u(t) = P_c(t)$ is the control input and $d(t) = \Delta P_l(t)$ is the load disturbance. As in the LFC of power systems, to maintain zero steady-state value of the frequency deviations, Δf, when the system is subjected to any step load disturbance, ΔP_l, it is necessary to include an integral controller. Thus, the state variables for the system are defined as follows

$$x(t) = [\Delta f(t)\ \Delta P_g(t) \int \Delta f(t)dt\ \Delta P_e(t)\ \Delta P_r(t)\ \Delta X_g(t)]^T.$$

Matrices $A \in \mathbb{R}^{6\times 6}$, $B \in \mathbb{R}^{6\times 1}$ and $D \in \mathbb{R}^{6\times 1}$ are as given below

$$A = \begin{bmatrix} -\frac{D}{M} & \frac{1}{M} & 0 & \frac{1}{M} & 0 & 0 \\ 0 & -\frac{1}{T_r} & 0 & 0 & \frac{T_t - K_r}{T_r T_t} & \frac{K_r K_t}{T_r T_t} \\ 1 & 0 & 0 & 0 & 0 & 0 \\ 0 & 0 & 0 & -\frac{1}{T_e} & 0 & 0 \\ 0 & 0 & 0 & 0 & -\frac{1}{T_t} & \frac{K_t}{T_t} \\ -\frac{K_g}{RT_g} & 0 & 0 & 0 & 0 & -\frac{1}{T_g} \end{bmatrix},$$

$$B = \begin{bmatrix} 0\ 0\ 0\ \frac{\alpha_e K_e}{T_e}\ 0\ \frac{\alpha_g K_g}{T_g} \end{bmatrix}^T, \text{ and } D = \begin{bmatrix} -\frac{1}{M}\ 0\ 0\ 0\ 0\ 0 \end{bmatrix}^T.$$

The data for the model is taken from [6], where $D = 0.0083$, $M = 0.1667$, $\alpha_g = 0.9$, $\alpha_e = 0.1$, $K_g = 1$, $T_g = 0.08$, $K_t = 1$, $T_t = 0.3$, $K_r = 0.5$, $T_r = 10$, $R = 2.4$, $K_e = 1$ and $T_e = 1$. From the above state-space model, we have

$$d(t) = -M\Delta\dot{f}(t) - D\Delta f(t) + \Delta P_g(t) + \Delta P_e(t). \tag{3.17}$$

Thus, for the trivial case where the state variables, $\Delta f(t)$, $\Delta P_g(t)$ and $\Delta P_e(t)$ are assumed to be available, the unknown input $d(t)$ can be directly obtained from them (see, (3.17)) where the first-derivative of $\Delta f(t)$ is required. To avoid this trivial case, let us now consider the following outputs for feedback control

$$y(t) = \begin{bmatrix} \Delta f(t) \\ \Delta P_g(t) \\ \int \Delta f(t)dt \end{bmatrix} = Cx(t),$$

where $C = \begin{bmatrix} 1\,0\,0\,0\,0\,0 \\ 0\,1\,0\,0\,0\,0 \\ 0\,0\,1\,0\,0\,0 \end{bmatrix}$. In fact, for the above outputs, we only use two output measurements $\Delta f(t)$ and $\Delta P_g(t)$ since $\int \Delta f(t)dt$ is obtained by feeding $\Delta f(t)$ into an input of an integrator. On the other hand, an integral controller is required to ensure zero steady-state value of $\Delta f(t)$ in the closed-loop control system for any step load change in $\Delta P_l(t)$.

Based on the above available output measurements, we consider a static output feedback controller, $u(t) = -F_o y(t)$, to stabilize the system, where

$$F_o = \begin{bmatrix} 5\ 25\ 5 \end{bmatrix}.$$

Thus, the following stable closed-loop system is obtained

$$\dot{x}(t) = A_{cl}x(t) + Dd(t), \tag{3.18}$$
$$y(t) = Cx(t), \tag{3.19}$$

where $A_{cl} = A - BF_oC$. Figure 3.5 shows the closed-loop responses of $\Delta f(t)$, $\Delta P_g(t)$ and $\Delta P_e(t)$, where a 0.1 p.u. step load change occurred in $\Delta P_l(t)$. It is clear from the figure that zero steady-state value of $\Delta f(t)$ is achieved and that the thermal power generator and EVs contribute their power according to the participation factors of 0.9 and 0.1, respectively.

Next, we design an unknown input observer to estimate the unknown input, $d(t) = \Delta P_l(t)$, for system (3.18)–(3.19). For this, Algorithm 1 presented in Sect. 2.2 can be used to design a reduced-order unknown input observer for $d(t)$. Alternatively, from (3.17), the unknown input $d(t)$ can be estimated as follows

$$\hat{d}(t) = -M\Delta\dot{f}(t) - D\Delta f(t) + \Delta P_g(t) + \Delta\hat{P}_e(t). \tag{3.20}$$

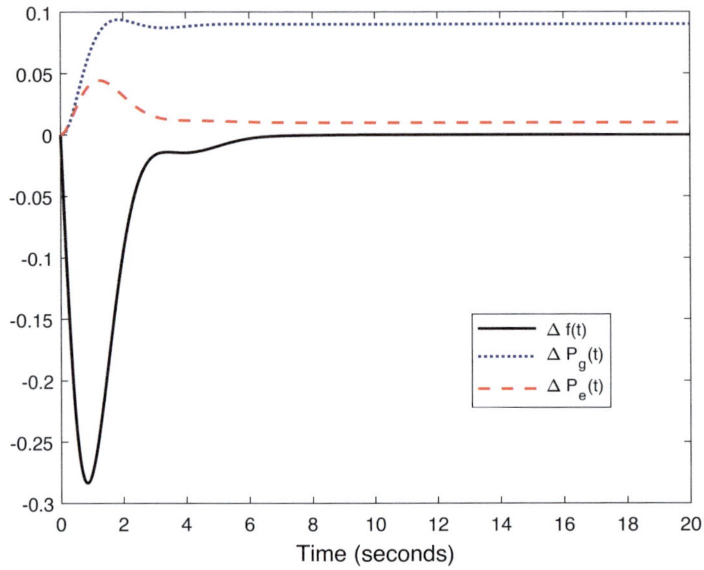

Fig. 3.5 $\Delta f(t)$, $\Delta P_g(t)$ and $\Delta P_e(t)$ responses to a 0.1 p.u. step load change in $\Delta P_l(t)$

Thus, the task here is to design a reduced-order observer to estimate just the state variable $\Delta P_e(t)$. Denoting $z(t) = \Delta P_e(t) = Fx(t)$, where $F = [0\,0\,0\,1\,0\,0]$, a scalar linear function to be estimated from system (3.18)–(3.19). We can design a reduced-order disturbance-free functional observer $\hat{z}(t)$, where

$$\hat{z}(t) = Kw(t) + My(t) \tag{3.21}$$

$$\dot{w}(t) = Nw(t) + Gy(t). \tag{3.22}$$

As such, from [7], a functional observer (3.21)–(3.22) can be designed by solving the following equations

$$NL + GC - LA_{cl} = 0 \tag{3.23}$$

$$LD = 0 \tag{3.24}$$

$$F - KL - MC = 0 \tag{3.25}$$

for the unknown constant matrices N (where N is Hurwitz), and L, G, K and M of appropriate dimensions. Based on a design method as presented in Sect. 5.4 in [7], we can design a stable first-order functional observer, where

$$N = -1,\ L = [0\,0\,0\,1\,0\,0],\ K = 1,\ M = [0\,0\,0],\ G = [-0.5\ \ -2.5\ \ -0.5].$$

With these parameters, it is easy to check that (3.23)–(3.25) are satisfied. Thus, the following unknown input observer is obtained

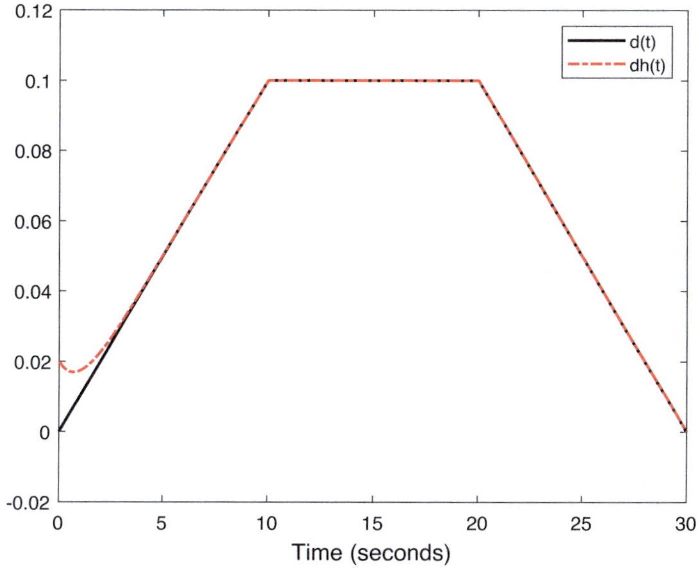

Fig. 3.6 Trajectories of $d(t)$ and $\hat{d}(t)$ (A practical example, Chap. 3)

$$\hat{d}(t) = -M\Delta\dot{f}(t) - D\Delta f(t) + \Delta P_g(t) + w(t), \tag{3.26}$$

$$\dot{w}(t) = Nw(t) + Gy(t). \tag{3.27}$$

Figure 3.6 shows the responses of $d(t)$ and $\hat{d}(t)$ with load disturbances in the form of ramp/step inputs. Observe that asymptotic tracking of the unknown input is achieved. It should be noted here that while we can design a stable first-order functional observer to estimate $z(t)$, the pole of the observer is fixed at $N = -1$. Therefore, the convergence time is a bit slow at around 4 s.

3.2.7 Design of Distributed/decentralized Unknown Input Observers

As a direct application of the result presented in this chapter, we consider an interconnected dynamical system comprising two subsystems, namely, S_1 and S_2, as described below

$$S_1 : \dot{x}_1(t) = A_{11}x_1(t) + A_{12}x_2(t) + D_1d_1(t), \tag{3.28}$$

$$y_1(t) = C_1x_1(t), \tag{3.29}$$

$$S_2 : \dot{x}_2(t) = A_{22}x_2(t) + A_{21}x_1(t) + D_2 d_2(t), \tag{3.30}$$

$$y_2(t) = C_2 x_2(t), \tag{3.31}$$

where for $i = 1, 2$, $x_i(t) \in \mathbb{R}^{n_i}$, $d_i(t) \in \mathbb{R}^{q_i}$ and $y_i(t) \in \mathbb{R}^{p_i}$ are, respectively, the state, unknown input and the output vectors of the local ith subsystem. Without loss of generality, it is assumed that $\text{rank}(C_i) = p_i$ and $\text{rank}(D_i) = q_i$. The terms $A_{12}x_2(t)$ and $A_{21}x_1(t)$ represent the couplings between the two subsystems S_1 and S_2.

Here, our objective is to design two unknown input observers, one at each subsystem. Therefore, at subsystem S_1, the designed unknown input observer is required to estimate $d_1(t)$, and it should be insensitive to $d_2(t)$ from subsystem S_2. Vice versa for subsystem S_2. The two designed unknown observers are dynamically decoupled from each other.

In the literature, there are many results available for the design of distributed and decentralized observers for interconnected dynamical systems (see, for example, [6, 8–11]). However, the design of distributed/decentralized unknown input observers to asymptotically estimate local unknown input vectors has not been widely available.

3.2.8 *Numerical Example 4: Distributed Unknown Input Observers*

In the following, we show how to design two unknown input observers for S_1 and S_2. For this example, we let A_{11}, A_{12}, A_{21}, A_{22}, D_1, D_2, C_1 and C_2 in (3.28)–(3.31) be as follows

$$A_{11} = \begin{bmatrix} -7 & -1 & -2 \\ 6 & -12 & -2 \\ -2 & 1 & -1 \end{bmatrix}, A_{12} = \begin{bmatrix} 0 & 0 & -1 \\ 0 & 1 & -2 \\ 0 & 0 & 1 \end{bmatrix}, A_{21} = \begin{bmatrix} -5 & 1 & 0 \\ 8 & -4 & -1 \\ 0 & -2 & -4 \end{bmatrix},$$

$$A_{22} = \begin{bmatrix} -5 & 3 & 4 \\ 1 & -8 & 2 \\ 0 & 2 & -6 \end{bmatrix}, C_1 = C_2 = \begin{bmatrix} I_2 & 0_{2\times 1} \end{bmatrix}, D_1 = D_2 = \begin{bmatrix} 1 \\ 2 \\ 3 \end{bmatrix}.$$

Each of the above subsystems has three state variables, two output measurements and one unknown input.

Let us now design a distributed unknown input observer to estimate $d_1(t)$ at S_1 (the same can be done for S_2). For this, we discuss the case where the output vector $y_2(t) \in \mathbb{R}^2$ from S_2 is sent over to S_1 for the estimation of $d_1(t)$ (note that other scenarios, such as no output from S_2 being used, can be considered based on the discussions below).

Now, as $y_2(t) \in \mathbb{R}^2$ from S_2 is used, we can augment S_1 and S_2 to form the following global system

$$\begin{bmatrix} \dot{x}_1(t) \\ \dot{x}_2(t) \end{bmatrix} = \begin{bmatrix} A_{11} & A_{12} \\ A_{21} & A_{22} \end{bmatrix} \begin{bmatrix} x_1(t) \\ x_2(t) \end{bmatrix} + \begin{bmatrix} D_1 & 0_{3\times1} \\ 0_{3\times1} & D_2 \end{bmatrix} \begin{bmatrix} d_1(t) \\ d_2(t) \end{bmatrix}, \tag{3.32}$$

$$y(t) = \begin{bmatrix} y_1(t) \\ y_2(t) \end{bmatrix} = \begin{bmatrix} C_1 & 0_{2\times3} \\ 0_{2\times3} & C_2 \end{bmatrix} \begin{bmatrix} x_1(t) \\ x_2(t) \end{bmatrix}. \tag{3.33}$$

Our result as presented in Design Approach 1 (see, Sect. 3.2.1 of this chapter) can now be readily applied to design an unknown input observer to estimate $d_1(t)$. The designed observer is insensitive to $d_2(t)$.

By letting $A = \begin{bmatrix} A_{11} & A_{12} \\ A_{21} & A_{22} \end{bmatrix}$, $D = [\bar{D}_1\ \bar{D}_2] = \begin{bmatrix} D_1 & 0_{3\times1} \\ 0_{3\times1} & D_2 \end{bmatrix}$ and $C = \begin{bmatrix} C_1 & 0_{2\times3} \\ 0_{2\times3} & C_2 \end{bmatrix}$. We then follow the steps as presented in Algorithm 2.

Firstly, we have $CD = \begin{bmatrix} 1 & 0 \\ 2 & 0 \\ 0 & 1 \\ 0 & 2 \end{bmatrix}$. Hence, let us pick the first output of subsystem S_1, and use its first-derivative for the estimation of $d_1(t)$. Without any confusion in the notation, we let the first output of S_1 be denoted as $y_{11}(t)$, i.e., $y_1(t) = \begin{bmatrix} y_{11}(t) \\ y_{12}(t) \end{bmatrix}$ (likewise $y_2(t) = \begin{bmatrix} y_{21}(t) \\ y_{22}(t) \end{bmatrix}$). Next, from Steps 1-2 of Algorithm 2, we obtain

$$P = \begin{bmatrix} 0_{1\times3} \\ I_3 \end{bmatrix}, \qquad V = \begin{bmatrix} C\bar{D}_1 & P \end{bmatrix} = \begin{bmatrix} 1&0&0&0 \\ 2&1&0&0 \\ 0&0&1&0 \\ 0&0&0&1 \end{bmatrix}, \qquad V_1 = \begin{bmatrix} 1&0&0&0 \end{bmatrix},$$

$$V_2 = \begin{bmatrix} -2&1&0&0 \\ 0&0&1&0 \\ 0&0&0&1 \end{bmatrix}, E = \begin{bmatrix} 0_{1\times5} \\ I_5 \end{bmatrix}, T = [E\ \bar{D}_1] = \begin{bmatrix} 0&0&0&0&0&1 \\ 1&0&0&0&0&2 \\ 0&1&0&0&0&3 \\ 0&0&1&0&0&0 \\ 0&0&0&1&0&0 \\ 0&0&0&0&1&0 \end{bmatrix},$$ and the following reduced-order system as defined by (3.4)–(3.5), where

$$\dot{\bar{x}}_1(t) = \begin{bmatrix} -10 & 2 & 0 & 1 & 0 \\ 4 & 5 & 0 & 0 & 4 \\ 1 & 0 & -5 & 3 & 4 \\ -4 & -1 & 1 & -8 & 2 \\ -2 & -4 & 0 & 2 & -6 \end{bmatrix} \bar{x}_1(t) + \begin{bmatrix} 6 \\ 42 \\ -3 \\ -3 \\ -16 \end{bmatrix} y_{11}(t) + \begin{bmatrix} 0 \\ 0 \\ 1 \\ 2 \\ 3 \end{bmatrix} d_2(t), \tag{3.34}$$

$$\bar{y}(t) = V_2 C E \bar{x}_1(t) = \begin{bmatrix} 1&0&0&0&0 \\ 0&0&1&0&0 \\ 0&0&0&1&0 \end{bmatrix} \bar{x}_1(t), \tag{3.35}$$

where $\bar{y}(t) = V_2 y(t)$. Also, from (3.1), the unknown input $d_1(t)$ is obtained as follows

$$d_1(t) = \dot{y}_{11}(t) - \bar{A}_{22} y_{11}(t) - \bar{A}_{21} \bar{x}_1(t), \tag{3.36}$$

where $\bar{A}_{22} = -15$ and $\bar{A}_{21} = \begin{bmatrix} -1 & -2 & 0 & 0 & -1 \end{bmatrix}$. Hence, by designing a reduced-order functional observer, $\hat{z}(t)$, to estimate the following function

$$z(t) = -\bar{A}_{21} \bar{x}_1(t) = \begin{bmatrix} 1 & 2 & 0 & 0 & 1 \end{bmatrix} \bar{x}_1(t), \tag{3.37}$$

for system (3.34)–(3.35), the unknown input $d_1(t)$ can now be estimated as follows

$$\hat{d}_1(t) = \dot{y}_{11}(t) + 15 y_{11}(t) + \hat{z}(t). \tag{3.38}$$

Based on existing results in the literature (for example, [3]), a first-order functional observer, $\hat{z}(t)$, can be obtained for system (3.34)–(3.35), where we select $N = -5$, and the observer is given below

$$\begin{aligned} \hat{z}(t) &= w(t) + \begin{bmatrix} 9.4167 & 1.3333 & 0.8333 \end{bmatrix} \bar{y}(t), \\ \dot{w}(t) &= -5w(t) + \begin{bmatrix} 50.0833 & -0.8333 & -7.9167 \end{bmatrix} \bar{y}(t) + 24 y_{11}(t). \end{aligned}$$

Note that $\bar{y}(t) = V_2 y(t) = \begin{bmatrix} -2 & 1 & 0 & 0 \\ 0 & 0 & 1 & 0 \\ 0 & 0 & 0 & 1 \end{bmatrix} \begin{bmatrix} y_{11}(t) \\ y_{12}(t) \\ y_{21}(t) \\ y_{22}(t) \end{bmatrix}$, and thus the above observer can now be expressed in terms of $y(t)$, as follows

$$\hat{z}(t) = w(t) + \begin{bmatrix} -18.8333 & 9.4167 & 1.3333 & 0.8333 \end{bmatrix} y(t), \tag{3.39}$$

$$\dot{w}(t) = -5w(t) + \begin{bmatrix} -76.1667 & 50.0833 & -0.8333 & -7.9167 \end{bmatrix} y(t). \tag{3.40}$$

Finally, by substituting (3.39) into (3.38), $\hat{d}_1(t)$ is given below

$$\hat{d}_1(t) = \dot{y}_{11}(t) + \begin{bmatrix} -3.8333 & 9.4167 & 1.3333 & 0.8333 \end{bmatrix} y(t) + w(t), \tag{3.41}$$

$$\dot{w}(t) = -5w(t) + \begin{bmatrix} -76.1667 & 50.0833 & -0.8333 & -7.9167 \end{bmatrix} y(t). \tag{3.42}$$

Figure 3.7 shows the trajectories of the disturbances $d_i(t)$, $i = 1, 2$ and the estimate $\hat{d}_1(t)$. It is clear that $\hat{d}_1(t)$ asymptotically converges to $d_1(t)$, while being insensitive to the disturbance $d_2(t)$.

We can carry out the same steps as above to design an unknown input observer to estimate $d_2(t)$ for Subsystem S_2 (the detail is omitted here). Here, we obtain the following observer where first-derivative of the first output of S_2 is used

$$\hat{d}_2(t) = \dot{y}_{21}(t) + \begin{bmatrix} 10.3333 & -11.6667 & -2.3333 & 2.6667 \end{bmatrix} y(t) + w(t), \tag{3.43}$$

$$\dot{w}(t) = -5w(t) + \begin{bmatrix} 8.6667 & -53.6667 & 59.3333 & -13.6667 \end{bmatrix} y(t). \tag{3.44}$$

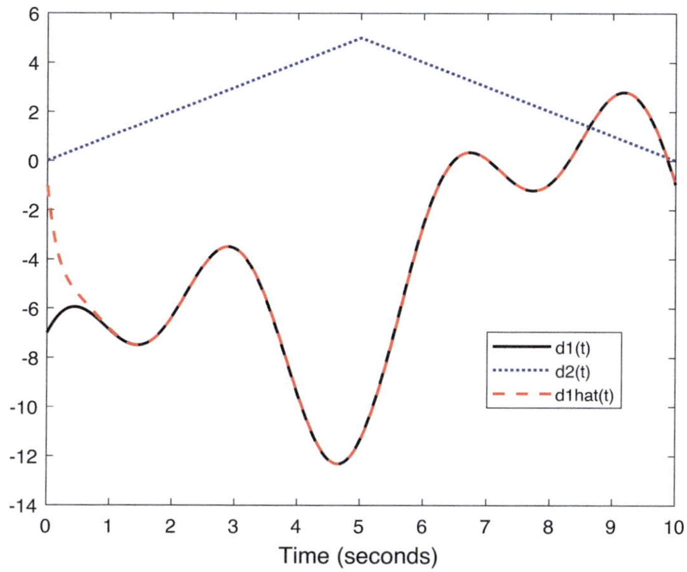

Fig. 3.7 Trajectories of $d_1(t)$, $d_2(t)$ and $\hat{d}_1(t)$ (Numerical Example 4: Distributed unknown input observer, Chap. 3)

Figure 3.8 shows the trajectories of the disturbances $d_i(t)$, $i = 1, 2$, and the estimate $\hat{d}_2(t)$. It is clear that $\hat{d}_2(t)$ asymptotically converges to $d_2(t)$, while being insensitive to the disturbance $d_1(t)$.

The reader can also refer to [11] where we discussed advantages of using distributed observation schemes over those centralized schemes.

Remark 5 Another way to design a distributed unknown input observer for S_1 (or S_2) as suggested by [7] is to decompose the interconnected term, i.e., $A_{12}x_2(t)$, as follows

$$A_{12}x_2(t) = H_1C_2x_2(t) + \tilde{A}_{12}x_2(t) = H_1y_2(t) + \tilde{D}_1\tilde{d}_1(t). \tag{3.45}$$

Thus, for the above decomposition, by utilizing the output measurement, $y_2(t)$, of Subsystem S_2, the interconnected term is split into two parts. The first part is $H_1y_2(t)$ and the second part, $(A_{12}x_2(t) - H_1y_2(t))$, is expressed in a compact form, which is then treated as an unknown input term, where $\tilde{D}_1$ is a full-column rank matrix. Thus, in this example, we can now express $A_{12}x_2(t)$ as follows

$$\begin{aligned} A_{12}x_2(t) &= \begin{bmatrix} 0 & 0 & -1 \\ 0 & 1 & -2 \\ 0 & 0 & 1 \end{bmatrix} x_2(t) \\ &= \begin{bmatrix} 0 \\ 1 \\ 0 \end{bmatrix} y_{22}(t) + \begin{bmatrix} -1 \\ -2 \\ 1 \end{bmatrix} \tilde{d}_1(t). \end{aligned} \tag{3.46}$$

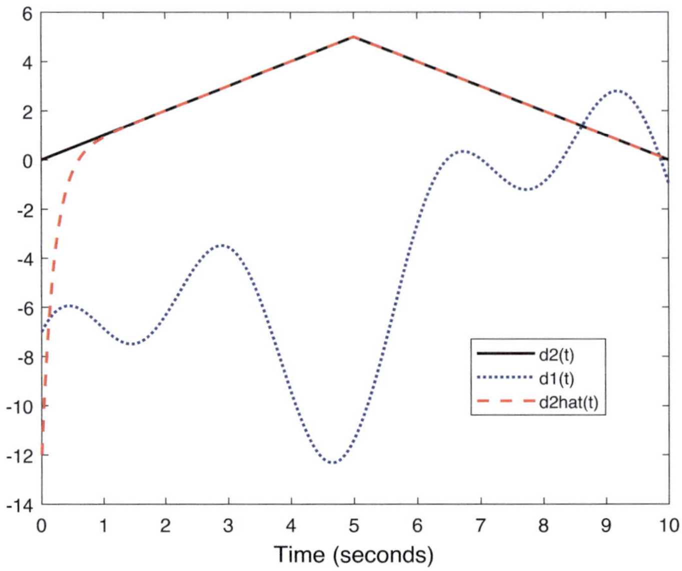

Fig. 3.8 Trajectories of $d_1(t)$, $d_2(t)$ and $\hat{d}_2(t)$ (Numerical Example 4: Distributed unknown input observer, Chap. 3)

By substituting (3.46) into (3.28)–(3.29), we obtain the following system

$$S_1 : \dot{x}_1(t) = A_{11}x_1(t) + \bar{H}_1 y_{22}(t) + \bar{D}\bar{d}(t), \tag{3.47}$$

$$y_1(t) = C_1 x_1(t), \tag{3.48}$$

where $\bar{H}_1 = \begin{bmatrix} 0 \\ 1 \\ 0 \end{bmatrix}$, $\bar{D} = [D_1 \ \ \tilde{D}_1] = \begin{bmatrix} 1 & -1 \\ 2 & -2 \\ 3 & 1 \end{bmatrix}$, and $\bar{d}(t) = \begin{bmatrix} d_1(t) \\ \tilde{d}_1(t) \end{bmatrix}$.

However, since $\tilde{D}_1(1, 1) \neq 0$, we cannot employ Design Approach 1 as discussed in Sect. 3.2.1 to decouple $d_1(t)$ from $\tilde{d}_1(t)$. We also have $\text{rank}(C_1\bar{D}) \neq \text{rank}(\bar{D})$, and thus Design Approach 2 in Sect. 3.2.4 is not applicable. Therefore, our approach based on the global system (3.32)–(3.33) is a workable option for this example.

3.2.9 Numerical Example 5: Decentralized Unknown Input Observers

For Numerical Example 4, let us now change D_1 and D_2 to $D_1 = \begin{bmatrix} 1 \\ 0 \\ -1 \end{bmatrix}$ and $D_2 = \begin{bmatrix} 1 \\ 0 \\ 0 \end{bmatrix}$. Let us design an unknown input observer to estimate $d_1(t)$ at Subsystem S_1. We consider the case where there is no output measurement from S_2 that is sent over to S_1. Thus, only $y_1(t)$ is used, i.e.,

$$y(t) = y_1(t) = \begin{bmatrix} C_1 \; 0_{2\times 3} \end{bmatrix} \begin{bmatrix} x_1(t) \\ x_2(t) \end{bmatrix}.$$

Note that we now have

$$\bar{y}(t) = V_2 C E \bar{x}_1(t) = \begin{bmatrix} 1 \; 0 \; 0 \; 0 \; 0 \end{bmatrix} \bar{x}_1(t), \tag{3.49}$$

where $V_2 = \begin{bmatrix} 0 \; 1 \end{bmatrix}$. In this case, a second-order functional observer is required to estimate $z(t)$. By following the same steps as in Numerical Example 4, $\hat{d}_1(t)$ is obtained as below

$$\hat{d}_1(t) = \dot{y}_{11}(t) + \begin{bmatrix} 5 \; 3 \end{bmatrix} y_1(t) + w_1(t), \tag{3.50}$$

$$\begin{bmatrix} \dot{w}_1(t) \\ \dot{w}_2(t) \end{bmatrix} = \begin{bmatrix} -2 & -2 \\ 0 & -3 \end{bmatrix} \begin{bmatrix} w_1(t) \\ w_2(t) \end{bmatrix} + \begin{bmatrix} -24 & 18 \\ -6 & 0 \end{bmatrix} y_1(t). \tag{3.51}$$

Figure 3.9 shows the trajectories of the disturbances $d_i(t)$, $i = 1, 2$ and the estimate $\hat{d}_1(t)$. It is clear that $\hat{d}_1(t)$ asymptotically converges to $d_1(t)$, while being insensitive to the disturbance $d_2(t)$.

Finally, as for Subsystem S_2, we would like to leave the design problem to the reader to investigate whether that it is possible to reduce the number of outputs to be sent over from S_1 to S_2. The reader should also investigate the case where the sum of two outputs, i.e., $y_{sum_1}(t) = y_{11}(t) + y_{12}(t)$ is sent over to S_2.

3.3 Concluding Remark

In this chapter, we have considered the problem of estimating an unknown input in the presence of multiple unknown inputs in the system. The designed unknown input observer asymptotically estimates a targeted unknown input, while it is insensitive to the rest of the unknown inputs. We have provided a detailed analysis of two design approaches (Design Approach 1 and Design Approach 2) and discussed

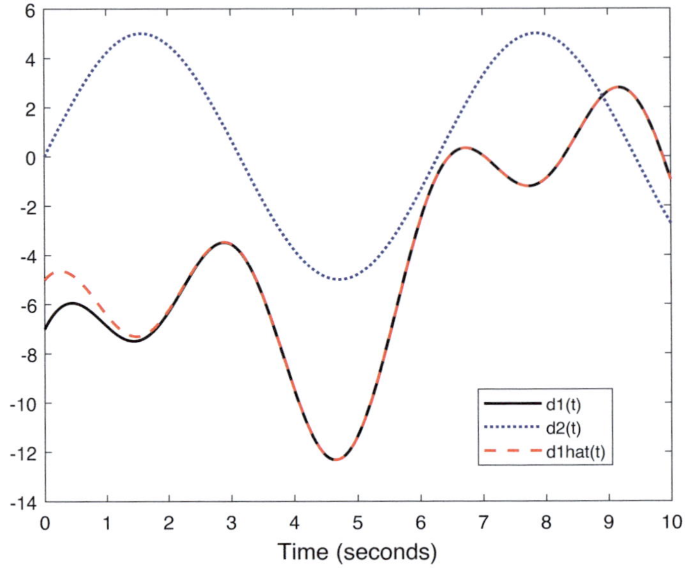

Fig. 3.9 Trajectories of $d_1(t)$, $d_2(t)$ and $\hat{d}_1(t)$ (Numerical Example 5: Decentralized unknown input observer, Chap. 3)

their advantages and drawbacks. Together, they offer the designer a viable tool for designing an unknown input observer with the least number of output differentiators and the simplest functional observers. Using the results derived in this chapter, we have also looked at the design of distributed/decentralized unknown input observers to asymptotically estimate local unknown input vectors of interconnected systems.

References

1. Petersen, I.R., Tempo, R.: Robust control of uncertain systems: Classical results and recent developments. Automatica **50**(5), 1315–1335 (2014)
2. Trinh, H., Tran, T.D., Fernando, T.: Disturbance decoupled observers for systems with unknown inputs. IEEE Trans. Automat. Control **53**(10), 2397–2402 (2008)
3. Trinh, H.M., Teh, P.S., Fernando, T.L.: Time-delay systems: design of delay-free and low-order observers. IEEE Trans. Automat. Control **55**(10), 2434–2438 (2010)
4. Darouach, M., Zasadzinski, M., Xu, S.J.: Full-order observers for linear systems with unknown inputs. IEEE Trans. Automat. Control **39**(3), 606–609 (1994)
5. Hou, M., Müller, P.C.: Design of observers for linear systems with unknown inputs. IEEE Trans. Automat. Control **37**(6), 871–875 (1992)
6. Pham, T.N., Trinh, H., Hien, L.V.: Load frequency control of power systems with electric vehicles and diverse transmission links using distributed functional observers. IEEE Trans. Smart Grid **7**(1), 238–252 (2016)
7. Trinh, H., Fernando, T.: Functional Observers for Dynamical Systems. Springer, Berlin (2012)

8. Hou, M., Müller, P.C.: Design of decentralized linear state function observers. Automatica **30**(11), 1801–1805 (1994)
9. Leong, W.Y., Trinh, H., Hien, L.V.: An LMI-based functional estimation scheme of large-scale time-delay systems with strong interconnections. J. Franklin Inst. **353**(11), 2482–2510 (2016)
10. Saif, M., Guan, Y.: Decentralized state estimation in large-scale interconnected dynamical systems. Automatica **28**(1), 215–219 (1992)
11. Trinh, H., Fernando, T., Iu, H.H.C., Wong, K.P.: Quasi-decentralized functional observers for the LFC of interconnected power systems. IEEE Trans. Power Syst. **28**(3), 3513–3514 (2013)

Chapter 4
Unknown Input Observers Design for Time-Delay Systems

4.1 System Description and Preliminaries

In this chapter, we discuss the design of unknown input observers for linear time-delay systems. We consider system (1.1)–(1.2) that is now subject to a constant time delay in the state vector as described below

$$\dot{x}(t) = Ax(t) + A_d x(t-\tau) + Dd(t),\ t \geq 0, \tag{4.1}$$

$$y(t) = Cx(t), \tag{4.2}$$

where $A_d \in \mathbb{R}^{n \times n}$ is a known constant matrix. The positive constant τ is a known time delay and $x(\theta) = \phi(\theta),\ \forall\theta \in [-\tau, 0]$ is a continuous initial function.

Based on the matrix decomposition technique presented in Sect. 2.2, we first obtain matrices V, T and E by following Steps 3-4 of Algorithm 1. Also, let us partition A_d and $\bar{A}_d$ as follows: $A_d = \begin{bmatrix} A_{d11} & A_{d12} \\ A_{d21} & A_{d22} \end{bmatrix}$, where $A_{d11} \in \mathbb{R}^{1\times 1}$, $A_{d12} \in \mathbb{R}^{1\times(n-1)}$, $A_{d21} \in \mathbb{R}^{(n-1)\times 1}$, $A_{d22} \in \mathbb{R}^{(n-1)\times(n-1)}$, and $\bar{A}_d = T^{-1}A_d T = \begin{bmatrix} \bar{A}_{d11} & \bar{A}_{d12} \\ \bar{A}_{d21} & \bar{A}_{d22} \end{bmatrix}$, where $\bar{A}_{d11} \in \mathbb{R}^{(n-1)\times(n-1)}$, $\bar{A}_{d12} \in \mathbb{R}^{(n-1)\times 1}$, $\bar{A}_{d21} \in \mathbb{R}^{1\times(n-1)}$ and $\bar{A}_{d22} \in \mathbb{R}$. Then, by applying the state transformation $x(t) = T\bar{x}(t)$ to system (4.1)–(4.2), we obtain the following disturbance-free time-delay system

$$\begin{aligned} \dot{\bar{x}}_1(t) &= \bar{A}_{11}\bar{x}_1(t) + \bar{A}_{d11}\bar{x}_1(t-\tau) \\ &\quad + \bar{A}_{12}V_1 y(t) + \bar{A}_{d12}V_1 y(t-\tau),\ t \geq 0, \end{aligned} \tag{4.3}$$

$$\bar{y}(t) = \tilde{C}_1\bar{x}_1(t), \tag{4.4}$$

where $\bar{y}(t) = V_2 y(t)$ and $\tilde{C}_1 = V_2 C E = [I_{(p-1)}\ 0_{(p-1)\times(n-p)}]$. The unknown input $d(t)$ is obtained as follows

H. Trinh et al., *Unknown Inputs Estimation in Linear Time-Delay Systems Using Generalized Functional Observers*, Studies in Systems, Decision and Control 648,
https://doi.org/10.1007/978-3-032-11870-7_4

$$\begin{aligned} d(t) &= V_1 \dot{y}(t) - \bar{A}_{22} V_1 y(t) - \bar{A}_{d22} V_1 y(t-\tau) \\ &\quad - \bar{A}_{21} \bar{x}_1(t) - \bar{A}_{d21} \bar{x}_1(t-\tau), \ t \geq 0. \end{aligned} \tag{4.5}$$

As can be seen from (4.5), to estimate $d(t)$ we only need to use the first-derivative of one output measurement since V_1 is of the form (2.4). Examining (4.5) reveals that there are several options in which $d(t)$ can be estimated.

4.2 Design Option 1

In this option, we design a minimum-order state observer, $\hat{\bar{x}}_1(t)$, to asymptotically estimate $\bar{x}_1(t)$ of the disturbance-free time-delay system (4.3)–(4.4). Then, the unknown input $d(t)$ is estimated as follows

$$\begin{aligned} \hat{d}(t) &= V_1 \dot{y}(t) - \bar{A}_{22} V_1 y(t) - \bar{A}_{d22} V_1 y(t-\tau) \\ &\quad - \bar{A}_{21} \hat{\bar{x}}_1(t) - \bar{A}_{d21} \hat{\bar{x}}_1(t-\tau), \ t \geq 0. \end{aligned} \tag{4.6}$$

In this option, the designed state observer $\hat{\bar{x}}_1(t)$ estimates all of the unmeasurable $(n-p)$ state variables of $\bar{x}_1(t)$. Even though its order is the minimum, i.e., $(n-p)$, it might not produce a low-order unknown input observer, $\hat{d}(t)$, since there might exist scalar functional observers of order lower than $(n-p)$ that can be used to estimate $d(t)$.

Thus, in the next option, Option 2, we thoroughly look at several scenarios in which functional observers for time-delay system (4.3)–(4.4) are considered instead of reduced-order state observers.

4.3 Design Option 2

Let us further partition $\bar{A}_{21}$ and $\bar{A}_{d21}$ as follows

$$\bar{A}_{21} = \begin{bmatrix} \bar{A}_{21}^1 & \bar{A}_{21}^2 \end{bmatrix}, \tag{4.7}$$

$$\bar{A}_{d21} = \begin{bmatrix} \bar{A}_{d21}^1 & \bar{A}_{d21}^2 \end{bmatrix}, \tag{4.8}$$

where $\bar{A}_{21}^1, \bar{A}_{d21}^1 \in \mathbb{R}^{1\times(p-1)}$ and $\bar{A}_{21}^2, \bar{A}_{d21}^2 \in \mathbb{R}^{1\times(n-p)}$.

4.3.1 Scenario 1

First, we consider the simplest case where $\bar{A}^2_{21} = 0_{1\times(n-p)}$ and $\bar{A}^2_{d21} = 0_{1\times(n-p)}$. Then functions $-\bar{A}_{21}\bar{x}_1(t)$ and $-\bar{A}_{d21}\bar{x}_1(t-\tau)$ in (4.5) can be expressed as

$$-\bar{A}_{21}\bar{x}_1(t) = -\bar{A}^1_{21}V_2 y(t), \tag{4.9}$$

$$-\bar{A}_{d21}\bar{x}_1(t-\tau) = -\bar{A}^1_{d21}V_2 y(t-\tau). \tag{4.10}$$

Note that (4.9)–(4.10) are obtained based on the fact that $\bar{y}(t) = V_2 y(t)$ and $\tilde{C}_1 = [I_{(p-1)}\ 0_{(p-1)\times(n-p)}]$. Hence, the unknown input $d(t)$ as defined in (4.5) is reduced to the following equation

$$\begin{aligned} d(t) &= V_1\dot{y}(t) - \bar{A}_{22}V_1 y(t) - \bar{A}_{d22}V_1 y(t-\tau) \\ &\quad - \bar{A}^1_{21}V_2 y(t) - \bar{A}^1_{d21}V_2 y(t-\tau),\ t \geq 0. \end{aligned} \tag{4.11}$$

Thus, provided that the structural constraints $\bar{A}^2_{21} = 0_{1\times(n-p)}$ and $\bar{A}^2_{d21} = 0_{1\times(n-p)}$ are satisfied, we do not need to estimate both functions $-\bar{A}_{21}\bar{x}_1(t)$ and $-\bar{A}_{d21}\bar{x}_1(t-\tau)$ (also, a state observer as in Option 1 is not needed) as they are directly obtained from the output measurement (see, (4.9)–(4.10)). Thus, there is no need to design any observer to estimate the unknown input.

The above structural constraints can also be established based on the original matrices A and A_d. For this, let us partition matrix A as $A = \begin{bmatrix} A_{11} & A_{12} \\ A_{21} & A_{22} \end{bmatrix}$, where $A_{11} \in \mathbb{R}^{1\times 1}$, $A_{12} \in \mathbb{R}^{1\times(n-1)}$, $A_{21} \in \mathbb{R}^{(n-1)\times 1}$, $A_{22} \in \mathbb{R}^{(n-1)\times(n-1)}$, and sub-matrices A_{12} and A_{d12} as $A_{12} = \begin{bmatrix} A^1_{12} & A^2_{12} \end{bmatrix}$ and $A_{d12} = \begin{bmatrix} A^1_{d12} & A^2_{d12} \end{bmatrix}$.

Now, with $T = \begin{bmatrix} E & D \end{bmatrix} = \begin{bmatrix} 0_{1\times(n-1)} & \alpha_1 \\ I_{(n-1)} & D_{12} \end{bmatrix}$ and $T^{-1} = \begin{bmatrix} -\alpha_1^{-1}D_{12} & I_{(n-1)} \\ \alpha_1^{-1} & 0_{1\times(n-1)} \end{bmatrix}$, it is easy to show that the structural constraints $\bar{A}^2_{21} = 0_{1\times(n-p)}$ and $\bar{A}^2_{d21} = 0_{1\times(n-p)}$ are equivalent to the following

$$A^2_{12} = 0_{1\times(n-p)}, \tag{4.12}$$

$$A^2_{d12} = 0_{1\times(n-p)}. \tag{4.13}$$

Conditions (4.12)–(4.13) allow us to quickly establish whether the unknown input $d(t)$ can be directly obtained from the available output measurement without the need to design any state or functional observers.

4.3.2 Scenario 2

In this scenario, we design a scalar functional observer to estimate the function $z(t) = F\bar{x}_1(t)$, where $F \in \mathbb{R}^{1\times(n-1)} \neq 0$ is given below

$$F = \begin{bmatrix} 0_{1\times(p-1)} & -\bar{A}_{21}^2 \end{bmatrix}.$$

This scenario occurs when: (i) $\bar{A}_{21}^2 \neq 0$ and $\bar{A}_{d21}^2 = 0$, or (ii) $\bar{A}_{21}^2 \neq 0$ and rank $\begin{bmatrix} \bar{A}_{21}^2 \\ \bar{A}_{d21}^2 \end{bmatrix} = 1$, which implies that $\bar{A}_{d21}^2$ is linearly dependent on $\bar{A}_{21}^2$. Thus for both cases, to estimate $d(t)$ we only need to design a scalar functional observer to estimate the function $z(t)$.

To estimate the function $z(t) = F\bar{x}_1(t)$ of the disturbance-free time-delay system (4.3)–(4.4), let us consider the following observer with delayed output measurements

$$\hat{z}(t) = Kw(t) + M\bar{y}(t), \tag{4.14}$$
$$\dot{w}(t) = Nw(t) + G\bar{y}(t) + G_d\bar{y}(t-\tau) + Jy(t) + J_d y(t-\tau) \tag{4.15}$$

where $w(t) \in \mathbb{R}^m$, $1 \leq m \leq (n-p)$, with the initial condition $w(\theta) = \rho(\theta)$, $\forall\theta \in [-\tau, 0]$. Existence conditions of observer (4.14)–(4.15) can be derived. For this, let $L \in \mathbb{R}^{m\times(n-1)}$ be a full-row rank matrix, and the following error vectors are considered

$$e(t) = \hat{z}(t) - z(t), \tag{4.16}$$
$$\varepsilon(t) = w(t) - L\bar{x}_1(t). \tag{4.17}$$

Based on the approach proposed in [1, 2], we can show that observer (4.14)–(4.15) exists such that $\hat{z}(t)$ converges asymptotically to $z(t)$ for any initial condition $\rho(\theta)$ if the following conditions hold

$$N \text{ is Hurwitz}, \tag{4.18}$$
$$NL + G\tilde{C}_1 - L\bar{A}_{11} = 0, \tag{4.19}$$
$$G_d\tilde{C}_1 - L\bar{A}_{d11} = 0, \tag{4.20}$$
$$F - KL - M\tilde{C}_1 = 0, \tag{4.21}$$
$$J - L\bar{A}_{12}V_1 = 0, \tag{4.22}$$
$$J_d - L\bar{A}_{d12}V_1 = 0. \tag{4.23}$$

Necessary and sufficient conditions for the solvability of coupled equations (4.19)–(4.21) with N being Hurwitz can be derived. The reader can refer to Chap. 10 for a procedure to solve for the unknown constant matrices N, L, G, G_d, K and M of equations (4.19)–(4.21). Once L is found, J and J_d are obtained from (4.22) and (4.23), where $J = L\bar{A}_{12}V_1$ and $J_d = L\bar{A}_{d12}V_1$.

Observer (4.14)–(4.15) for time-delay system (4.3)–(4.4) has a slightly more complex structure than observer (2.10)–(2.11), where delayed output measurements (hence memory units) are needed for its implementation and realization. Here, we include $G_d\bar{y}(t-\tau)$ and $J_d y(t-\tau)$ in (4.15) in order to deal with the two delayed terms $\bar{A}_{d11}\bar{x}_1(t-\tau)$ and $\bar{A}_{d12}V_1 y(t-\tau)$ in (4.3). This results in an extra condition

(4.20) and thus makes the solvability conditions of the coupled equations (4.19)–(4.21) harder to satisfy than (2.12)–(2.13).

On the other hand, there is always a trade-off between simplicity of observer's structures and conservatism of their existence conditions. The reader can refer to [3], where an internal delay term $N_d w(t-\tau)$ is included in (4.15) to relax condition (4.20) to $N_d L + G_d \tilde{C}_1 - L\bar{A}_{d11} = 0$, and condition (4.18) is changed to the requirement that the following time-delay system is asymptotically stable [3]

$$\dot{\varepsilon}(t) = N\varepsilon(t) + N_d\varepsilon(t-\tau). \tag{4.24}$$

In contrast, in [4], by employing a general Luenberger-type observer which includes multiple delayed output measurements, a less conservative existence condition than the one reported in [3] can be obtained.

4.3.3 Numerical Example 1

Let us now reconsider Numerical Example 1 of Chap. 2, where the system is subject to a constant time delay of $\tau = 1s$ in the state vector $x(t)$ and matrix A_d is as given below

$$A_d = \begin{bmatrix} -1 & -2 & -3 & 0 & 0 & 0 \\ 2.5 & -7 & -5 & 0 & 0.5 & 0 \\ -1 & 1 & -5 & 0 & 0 & 1 \\ 4 & 0 & -2 & -2 & -2 & 1 \\ 3 & -2 & -6 & 6 & -4 & 0 \\ 7 & -4 & 2 & 0 & 1 & -2 \end{bmatrix}.$$

With A_d given above, we have $A_{d12} = \begin{bmatrix} A^1_{d12} & A^2_{d12} \end{bmatrix}$, where $A^1_{d12} = \begin{bmatrix} -2 & -3 \end{bmatrix}$ and $A^2_{d12} = \begin{bmatrix} 0 & 0 & 0 \end{bmatrix}$. Thus, as discussed in Scenario 2, we design a scalar functional observer to estimate the unknown input $d(t)$. Now, with matrices T, E, V_1 and V_2 as given in Numerical Example 1 (Chap. 2), a disturbance-free time-delay system (4.3)–(4.4) is obtained, where $\bar{A}_{d11}$, $\bar{A}_{d12}$, $\bar{A}_{d21}$ and $\bar{A}_{d22}$ are as given below $\bar{A}_{d11} = \begin{bmatrix} -3 & 1 & 0 & 0.5 & 0 \\ 1 & -5 & 0 & 0 & 1 \\ 0 & -2 & -2 & -2 & 1 \\ 0 & -3 & 6 & -4 & 0 \\ -4 & 2 & 0 & 1 & -2 \end{bmatrix}$, $\bar{A}_{d12} = \begin{bmatrix} -1 \\ 1 \\ 2 \\ 0 \\ 0 \end{bmatrix}$, $\bar{A}_{d21} = \begin{bmatrix} -2 & -3 & 0 & 0 & 0 \end{bmatrix}$ and $\bar{A}_{d22} = -5$.

Substituting $-\bar{A}_{22}V_1 y(t) = 9y_1(t)$, $-\bar{A}_{d22}V_1 y(t-\tau) = 5y_1(t-\tau)$, $-\bar{A}_{d21}\bar{x}_1(t-\tau) = \begin{bmatrix} -4 & 2 & 3 \end{bmatrix} y(t-\tau)$ and $z(t) = -\bar{A}_{21}\bar{x}_1(t)$ into (4.5), the unknown input $d(t)$ is obtained as follows

$$d(t) = \dot{y}_1(t) + 9y_1(t) + [1\ 2\ 3]y(t-\tau) + z(t),\ t \geq 0. \tag{4.25}$$

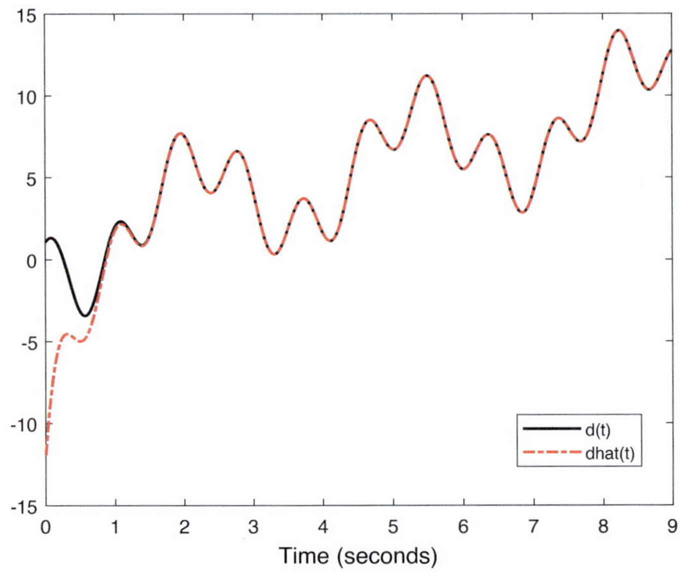

Fig. 4.1 Trajectories of $d(t)$ and $\hat{d}(t)$ (Numerical Example 1, Chap. 4)

From (4.25), we can design a scalar functional observer of the form (4.14)–(4.15) to estimate the function $z(t)$ (refer to Chap. 10 for a design procedure). A shortcut to this design problem is to use the existing matrix L in Numerical Example 1 (Chap. 2) and solve for the unknown matrix G_d (if it exists) so that condition (4.20) is satisfied. Note that with L, K, N, M and G as given in Numerical Example 1 (Chap. 2), conditions (4.18), (4.19) and (4.21) are already satisfied. Here, matrix G_d exists and condition (4.20) is satisfied with $G_d = \begin{bmatrix} 4 & -10 \end{bmatrix}$. Thus, with the designed functional observer, $\hat{z}(t)$, the unknown input $d(t)$ given in (4.25) can now be estimated as follows

$$\begin{aligned}\hat{d}(t) &= \dot{y}_1(t) + 9y_1(t) + [1\ 2\ 3]y(t-\tau) + \hat{z}(t),\\ \hat{z}(t) &= w(t) + \begin{bmatrix} -6 & 3 & 0 \end{bmatrix} y(t),\\ \dot{w}(t) &= -4w(t) + \begin{bmatrix} -28 & 12 & -2 \end{bmatrix} y(t) + \begin{bmatrix} -4 & 4 & -10 \end{bmatrix} y(t-\tau),\ t \geq 0.\end{aligned}$$

The above unknown input observer, $\hat{d}(t)$, has the simplest structure, which employs a first-order scalar functional observer, $\hat{z}(t)$, and first-derivative of only one output measurement. Figure 4.1 shows the responses of $d(t)$ and $\hat{d}(t)$. It is clear from the figure that asymptotic tracking of the unknown input is achieved.

4.3.4 Numerical Example 2

In this example, we show the advantage of having a scalar functional observer over a reduced-order state observer in estimating an unknown input $d(t)$. For this, let us reconsider Numerical Example 1 with $\tau = 2s$ and the following changes, highlighted in bold, are made to matrices A and A_d

$$A = \begin{bmatrix} -7 & -1 & -2 & 0 & 0 & -1 \\ 6 & -12 & -2 & 0 & 1 & -2 \\ -2 & 1 & -1 & 0 & 0 & 1 \\ -5 & 1 & 0 & -5 & 3 & 4 \\ 8 & -4 & -1 & \mathbf{0} & -8 & 2 \\ \mathbf{1} & -2 & -4 & 0 & \mathbf{1} & -6 \end{bmatrix},$$

and

$$A_d = \begin{bmatrix} -1 & -2 & -3 & 0 & 0 & 0 \\ 2.5 & -7 & -5 & 0 & 0.5 & 0 \\ -1 & 1 & -5 & \mathbf{1} & 0 & 1 \\ 4 & 0 & -2 & -2 & -2 & 1 \\ \mathbf{-1} & -2 & -6 & \mathbf{0} & \mathbf{0} & \mathbf{0} \\ \mathbf{8} & -4 & 2 & \mathbf{1} & \mathbf{0} & \mathbf{1} \end{bmatrix}.$$

With T, E, V_1 and V_2 as derived in Numerical Example 1 (Chap. 2), we obtain a disturbance-free time-delay system (4.3)–(4.4), where $\bar{A}_{11}$, $\bar{A}_{12}$, $\bar{A}_{d11}$ and $\bar{A}_{d12}$ are as given below

$$\bar{A}_{11} = \begin{bmatrix} -10 & 2 & 0 & 1 & 0 \\ 1 & -1 & 0 & 0 & 1 \\ 1 & 0 & -5 & 3 & 4 \\ -3 & 1 & 0 & -8 & 3 \\ -2 & -4 & 0 & 1 & -6 \end{bmatrix}, \; \bar{A}_{12} = \begin{bmatrix} 1 \\ 0 \\ 0 \\ 1 \\ -2 \end{bmatrix},$$

$$\bar{A}_{d11} = \begin{bmatrix} -3 & 1 & 0 & 0.5 & 0 \\ 1 & -5 & 1 & 0 & 1 \\ 0 & -2 & -2 & -2 & 1 \\ 0 & -3 & 0 & 0 & 0 \\ -4 & 2 & 1 & 0 & 1 \end{bmatrix}, \; \bar{A}_{d12} = \begin{bmatrix} -1 \\ 1 \\ 2 \\ 0 \\ 0 \end{bmatrix}.$$

Following from (4.5), the unknown input $d(t)$ is obtained as

$$d(t) = \dot{y}_1(t) + 9y_1(t) + [1\;2\;3]y(t-\tau) + z(t), \; t \geq 0, \tag{4.26}$$

where $z(t) = -\bar{A}_{21}\bar{x}_1(t) = F\bar{x}_1(t)$ and $F = \begin{bmatrix} 1 & 2 & 0 & 0 & 1 \end{bmatrix}$.

Thus to estimate the unknown input $d(t)$ as defined by (4.26), we design a functional observer to estimate the function $z(t)$ of the disturbance-free time-delay system (4.3)–(4.4), where $\bar{A}_{11}$, $\bar{A}_{12}$, $\bar{A}_{d11}$ and $\bar{A}_{d12}$ are as given above.

In the following, we follow the design procedure as given in Chap. 10 to design a functional observer. Accordingly, the following partitioned matrices are obtained

$F_1 = \begin{bmatrix}1\,2\end{bmatrix}, F_2 = \begin{bmatrix}0\,0\,1\end{bmatrix}, \bar{A}_{11}^{11} = \begin{bmatrix} -10 & 2 \\ 1 & -1 \end{bmatrix}, \bar{A}_{d11}^{11} = \begin{bmatrix} -3 & 1 \\ 1 & -5 \end{bmatrix}, \bar{A}_{11}^{12} = \begin{bmatrix} 0 & 1 & 0 \\ 0 & 0 & 1 \end{bmatrix},$
$\bar{A}_{d11}^{12} = \begin{bmatrix} 0 & 0.5 & 0 \\ 1 & 0 & 1 \end{bmatrix}, \bar{A}_{11}^{21} = \begin{bmatrix} 1 & 0 \\ -3 & 1 \\ -2 & -4 \end{bmatrix}, \bar{A}_{d11}^{21} = \begin{bmatrix} 0 & -2 \\ 0 & -3 \\ -4 & 2 \end{bmatrix}, \bar{A}_{11}^{22} = \begin{bmatrix} -5 & 3 & 4 \\ 0 & -8 & 3 \\ 0 & 1 & -6 \end{bmatrix}$
and $\bar{A}_{d11}^{22} = \begin{bmatrix} -2 & -2 & 1 \\ 0 & 0 & 0 \\ 1 & 0 & 1 \end{bmatrix}$.

First, we start with a first-order functional observer, i.e., $m = 1$. For this, condition (10.16) is checked with $F_2 = \begin{bmatrix}0 & 0 & 1\end{bmatrix}$ and it is found to be satisfied, where rank $\begin{bmatrix} \bar{A}_{d11}^{12} \\ F_2\bar{A}_{d11}^{22} \end{bmatrix} = 2$ and rank$\begin{bmatrix}\bar{A}_{d11}^{12}\end{bmatrix} = 2$. We then check if condition (10.28) is satisfied or not. For this, we have

$$\text{rank} \begin{bmatrix} F_2\bar{A}_{11}^{22} & F_2\bar{A}_{d11}^{22} \\ \bar{A}_{11}^{12} & \bar{A}_{d11}^{12} \\ F_2 & 0 \end{bmatrix} = 4,$$

but

$$\text{rank} \begin{bmatrix} \bar{A}_{11}^{12} & \bar{A}_{d11}^{12} \\ F_2 & 0 \end{bmatrix} = 3.$$

Condition (10.28) is not satisfied and thus a first-order functional observer of the form (4.14)–(4.15) does not exist.

We then seek a higher-order functional observer with $m = 2$. Since $\bar{A}_{d11}^{12}$ is a full-row rank matrix, we follow the steps as presented in Case 1 of Chap. 10.

In Step 1 of Chap. 10 and by following its Remark 1, a row vector $\bar{F}_{ex} = \begin{bmatrix}0 & 1 & 0\end{bmatrix}$ is obtained, where $\bar{F}_{ex}$ is an orthogonal basis of the null-space of $\bar{A}_{d11}^{22}$. Note that we can only find one row vector $\bar{F}_{ex}$ that can satisfy condition (10.37) (see Remark 1, Chap. 10).

With the found $\bar{F}_{ex}$, we have rank $\begin{bmatrix} F_2 \\ \bar{F}_{ex} \end{bmatrix} = 2$ and condition (10.37) is satisfied.

In Step 2, we check condition (10.38) and it is found to be satisfied. Thus, from (10.25) and (10.26), N_1 and N_2, respectively, are obtained as $N_1 = \begin{bmatrix} -7 & 1 \\ 3 & -8 \end{bmatrix}$ and $N_2 = 0_{2\times 2}$. In this case, the pair (N_1, N_2) is not observable but it is detectable. By letting $Z = 0$, matrix $N = N_1$ is stable with two eigenvalues at $s_1 = -5.6972$ and $s_2 = -9.3028$.

The rest of the observer's parameters can be obtained as follows. From (10.23), (10.10), (10.12), and (10.14), matrices L_1, M, G and G_d, respectively, are obtained, where $L_1 = \begin{bmatrix} 0 & -1 \\ 0 & 0 \end{bmatrix}$, $M = \begin{bmatrix} 1 & 3 \\ 0 & 0 \end{bmatrix}$, $G = \begin{bmatrix} -3 & -10 \\ -3 & 4 \end{bmatrix}$ and $G_d = \begin{bmatrix} -5 & 7 \\ 0 & -3 \end{bmatrix}$. With $L = \begin{bmatrix} L_1 & F_a^2 \end{bmatrix} = \begin{bmatrix} 0 & -1 & 0 & 0 & 1 \\ 0 & 0 & 0 & 1 & 0 \end{bmatrix}$, we obtain $J = L\bar{A}_{12}V_1 = \begin{bmatrix} -2 & 0 & 0 \\ 1 & 0 & 0 \end{bmatrix}$ and $J_d =$

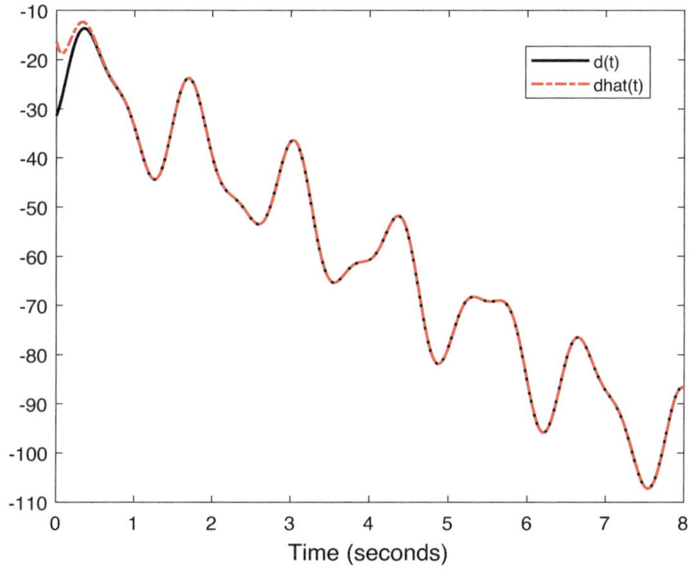

Fig. 4.2 Trajectories of $d(t)$ and $\hat{d}(t)$ (Numerical Example 2, Chap. 4)

$L\bar{A}_{d12}V_1 = \begin{bmatrix} -1 & 0 & 0 \\ 0 & 0 & 0 \end{bmatrix}$. It is easy to check that with the above obtained matrices, conditions (10.7)–(10.9) are satisfied, thus verifying the correctness of our design procedure.

Finally, the unknown input $d(t)$ as defined by (4.26) can now be estimated by using a second-order functional observer, where

$$\begin{aligned}
\hat{d}(t) &= \dot{y}_1(t) + 9y_1(t) + [1\ 2\ 3]y(t-\tau) + \hat{z}(t), \\
\hat{z}(t) &= w_1(t) + \begin{bmatrix} -2 & 1 & 3 \end{bmatrix} y(t), \\
\begin{bmatrix} \dot{w}_1(t) \\ \dot{w}_2(t) \end{bmatrix} &= \begin{bmatrix} -7 & 1 \\ 3 & -8 \end{bmatrix} \begin{bmatrix} w_1(t) \\ w_2(t) \end{bmatrix} \\
&+ \begin{bmatrix} 4 & -3 & -10 \\ 7 & -3 & 4 \end{bmatrix} y(t) + \begin{bmatrix} 9 & -5 & 7 \\ 0 & 0 & -3 \end{bmatrix} y(t-\tau).
\end{aligned}$$

Note that if Design Option 1 as presented in Sect. 4.2 was to be adopted, then we would find that condition (10.27) is, however, not satisfied. This implies that a third-order state observer does not exist. On the other hand, a second-order functional observer exists that can be used to estimate the required function $z(t) = F\bar{x}_1(t)$ instead of $\bar{x}_1(t)$.

Figure 4.2 shows the responses of $d(t)$ and $\hat{d}(t)$. It is clear from the figure that asymptotic tracking of the unknown input is achieved.

4.3.5 Numerical Example 3

In this example, we illustrate the design of an unknown input observer, where it is necessary to employ an observer with an internal delay as presented in Chap. 10. For this, we reconsider Numerical Example 1 but change the element in $A(6, 6)$ from -6 to -1.9, i.e., $A(6, 6) = -1.9$. With this change, it is found that a first-order functional observer (10.1)–(10.2) does not exist. Also, as explained in Chap. 10 (Case 1), increasing the order of the observer does not work, since there is an unstable fixed pole at $s = 0.1$.

As observer (10.1)–(10.2) does not exist, we therefore look at an observer with an internal delay, i.e., observer (10.48)-(10.49). Following the procedure as given in Chap. 10, condition (10.60) is satisfied. This is as expected since condition (10.60) is less conservative than condition (10.28). From (10.64)–(10.67), we obtain $\bar{N}_1 = -1.9$, $\bar{N}_2 = \begin{bmatrix} 0 \\ -1 \end{bmatrix}$, $\bar{N}_{d1} = -2$, and $\bar{N}_{d2} = \begin{bmatrix} 0 \\ -1 \end{bmatrix}$. By letting $\bar{Z} = \begin{bmatrix} 0 \ \bar{z} \end{bmatrix}$, we now need to find a $\bar{z}$ such that the following time-delay is asymptotically stable for any initial condition $\varepsilon(\theta)$, $\forall\theta \in [-\tau, 0]$

$$\dot{\varepsilon}(t) = (\bar{z} - 1.9)\varepsilon(t) + (\bar{z} - 2)\varepsilon(t - \tau). \tag{4.27}$$

For the stability of the above one-dimensional time-delay system, an *exact* stability condition was derived by Mori [5] where it is required that $(\bar{z} - 1.9) + (\bar{z} - 2) < 0$, with $(\bar{z} - 2) \geq -\frac{1}{\tau}$. Thus, with $\tau = 1s$ in this example, its stability can be satisfied for $1 \leq \bar{z} < 1.95$. We can now see the significance of having an internal delay term $N_d w(t - \tau)$ in observer (10.48)–(10.49), as the unstable pole at $s = 0.1$ has now been removed and replaced by an asymptotic stable error system (4.27). Thus, $\bar{z}$ can now be selected from the range $[1, 1.95)$. Figure 4.3 shows the trajectories of $\varepsilon(t)$ for some values of $\bar{z}$. Here, we select $\bar{z} = 1.6$ to achieve a faster convergence rate. Note also that with $\bar{z} = 1.6$, the eigenvalues of (4.27) can be computed by using the Matlab package developed by Wu and Michiels [6]. It is found that there is a pair of dominant poles at $-1.042 \pm j0.8575$.

Thus, with $\bar{z} = 1.6$, we obtain $N = -0.3$, $N_d = -0.4$, $M = [3\ \ 0.4]$, $G = [19\ \ -9.12]$, $G_d = [2.8\ \ -7.36]$, $J = [-4\ \ 0\ \ 0]$, and $J_d = [3.6\ \ 0\ \ 0]$. Thus, the unknown input $d(t)$ given in (4.25) can be estimated as follows

$$\begin{aligned}
\hat{d}(t) &= \dot{y}_1(t) + 9y_1(t) + [1\ \ 2\ \ 3]y(t - \tau) + \hat{z}(t), \\
\hat{z}(t) &= w(t) + [-6\ \ 3\ \ 0.4]y(t), \\
\dot{w}(t) &= -0.3w(t) - 0.4w(t - \tau) + [-42\ \ 19\ \ -9.12]y(t) \\
&\quad + [-2\ \ 2.8\ \ -7.36]y(t - \tau), \ t \geq 0.
\end{aligned}$$

Figure 4.4 shows the responses of $d(t)$ and $\hat{d}(t)$.

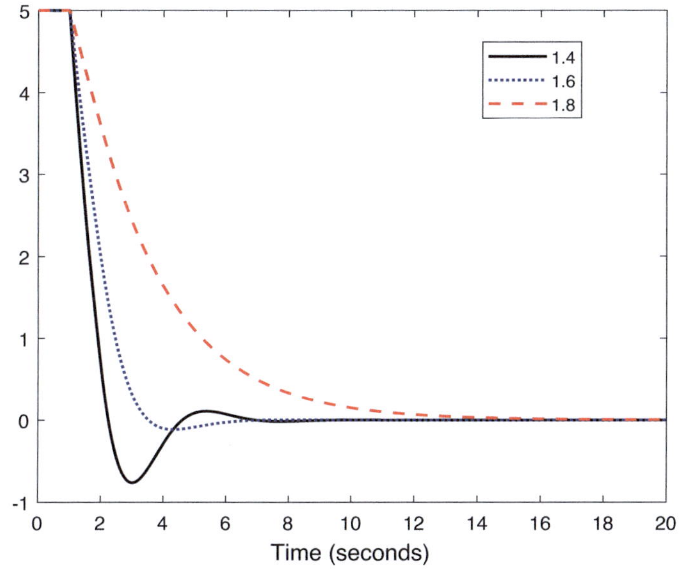

Fig. 4.3 Trajectories of $\varepsilon(t)$ for $\bar{z} = 1.4, 1.6, 1.8$ (Numerical Example 3, Chap. 4)

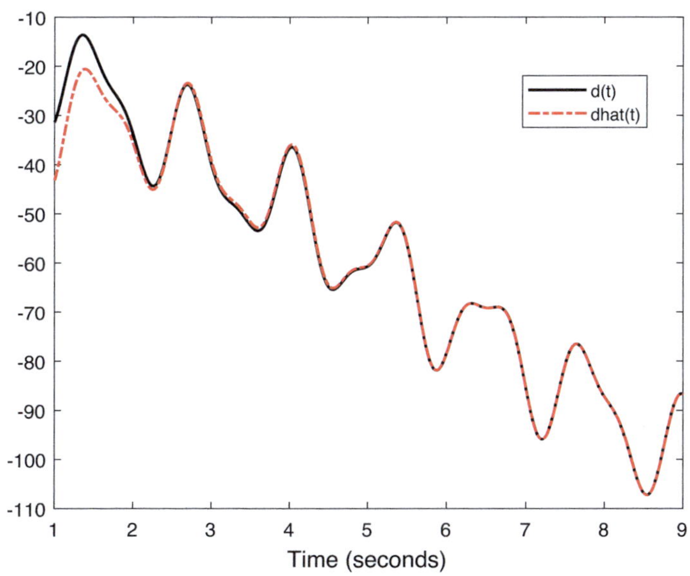

Fig. 4.4 Trajectories of $d(t)$ and $\hat{d}(t)$ (Numerical Example 3, Chap. 4)

4.3.6 Numerical Example 4

In this example, we illustrate the design of an unknown input observer, where we employ a first-order scalar functional observer with an internal delay and multiple delayed output measurements as presented in Chap. 10 (see, observer (10.72)–(10.73)). Let us reconsider Numerical Example 2 but with the following changes, highlighted in bold, made to matrices A and A_d

$$A = \begin{bmatrix} -7 & -1 & -2 & 0 & 0 & -1 \\ 6 & -12 & -2 & \mathbf{1} & 1 & -2 \\ \mathbf{-3} & 1 & -1 & 0 & \mathbf{1} & 1 \\ -5 & 1 & 0 & -5 & 3 & 4 \\ 8 & -4 & -1 & 0 & -8 & 2 \\ \mathbf{2} & -2 & -4 & 0 & \mathbf{0} & -6 \end{bmatrix},$$

and

$$A_d = \begin{bmatrix} -1 & -2 & -3 & 0 & 0 & 0 \\ \mathbf{3} & -7 & -5 & 0 & \mathbf{0} & 0 \\ -1 & 1 & -5 & \mathbf{0} & 0 & \mathbf{0} \\ 4 & 0 & -2 & -2 & -2 & 1 \\ -1 & -2 & -6 & 0 & 0 & 0 \\ 8 & -4 & 2 & 1 & 0 & 1 \end{bmatrix}.$$

With T, E, V_1 and V_2 as derived in Numerical Example 1 (Chap. 2), we obtain a disturbance-free time-delay system (4.3)–(4.4), where $\bar{A}_{11}$, $\bar{A}_{12}$, $\bar{A}_{d11}$ and $\bar{A}_{d12}$ are as given below

$$\bar{A}_{11} = \begin{bmatrix} -10 & 2 & 1 & 1 & 0 \\ 1 & -1 & 0 & 1 & 1 \\ 1 & 0 & -5 & 3 & 4 \\ -3 & 1 & 0 & -8 & 3 \\ -2 & -4 & 0 & 0 & -6 \end{bmatrix}, \bar{A}_{12} = \begin{bmatrix} 1 \\ 0 \\ 0 \\ 1 \\ -2 \end{bmatrix},$$

$$\bar{A}_{d11} = \begin{bmatrix} -3 & 1 & 0 & 0 & 0 \\ 1 & -5 & 0 & 0 & 0 \\ 0 & -2 & -2 & -2 & 1 \\ 0 & -3 & 0 & 0 & 0 \\ -4 & 2 & 1 & 0 & 1 \end{bmatrix}, \bar{A}_{d12} = \begin{bmatrix} -1 \\ 1 \\ 2 \\ 0 \\ 0 \end{bmatrix}.$$

The unknown input $d(t)$ is as given in (4.26), where $z(t) = F\bar{x}_1(t)$, $F = \begin{bmatrix} 1 & 2 & 0 & 0 & 1 \end{bmatrix}$. Thus to estimate the unknown input $d(t)$ as defined by (4.26), we need to estimate $z(t)$ of the disturbance-free time-delay system (4.3)–(4.4), where $\bar{A}_{11}$, $\bar{A}_{12}$, $\bar{A}_{d11}$ and $\bar{A}_{d12}$ are given above.

With $\bar{A}_{11}$, $\bar{A}_{d11}$ and F given above, we obtain the following partitioned matrices

$F_1 = \begin{bmatrix}1\,2\end{bmatrix}, F_2 = \begin{bmatrix}0\,0\,1\end{bmatrix}, \bar{A}_{11}^{11} = \begin{bmatrix} -10 & 2 \\ 1 & -1 \end{bmatrix}, \bar{A}_{d11}^{11} = \begin{bmatrix} -3 & 1 \\ 1 & -5 \end{bmatrix}, \bar{A}_{11}^{12} = \begin{bmatrix} 1 & 1 & 0 \\ 0 & 1 & 1 \end{bmatrix},$
$\bar{A}_{d11}^{12} = \begin{bmatrix} 0 & 0 & 0 \\ 0 & 0 & 0 \end{bmatrix}, \bar{A}_{11}^{21} = \begin{bmatrix} 1 & 0 \\ -3 & 1 \\ -2 & -4 \end{bmatrix}, \bar{A}_{d11}^{21} = \begin{bmatrix} 0 & -2 \\ 0 & -3 \\ -4 & 2 \end{bmatrix}, \bar{A}_{11}^{22} = \begin{bmatrix} -5 & 3 & 4 \\ 0 & -8 & 3 \\ 0 & 0 & -6 \end{bmatrix}$
and $\bar{A}_{d11}^{22} = \begin{bmatrix} -2 & -2 & 1 \\ 0 & 0 & 0 \\ 1 & 0 & 1 \end{bmatrix}$.

We can find that a functional observer (10.1)–(10.2) does not exist even for the case where its order has reached $(n - p)$. This is due to the fact that neither condition (10.16) nor condition (10.27) hold.

As for observer (10.48)–(10.49), it is found that condition (10.60) is not satisfied, and thus a first-order scalar functional observer (10.48)–(10.49) does not exist. Note that we can follow similar logic as in Step 1 of Chap. 10 by adding an additional row vector into F_2 and then repeat the design process for a second-order functional observer. However, in this example, we cannot find an additional row vector to be added into F_2 such that condition (10.60) is satisfied. On the other hand, by letting $F_2 = I_3$, a third-order state observer (10.48)–(10.49) exists since condition (10.60) is satisfied and L_1 can be found to ensure that the time-delay system (10.71) is asymptotically stable. However, a third-order state observer does not provide a simple observer scheme for estimating the unknown input $d(t)$, if indeed there might exist a simpler observer for $z(t)$.

Next, we look at observer (10.72)–(10.73) and design a first-order observer if it exists. For this, we first check condition (10.104) (or (10.103)) and it is found to be satisfied. From (10.108)–(10.111), we obtain $\tilde{N}_1 = -6$, $\tilde{N}_2 = 0_{4\times1}$, $\tilde{N}_{d1} = 2$ and $\tilde{N}_{d2} = 0_{4\times1}$. Since both pairs $(\tilde{N}_1, \tilde{N}_2)$ and $(\tilde{N}_{d1}, \tilde{N}_{d2})$ are not observable, we can let $\tilde{Z} = 0_{1\times4}$. Thus, the error time-delay system (10.112) is obtained, where

$$\dot{e}(t) = -6e(t) + 2e(t - \tau). \tag{4.28}$$

The above time-delay system is asymptotically stable independent of τ ([5]), i.e., for any value of τ.

From (10.105), (10.106) and (10.107), respectively, we obtain $M = \begin{bmatrix}1 & 2\end{bmatrix}$, $M_d = \begin{bmatrix}1 & -1\end{bmatrix}$, $N = -6$, $N_d = 2$. From (10.90), (10.92) and (10.94), respectively, we obtain $\tilde{T} = \begin{bmatrix}4 & 8\end{bmatrix}$, $\tilde{W} = \begin{bmatrix}5 & -5\end{bmatrix}$ and $\tilde{U} = \begin{bmatrix}4 & -6\end{bmatrix}$. Matrices G, G_d and G_{dd} are obtained from (10.84)–(10.86), where $G = \begin{bmatrix}-2 & -4\end{bmatrix}$, $G_d = \begin{bmatrix}1 & 5\end{bmatrix}$ and $G_{dd} = \begin{bmatrix}6 & -8\end{bmatrix}$. Finally, we obtain matrices J, J_d and J_{dd} from (10.78)–(10.80), respectively, where $J = \begin{bmatrix}-2 & 0 & 0\end{bmatrix}$, $J_d = \begin{bmatrix}-1 & 0 & 0\end{bmatrix}$ and $J_{dd} = \begin{bmatrix}2 & 0 & 0\end{bmatrix}$. With the above obtained matrices, conditions (10.81)–(10.83) are satisfied thus verifying the correctness of the design procedure.

Finally, the unknown input $d(t)$ can now be estimated by using a first-order functional observer (10.72)–(10.73), where

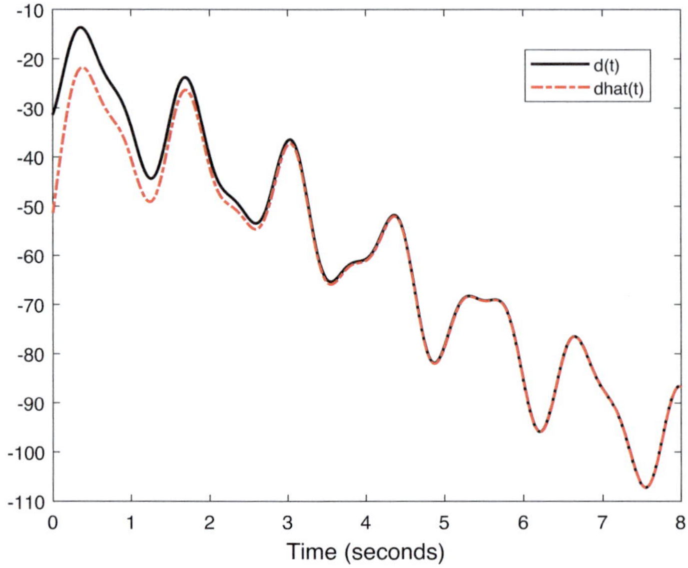

Fig. 4.5 Trajectories of $d(t)$ and $\hat{d}(t)$ (Numerical Example 4, Chap. 4)

$$
\begin{aligned}
\hat{d}(t) &= \dot{y}_1(t) + 9y_1(t) + [1 \ \ 2 \ \ 3]y(t-\tau) + \hat{z}(t), \\
\hat{z}(t) &= w(t) + [-2 \ \ 1 \ \ 2]y(t) + [-2 \ \ 1 \ \ -1]y(t-\tau), \\
\dot{w}(t) &= -6w(t) + 2w(t-\tau) + [2 \ \ -2 \ \ -4]y(t) \\
&\quad + [-3 \ \ 1 \ \ 5]y(t-\tau) + [-10 \ \ 6 \ \ -8]y(t-2\tau), \ t \geq 0.
\end{aligned}
$$

Figure 4.5 shows the responses of $d(t)$ and $\hat{d}(t)$.

4.3.7 Scenario 3

So far, as for Design Option 2, we have covered two scenarios based on the structure of submatrices $\bar{A}_{21}$ and $\bar{A}_{d21}$. In this scenario, we discuss the design of an unknown input $d(t)$ by considering the case where $\bar{A}^2_{d21}$ is linearly *independent* of $\bar{A}^2_{21}$, i.e., rank $\begin{bmatrix} \bar{A}^2_{21} \\ \bar{A}^2_{d21} \end{bmatrix} = 2$. In this scenario, apart from Design Option 1 which employs a reduced-order state observer to estimate $(n-p)$ unmeasurable state variables of $\bar{x}_1(t)$, we design a linear functional observer to estimate two scalar functions, namely, $z_1(t) = -\bar{A}_{21}\bar{x}_1(t)$ and $z_2(t) = -\bar{A}_{d21}\bar{x}_1(t)$. Thus, if a functional observer of order less than $(n-p)$ exists, then it will be used to estimate the unknown input. In such a case, the unknown input as defined in (4.5) can be estimated as follows

$$\hat{d}(t) = V_1\dot{y}(t) - \bar{A}_{22}V_1y(t) - \bar{A}_{d22}V_1y(t-\tau) \\ + \hat{z}_1(t) + \hat{z}_2(t-\tau), \; t \geq \tau. \tag{4.29}$$

By letting $\zeta(t) = \begin{bmatrix} z_1(t) \\ z_2(t) \end{bmatrix} = -\begin{bmatrix} \bar{A}_{21} \\ \bar{A}_{d21} \end{bmatrix}\bar{x}_1(t) \triangleq F_a\bar{x}_1(t)$, i.e., $\zeta(t) = F_a\bar{x}_1(t)$, $F_a \in \mathbb{R}^{2\times(n-1)}$, we design a low-order functional observer to estimate $\zeta(t)$, and hence $d(t)$ can be estimated as in (4.29).

4.3.8 Numerical Example 5

This numerical example is given to illustrate Scenario 3. We reconsider Numerical Example 2 but change the element in $A_d(1,5)$ from 0 to -1, i.e., $A_d(1,5) = -1$. With this change in matrix A_d, we obtain the following submatrices

$$\bar{A}_{d11} = \begin{bmatrix} -3 & 1 & 0 & 2.5 & 0 \\ 1 & -5 & 1 & 0 & 1 \\ 0 & -2 & -2 & -2 & 1 \\ 0 & -3 & 0 & 1 & 0 \\ -4 & 2 & 1 & 0 & 1 \end{bmatrix}, \; \bar{A}_{d12} = \begin{bmatrix} 1 \\ 1 \\ 2 \\ 1 \\ 0 \end{bmatrix},$$

$\bar{A}_{d21} = \begin{bmatrix} -2 & -3 & 0 & -1 & 0 \end{bmatrix}$ and $\bar{A}_{d22} = -6$.

We proceed to design a second-order functional observer (10.1)–(10.2) to estimate $\zeta(t) = F_a\bar{x}_1(t)$, where $F_a = \begin{bmatrix} 1 & 2 & 0 & 0 & 1 \\ 2 & 3 & 0 & 1 & 0 \end{bmatrix}$. Now, F_a is partitioned as $F_a = [F_1 \;\; F_2]$, where $F_1 = \begin{bmatrix} 1 & 2 \\ 2 & 3 \end{bmatrix}$ and $F_2 = \begin{bmatrix} 0 & 0 & 1 \\ 0 & 1 & 0 \end{bmatrix}$.

First, we check condition (10.22) and it is found to be satisfied. From (10.25) and (10.26), we obtain $N_1 = \begin{bmatrix} -7 & 1 \\ 3 & -8.4 \end{bmatrix}$ and $N_2 = 0_{2\times 2}$. By letting $Z = 0$, matrix $N = N_1$ is stable with two eigenvalues at $s_1 = -5.8318$ and $s_2 = -9.5682$. Then the rest of the observer's parameters can be obtained as follows. From (10.23), (10.10), (10.12), and (10.14), matrices L_1, M, G and G_d, respectively, are obtained, where $L_1 = \begin{bmatrix} 0 & -1 \\ -0.4 & 0 \end{bmatrix}$, $M = \begin{bmatrix} 1 & 3 \\ 2.4 & 3 \end{bmatrix}$, $G = \begin{bmatrix} -2.6 & -10 \\ -2.36 & 3.2 \end{bmatrix}$ and $G_d = \begin{bmatrix} -5 & 7 \\ 1.2 & -3.4 \end{bmatrix}$. With $L = \begin{bmatrix} L_1 & F_2 \end{bmatrix} = \begin{bmatrix} 0 & -1 & 0 & 0 & 1 \\ -0.4 & 0 & 0 & 1 & 0 \end{bmatrix}$, we obtain $J = L\bar{A}_{12}V_1 = \begin{bmatrix} -2 & 0 & 0 \\ 0.6 & 0 & 0 \end{bmatrix}$ and $J_d = L\bar{A}_{d12}V_1 = \begin{bmatrix} -1 & 0 & 0 \\ 0.6 & 0 & 0 \end{bmatrix}$. We can verify that with the above obtained matrices, conditions (10.7)–(10.9) are satisfied. This completes the design of a second-order functional observer (10.1)–(10.2) to estimate the function $\zeta(t)$. Accordingly, $\hat{\zeta}(t) = \begin{bmatrix} \hat{z}_1(t) \\ \hat{z}_2(t) \end{bmatrix}$ is obtained as below

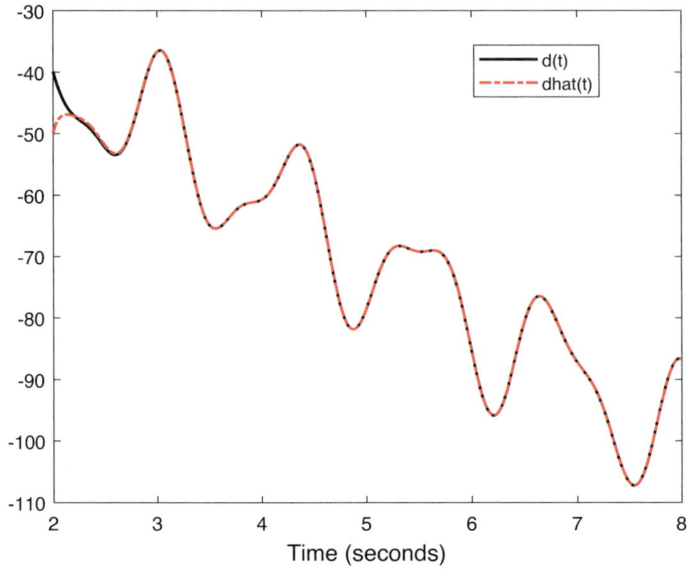

Fig. 4.6 Trajectories of $d(t)$ and $\hat{d}(t)$ (Numerical Example 5, Chap. 4)

$$\begin{bmatrix} \hat{z}_1(t) \\ \hat{z}_2(t) \end{bmatrix} = \begin{bmatrix} w_1(t) \\ w_2(t) \end{bmatrix} + \begin{bmatrix} -2 & 1 & 3 \\ -4.8 & 2.4 & 3 \end{bmatrix} y(t), \tag{4.30}$$

$$\begin{bmatrix} \dot{w}_1(t) \\ \dot{w}_2(t) \end{bmatrix} = \begin{bmatrix} -7 & 1 \\ 3 & -8.4 \end{bmatrix} \begin{bmatrix} w_1(t) \\ w_2(t) \end{bmatrix} + \begin{bmatrix} 3.2 & -2.6 & -10 \\ 5.32 & -2.36 & 3.2 \end{bmatrix} y(t) + \begin{bmatrix} 9 & -5 & 7 \\ -1.8 & 1.2 & -3.4 \end{bmatrix} y(t-\tau). \tag{4.31}$$

Thus, from (4.29), the unknown input can now be estimated with $\hat{z}_1(t)$ and $\hat{z}_2(t-\tau)$ as given in (4.30)–(4.31). Figure 4.6 shows the responses of $d(t)$ and $\hat{d}(t)$.

4.3.9 Scenario 4

In this scenario, thanks to the generalized functional estimation problem presented in Chap. 13, we estimate the unknown input $d(t)$ by designing a functional observer to estimate directly the following linear function

$$\bar{z}(t) = -\bar{A}_{21}\bar{x}_1(t) - \bar{A}_{d21}\bar{x}_1(t-\tau) \tag{4.32}$$

for the reduced-order disturbance-free time-delay system (4.3)–(4.4).

The distinctive feature of the above estimation problem is that $\bar{z}(t)$ contains linear combinations of both the instantaneous and delayed state vectors, i.e., $-\bar{A}_{21}\bar{x}_1(t)$ and $-\bar{A}_{d21}\bar{x}_1(t-\tau)$. Please refer to Chap. 13 for their design and several numerical examples showing their salient features. Thus, if a functional observer, $\hat{\bar{z}}(t)$, can be designed such that $\hat{\bar{z}}(t)$ converges asymptotically to $\bar{z}(t)$, then from (4.5), the unknown input $d(t)$ can be estimated according to the following equation

$$\hat{d}(t) = V_1\dot{y}(t) - \bar{A}_{22}V_1y(t) - \bar{A}_{d22}V_1y(t-\tau) + \hat{\bar{z}}(t). \tag{4.33}$$

4.3.10 Numerical Example 6

This numerical example is given to illustrate Scenario 4, where to estimate the unknown input $d(t)$, we need to employ a delayed functional observer. We consider the time-delay system (4.1)–(4.2) with the following matrices

$$A = \begin{bmatrix} -10 & 0 & 0 \\ -21 & 0 & 0 \\ -31 & 0 & 1 \end{bmatrix},\ A_d = \begin{bmatrix} -8 & 0 & 1 \\ -26 & -1 & 4 \\ -34 & 0 & 3 \end{bmatrix},\ D = \begin{bmatrix} 1 \\ 2 \\ 3 \end{bmatrix} \text{ and } C = \begin{bmatrix} 1 & 0 & 0 \\ 0 & 1 & 0 \end{bmatrix}.$$

Our aim is to design an observer to estimate the unknown input $d(t)$. Based on the matrix decomposition technique presented in Sect. 2.2, we obtain the following matrices

$$P = \begin{bmatrix} 0 \\ 1 \end{bmatrix},\ E = \begin{bmatrix} 0 & 0 \\ 1 & 0 \\ 0 & 1 \end{bmatrix},\ V = [CD\ \ P] = \begin{bmatrix} 1 & 0 \\ 2 & 1 \end{bmatrix},\ V_1 = [1\ \ 0],\ V_2 = [-2\ \ 1] \text{ and}$$

$$T = [E\ \ D] = \begin{bmatrix} 0 & 0 & 1 \\ 1 & 0 & 2 \\ 0 & 1 & 3 \end{bmatrix}.$$

With the state transformation, $x(t) = T\bar{x}(t)$, we obtain the disturbance-free time-delay system (4.3)–(4.4), where

$$\bar{A}_{11} = \begin{bmatrix} 0 & 0 \\ 0 & 1 \end{bmatrix},\ \bar{A}_{d11} = \begin{bmatrix} -1 & 2 \\ 0 & 0 \end{bmatrix},\ \bar{A}_{12} = \begin{bmatrix} -1 \\ 2 \end{bmatrix},\ \bar{A}_{d12} = \begin{bmatrix} -6 \\ -10 \end{bmatrix} \text{ and } \tilde{C}_1 = V_2CE$$
$$= [1\ \ 0].$$

With the state transformation matrix T, we can obtain the following sub-matrices

$\bar{A}_{21} = [0\ \ 0]$, $\bar{A}_{d21} = [0\ \ 1]$, $\bar{A}_{22} = -10$ and $\bar{A}_{d22} = -5$.

With the above sub-matrices and from (4.5), the unknown input $d(t)$ is

$$d(t) = \dot{y}_1(t) + 10y_1(t) + 5y_1(t-\tau) - \bar{A}_{d21}\bar{x}_1(t-\tau),\ t \geq 0. \tag{4.34}$$

Thus, to estimate the unknown input as given in (4.34), we design an observer to estimate the function

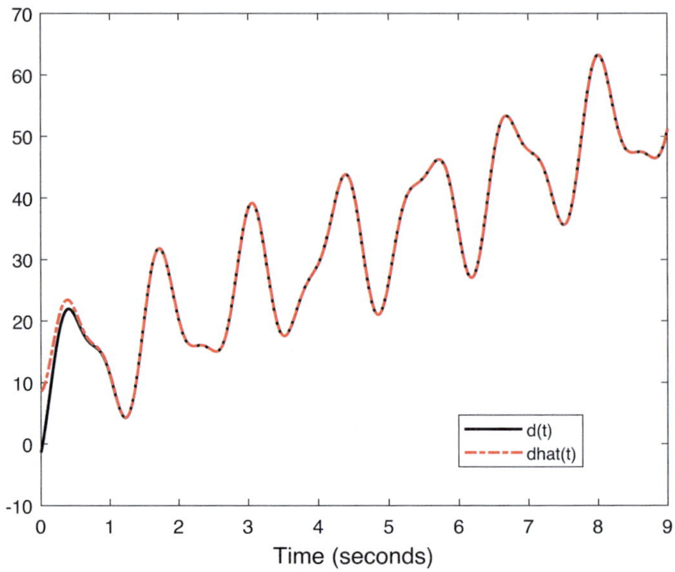

Fig. 4.7 Trajectories of $d(t)$ and $\hat{d}(t)$ (Numerical Example 6, Chap. 4)

$$\bar{z}(t) = \bar{A}_{d21}\bar{x}_1(t-\tau).$$

Please refer to Numerical Example 1 (Chap. 13) where we discuss the advantages and design of a first-order observer, $\hat{\bar{z}}(t)$, to estimate $\bar{z}(t)$. Hence, $d(t)$ can now be estimated based on the following equations

$$\begin{aligned}
\hat{d}(t) &= \dot{y}_1(t) + 10y_1(t) + 5y_1(t-\tau) - \hat{\bar{z}}(t),\\
\hat{\bar{z}}(t) &= w(t) - 6y_1(t) + 3y_2(t),\\
\dot{w}(t) &= -5w(t) + 33y_1(t) - 15y_2(t) + 14y_1(t-\tau)\\
&\quad + 3y_2(t-\tau) - 10y_1(t-2\tau).
\end{aligned}$$

Figure 4.7 shows the responses of $d(t)$ and $\hat{d}(t)$, where asymptotic tracking is achieved.

4.4 Concluding Remark

In this chapter, we have presented the first result on the design of unknown input observers for a class of time-delay systems, where a constant time delay τ occurs in the state vector. Based on the matrix decomposition technique presented in Chap. 2, we can effectively obtain a reduced-order disturbance-free time-delay system. The derived unknown input is in the simplest form that contains only one output measurement, its delayed output measurement, first-derivative of one output measurement, and a linear function of the state vector which contains a state-delayed term. We have looked at some possible ways to realize the estimation of the generalized functional by examining two design options, where Design Option 2 comprises four scenarios. Scenario 4 of Design Option 2 points to the need to estimate the generalized functional directly as this is an effective and economical way. Thus, the reader can refer to later chapters (see, Chap. 13) in the book for some results and discussions on the significance of estimating (4.32) directly.

Finally, we conclude this chapter by pointing out that, based on the two design approaches discussed in Chap. 3, we can design an unknown input observer in the presence of multiple unknown inputs for time-delay system (4.1)–(4.2). For example, as in Design Approach 1, if $D_{r1} = 0_{1\times(q-1)}$, then $\hat{d}_1(t)$ can be obtained as follows

$$\hat{d}_1(t) = V_1\dot{y}(t) - \bar{A}_{22}V_1y(t) - \bar{A}_{d22}V_1y(t-\tau) + \hat{\bar{z}}(t), \tag{4.35}$$

where we design an observer $\hat{\bar{z}}(t)$ to estimate $\bar{z}(t)$ as defined by (4.32) for the following time-delay system with an unknown input vector $d_r(t)$

$$\begin{aligned} \dot{\bar{x}}_1(t) &= \bar{A}_{11}\bar{x}_1(t) + \bar{A}_{d11}\bar{x}_1(t-\tau) + D_{r2}d_r(t) \\ &\quad + \bar{A}_{12}V_1y(t) + \bar{A}_{d12}V_1y(t-\tau), \ t \geq 0, \qquad (4.36) \\ \bar{y}(t) &= \tilde{C}_1\bar{x}_1(t). \qquad (4.37) \end{aligned}$$

We would like to leave this problem for the reader to investigate.

References

1. Trinh, H.: Linear functional state observers for time-delay systems. Int. J. Control **72**(18), 1642–1658 (1999)
2. Trinh, H., Fernando, T.: Functional Observers for Dynamical Systems. Springer, Berlin (2012)
3. Darouach, M.: Linear functional observers for systems with delays in state variables. IEEE Trans. Automat. Control **46**(3), 491–496 (2001)
4. Nam, P.T., Pathirana, P.N., Trinh, H.: ε-bounded state estimation for time-delay systems with bounded disturbances. Int. J. Control **87**(9), 1747–1756 (2014)
5. Mori, T.: Criteria for asymptotic stability of linear time-delay systems. IEEE Trans. Automat. Control **30**(2), 158–161 (1985)
6. Wu, Z., Michiels, W.: Reliably computing all characteristic roots of delay differential equations in a given right half plane using a spectral method. J. Comput. Appl. Math. **236**(9), 2499–2514 (2012)

Chapter 5
Unknown Input Observers Design: A Time Delay in Both State and Output Vectors

In this chapter, we discuss the design of unknown input observers for linear time-delay systems, where both the state and output vectors are subject to constant delays. In this class of time-delay systems, recently, the authors in [1] addressed the long-standing problem of estimating both the states and unknown inputs (faults) using sliding mode observers. By the $\mathscr{H}_\infty$ filtering approach, $\mathscr{H}_\infty$ asymptotic estimation for both states and faults can be achieved. Here, our objective is to design an unknown input observer which uses first-derivative of only one output measurement, $y_i(t)$, together with a state/functional observer to asymptotically estimate the unknown input. With our matrix decomposition technique as presented in Chap. 2, however, the presence of a time delay in the output measurements poses some difficulties when decomposing the system into a disturbance-free time-delay system. To overcome this challenge, we impose some additional constraints on the output matrices, to be detailed in this chapter.

5.1 System Description and Preliminaries

Let us now consider system (4.1)–(4.2) where the output vector is further subject to a constant time delay, i.e.,

$$\dot{x}(t) = Ax(t) + A_d x(t-\tau) + Dd(t), \ t \geq 0, \tag{5.1}$$

$$y(t) = Cx(t) + C_d x(t-\tau), \tag{5.2}$$

where $C_d \in \mathbb{R}^{p\times n}$ is a known constant matrix. For ease of presentation, we assume that matrix C has full-row rank and without loss of generality, it takes the form $C = [I_p \ 0_{p\times(n-p)}]$. The case where C is not a full-row rank matrix will be discussed in a numerical example later on in this chapter (see, Numerical Example 4).

H. Trinh et al., *Unknown Inputs Estimation in Linear Time-Delay Systems Using Generalized Functional Observers*, Studies in Systems, Decision and Control 648,
https://doi.org/10.1007/978-3-032-11870-7_5

Based on the matrix decomposition technique as presented in Sect. 2.2 (see Algorithm 1, Steps 1-4), we first select an output measurement, $y_i(t)$, where its first-derivative will be used to estimate the unknown input. We then obtain matrices V, T and E such that V_1 takes the form (2.4), condition (2.1) is satisfied, i.e., $V_1CE = 0_{1\times(n-1)}$, and $T^{-1}D = \begin{bmatrix} 0_{n-1} \\ 1 \end{bmatrix}$. We also partition matrix C_d as $C_d = \begin{bmatrix} C_{di} \\ C_{dr} \end{bmatrix}$, where $C_{di} \in \mathbb{R}^{1\times n}$ and $C_{dr} \in \mathbb{R}^{(p-1)\times n}$.

Firstly, by applying the state transformation $x(t) = T\bar{x}(t)$ to (5.1)–(5.2), we obtain the following equations

$$\dot{\bar{x}}_1(t) = \bar{A}_{11}\bar{x}_1(t) + \bar{A}_{12}\bar{x}_2(t) + \bar{A}_{d11}\bar{x}_1(t-\tau) + \bar{A}_{d12}\bar{x}_2(t-\tau), \tag{5.3}$$

$$\dot{\bar{x}}_2(t) = \bar{A}_{21}\bar{x}_1(t) + \bar{A}_{22}\bar{x}_2(t) + \bar{A}_{d21}\bar{x}_1(t-\tau) + \bar{A}_{d22}\bar{x}_2(t-\tau) + d(t), \tag{5.4}$$

$$y(t) = [CE \;\; CD]\begin{bmatrix} \bar{x}_1(t) \\ \bar{x}_2(t) \end{bmatrix} + [C_dE \;\; C_dD]\begin{bmatrix} \bar{x}_1(t-\tau) \\ \bar{x}_2(t-\tau) \end{bmatrix}. \tag{5.5}$$

Next, we pre-multiply both sides of (5.5) by V^{-1} (where $V^{-1} = \begin{bmatrix} V_1 \\ V_2 \end{bmatrix}$) and by noting that $V_1CE = 0$, $V_1CD = 1$, $V_2CD = 0$, we obtain the following two equations

$$V_1y(t) = \bar{x}_2(t) + V_1C_dE\bar{x}_1(t-\tau) + V_1C_dD\bar{x}_2(t-\tau), \tag{5.6}$$

$$V_2y(t) = V_2CE\bar{x}_1(t) + V_2C_dE\bar{x}_1(t-\tau) + V_2C_dD\bar{x}_2(t-\tau). \tag{5.7}$$

5.2 Structural Constraint on C_d

It is clear from (5.6) that the presence of the state-delayed term, $V_1C_dD\bar{x}_2(t-\tau)$, in the output measurement $y_i(t)$, i.e., $V_1y(t)$, makes it not possible to express $\bar{x}_2(t)$ in terms of $y(t)$ and $\bar{x}_1(t-\tau)$. To overcome this problem, we impose a constraint on the output matrix C_d such that the following condition holds

$$V_1C_dD = 0. \tag{5.8}$$

With the structural constraint of V_1 in the form (2.4), we can see that condition (5.8) holds if either one of the following conditions holds:

$$V_1C_d = 0_{1\times n}, \tag{5.9}$$

$$C_{di}D = 0. \tag{5.10}$$

5.2.1 Structural Constraint 1: $V_1 C_d = 0_{1\times n}$

For condition (5.9) to hold, we need the first row of C_d to be zeros, i.e., $C_{di} = 0$. The implication of this requirement is that the selected output measurement, $y_i(t) = C_i x(t) + C_{di} x(t-\tau)$, where its first-derivative is to be used for the estimation of the unknown input, must not contain any state-delayed term, i.e., we require $C_{di} x(t-\tau) = 0$ and hence $C_{di} = 0$. Also, note that condition (5.9) implies $V_1 C_d E = 0_{1\times(n-1)}$, and thus from (5.6), we obtain

$$\bar{x}_2(t) = V_1 y(t). \tag{5.11}$$

5.2.2 Structural Constraint 2: $C_{di} D = 0$

On the other hand, as for condition (5.10) to hold, we can have $C_{di} \neq 0$ such that $C_{di} \subset \mathscr{D}_{null}$ where $\mathscr{D}_{null} \in \mathbb{R}^{(n-1)\times n}$ is an orthogonal basis of the left null-space of D (i.e., $\mathscr{D}_{null} D = 0_{(n-1)\times 1}$). Thus, we can allow time delay to enter into the output measurement, $y_i(t)$, but it needs to subject to the constraint (5.10). Hence, from (5.6), we obtain

$$\bar{x}_2(t) = V_1 y(t) - V_1 C_d E \bar{x}_1(t-\tau). \tag{5.12}$$

It should be noted here that condition (5.10) covers a wider class of time-delay systems than condition (5.9). However, decomposition (5.12) will result in a more complex disturbance-free time-delay system than decomposition (5.11), which in turn requires a more complex functional observer with multiple internal delays, to be detailed in the subsequent.

5.3 Unknown Input Observer Design Subject to Constraint 1

For each of the decompositions (5.11) and (5.12), we can derive a disturbance-free time-delay system and obtain a closed form expression for the unknown input, $d(t)$. Firstly, with the decomposition (5.11), and from (5.3) and (5.7), we obtain the following disturbance-free time-delay system

$$\begin{aligned}\dot{\bar{x}}_1(t) &= \bar{A}_{11}\bar{x}_1(t) + \bar{A}_{d11}\bar{x}_1(t-\tau)\\ &\quad + \bar{A}_{12}V_1 y(t) + \bar{A}_{d12}V_1 y(t-\tau),\ t \geq 0, \end{aligned} \tag{5.13}$$

$$\tilde{y}(t) = \tilde{C}_1\bar{x}_1(t) + \tilde{C}_{d1}\bar{x}_1(t-\tau), \tag{5.14}$$

where

$$\tilde{y}(t) = V_2 y(t) - V_2 C_d D V_1 y(t-\tau), \tag{5.15}$$

$$\tilde{C}_{d1} = V_2 C_d E. \tag{5.16}$$

By substituting (5.11) into (5.4), the unknown input, $d(t)$, is given in the form (4.5) as follows

$$\begin{aligned} d(t) &= V_1 \dot{y}(t) - \bar{A}_{22} V_1 y(t) - \bar{A}_{d22} V_1 y(t-\tau) \\ &- \bar{A}_{21} \bar{x}_1(t) - \bar{A}_{d21} \bar{x}_1(t-\tau),\ t \geq 0. \end{aligned}$$

Thus, the design of the unknown input, $d(t)$, can be carried out by designing a reduced-order state/functional observer for system (5.13)–(5.14). Note that for the time-delay system (5.13)–(5.14), the output vector, $\tilde{y}(t)$, contains a state-delayed term, $\tilde{C}_{d1}\bar{x}_1(t-\tau)$, and the reader can refer to Chap. 11 for an observer design procedure to estimate a linear function of the state vector i.e., $z(t) = F\bar{x}_1(t)$, or Chap. 13 for a design procedure to estimate both the instantaneous and delayed functions of the state vector, i.e., $\bar{z}(t) = F\bar{x}_1(t) + F_d\bar{x}_1(t-\tau)$.

Before we proceed to look at decomposition (5.12), let us consider a few numerical examples, where we design unknown input observers by using functional observers for system (5.13)–(5.14).

5.3.1 Numerical Example 1

In this example, we consider time-delay system (5.1)–(5.2), where $\tau = 0.5s$, matrices A, A_d, D and C are as given in Numerical Example 6 (Chap. 4). In this example, we let the second output of the system be further subject to a constant time-delay in the state vector, where $C_d = \begin{bmatrix} 0 & 0 & 0 \\ 1 & 1 & 1 \end{bmatrix}$. Our aim is to design an observer to estimate the unknown input $d(t)$.

With C_d as given above, we can verify that condition (5.9) is satisfied and hence decomposition (5.11) is achieved, which results in the disturbance-free time-delay system (5.13)–(5.14), where $\bar{A}_{11}, \bar{A}_{d11}, \bar{A}_{12}, \bar{A}_{d12}$ and $\tilde{C}_1$ are as obtained in Numerical Example 6 (Chap. 4), $\tilde{C}_{d1} = V_2 C_d E = \begin{bmatrix} 1 & 1 \end{bmatrix}$, and $\tilde{y}(t) = V_2 y(t) - V_2 C_d D V_1 y(t - \tau) = \begin{bmatrix} -2 & 1 \end{bmatrix} y(t) - 6y_1(t-\tau)$.

The unknown input, $d(t)$, is as given in (4.34), where

$$d(t) = \dot{y}_1(t) + 10y_1(t) + 5y_1(t-\tau) - \begin{bmatrix} 0 & 1 \end{bmatrix}\bar{x}_1(t-\tau),\ t \geq 0.$$

We now design an observer to estimate the following function

$$\bar{z}(t) = F\bar{x}_1(t) + F_d\bar{x}_1(t-\tau)$$

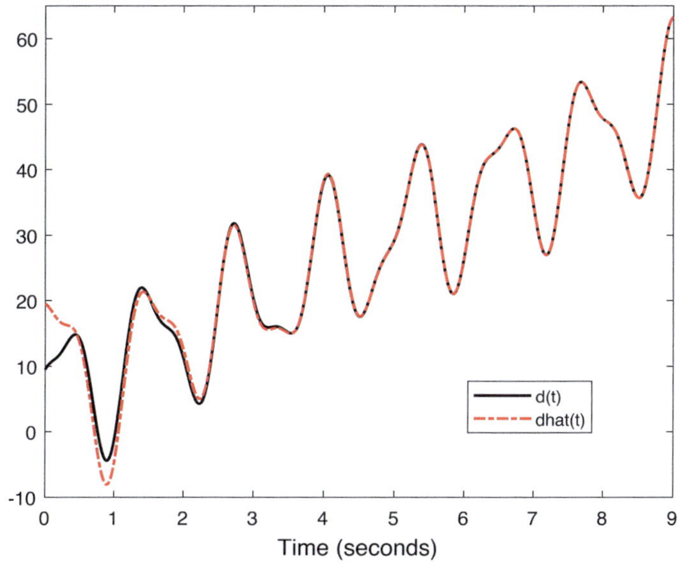

Fig. 5.1 Trajectories of $d(t)$ and $\hat{d}(t)$ (Numerical Example 1, Chap. 5)

where $F = \begin{bmatrix} 0 & 0 \end{bmatrix}$ and $F_d = \begin{bmatrix} 0 & 1 \end{bmatrix}$, and thus the unknown input can be estimated as follows

$$\hat{d}(t) = \dot{y}_1(t) + 10y_1(t) + 5y_1(t-\tau) - \hat{\bar{z}}(t). \tag{5.17}$$

Note that for this example, the designed functional observer directly estimates the delayed state variable, $\bar{x}_{12}(t-\tau)$, of the time-delay system (5.13)–(5.14). Please refer to Numerical Example 4 (Chap. 13) where we designed a first-order observer with an internal delay to estimate $\bar{z}(t)$. The observer is as follows

$$\hat{\bar{z}}(t) = w(t) - 1.2y_1(t) + 0.6y_2(t) - 3.6y_1(t-\tau), \tag{5.18}$$

$$\begin{aligned} \dot{w}(t) &= -0.8w(t) - 1.8w(t-\tau) + 1.56y_1(t) - 0.48y_2(t) \\ &+ 8.84y_1(t-\tau) - 0.48y_2(t-\tau) + 2.48y_1(t-2\tau). \end{aligned} \tag{5.19}$$

Figure 5.1 shows the trajectories of $d(t)$ and $\hat{d}(t)$, where asymptotic tracking is achieved.

Compared to the case of Numerical Example 6 (Chap. 4) (i.e., $C_d = 0$), asymptotic tracking is achieved, but it takes a bit longer for $\hat{d}(t)$ to converge to $\hat{d}(t)$ due to the stability requirement of the error time-delay system, which can only be stable for a range of time delay τ (see Numerical Example 4, Chap. 13, for further details).

5.3.2 Numerical Example 2

In this example, we consider time-delay system (5.1)–(5.2), where $\tau = 1s$, matrices A, A_d, D, C and C_d are given as follows $A = \begin{bmatrix} -10 & 0 & 1 \\ -21 & -10 & 0 \\ -31 & 0 & -4 \end{bmatrix}$, $A_d = \begin{bmatrix} -8 & 1 & 1 \\ -26 & -1 & 4 \\ -34 & 0 & -3 \end{bmatrix}$, $D = \begin{bmatrix} 1 \\ 2 \\ 3 \end{bmatrix}$, $C = \begin{bmatrix} 1 & 0 & 0 \\ 0 & 1 & 0 \end{bmatrix}$ and $C_d = \begin{bmatrix} 0 & 0 & 0 \\ 1 & 1 & 1 \end{bmatrix}$.

Our aim is to design an unknown observer to estimate the unknown input $d(t)$. With C_d as given, condition (5.9) is satisfied and the disturbance-free time-delay system (5.13)–(5.14) is obtained, where $\bar{A}_{11} = \begin{bmatrix} -10 & -2 \\ 0 & -7 \end{bmatrix}$, $\bar{A}_{d11} = \begin{bmatrix} -3 & 2 \\ -3 & -6 \end{bmatrix}$, $\bar{A}_{12} = \begin{bmatrix} -27 \\ -22 \end{bmatrix}$, $\bar{A}_{d12} = \begin{bmatrix} -10 \\ -34 \end{bmatrix}$, $V_1 = \begin{bmatrix} 1 & 0 \end{bmatrix}$, $\tilde{C}_1 = \begin{bmatrix} 1 & 0 \end{bmatrix}$ and $\tilde{C}_{d1} = \begin{bmatrix} 1 & 1 \end{bmatrix}$.

Note that $\tilde{y}(t)$ defined by (5.15) is related to $y(t)$ in (5.2) as $\tilde{y}(t) = V_2 y(t) - V_2 C_d D V_1 y(t-\tau)$, where $V_2 = [-2 \ \ 1]$ and thus, $\tilde{y}(t) = -2y_1(t) + y_2(t) - 6y_1(t-\tau)$.

With the state transformation $x(t) = T\bar{x}(t)$, we also obtain $\bar{A}_{22} = -7$, $\bar{A}_{d22} = -3$, $\bar{A}_{21} = [0 \ \ 1]$ and $\bar{A}_{d21} = [1 \ \ 1]$. Hence from (4.5), the unknown input, $d(t)$, is obtained as follows

$$\begin{aligned} d(t) &= \dot{y}_1(t) + 7y_1(t) + 3y_1(t-\tau) \\ &\quad - \begin{bmatrix} 0 & 1 \end{bmatrix}\bar{x}_1(t) - \begin{bmatrix} 1 & 1 \end{bmatrix}\bar{x}_1(t-\tau), \ t \geq 0. \end{aligned} \tag{5.20}$$

From (5.20), we design an observer to estimate the following function

$$\bar{z}(t) = F\bar{x}_1(t) + F_d\bar{x}_1(t-\tau), \tag{5.21}$$

where $F = \begin{bmatrix} 0 & 1 \end{bmatrix}$ and $F_d = \begin{bmatrix} 1 & 1 \end{bmatrix}$. Then the unknown input can be estimated as follows

$$\hat{d}(t) = \dot{y}_1(t) + 7y_1(t) + 3y_1(t-\tau) - \hat{\bar{z}}(t). \tag{5.22}$$

In Numerical Example 5 (Chap. 13), we discuss the design of functional observers to estimate $\bar{z}(t)$ as defined by (5.21) by exploring different available options via successively expanding the order of the designed observer. As a result, we obtained two observers based on Option (i) and Option (ii). In Option (i), we obtain a second-order observer with an internal delay; whereas in Option (ii), we obtain a third-order observer with no internal delay, and thus its stability is independent of the time delay, τ. Let us now use both observers to estimate the unknown input as defined by (5.22).

Option (i): Second-order functional observer (11.1)–(11.2). As detailed in Numerical Example 5 (Chap. 13), Option (i), to estimate $\bar{z}(t)$, we obtain a second-order observer as defined by (13.47) and (13.46). Thus, the unknown input, $d(t)$, can now be estimated according to the following equations

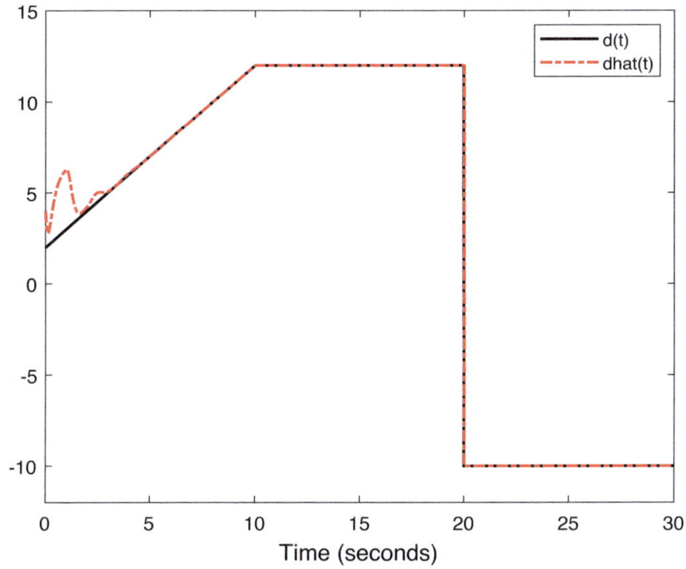

Fig. 5.2 Trajectories of $d(t)$ and $\hat{d}(t)$ (Numerical Example 2, second-order observer with an internal delay, Chap. 5)

$$\hat{d}(t) = \dot{y}_1(t) + 7y_1(t) + 3y_1(t-\tau) - \hat{\bar{z}}(t), \tag{5.23}$$

$$\hat{\bar{z}}(t) = w_1(t) + 0.1852\tilde{y}(t), \tag{5.24}$$

$$\begin{aligned}\begin{bmatrix}\dot{w}_1(t)\\ \dot{w}_2(t)\end{bmatrix} &= \begin{bmatrix}-6.6296 & 4.9259\\ -0.6914 & -5.7284\end{bmatrix}\begin{bmatrix}w_1(t)\\ w_2(t)\end{bmatrix} + \\ &\begin{bmatrix}0.9259 & -0.8148\\ 4.2716 & -3.3457\end{bmatrix}\begin{bmatrix}w_1(t-\tau)\\ w_2(t-\tau)\end{bmatrix} + \begin{bmatrix}-1.0786\\ -1.6046\end{bmatrix}\tilde{y}(t)\\ &+ \begin{bmatrix}-5.3617\\ 1.6019\end{bmatrix}\tilde{y}(t-\tau) + \begin{bmatrix}-17\\ -9.3333\end{bmatrix}y_1(t)\\ &+ \begin{bmatrix}-72.0741\\ 23.6049\end{bmatrix}y_1(t-\tau) + \begin{bmatrix}-35.8519\\ 52.7901\end{bmatrix}y_1(t-2\tau).\end{aligned} \tag{5.25}$$

To show how well $\hat{d}(t)$ tracks $d(t)$ using the above second-order observer, Fig. 5.2 shows the trajectories of $d(t)$ and $\hat{d}(t)$, where asymptotic tracking is achieved. As expected, asymptotic tracking is achieved within about four seconds (please refer to the discussion on the stability of the error time-delay system for this example in Numerical Example 5, Chap. 13, Option (i)).

Option (ii): For this option, we estimate the unknown input (5.22) by employing a third-order functional observer (11.1)–(11.2) without N_d, and thus the stability of the error $e(t)$ is independent of τ, which is its main attractive feature. As detailed in Numerical Example 5 (Chap. 13), Option (ii), to estimate $\bar{z}(t)$, we design a third-order observer, and thus the unknown input, $d(t)$, can now be estimated according to the following equations

$$
\begin{aligned}
&\hat{d}(t) = \dot{y}_1(t) + 7y_1(t) + 3y_1(t-\tau) - \hat{\dot{z}}(t),\\
&\hat{\dot{z}}(t) = w_1(t) + \tilde{y}(t),
\end{aligned} \tag{5.26}
$$

$$
\begin{aligned}
\begin{bmatrix} \dot{w}_1(t) \\ \dot{w}_2(t) \\ \dot{w}_3(t) \end{bmatrix} &= \begin{bmatrix} -7.5 & 4 & 2.5 \\ -4.5 & -10 & 10.5 \\ -2.5 & -1 & -6.5 \end{bmatrix} \begin{bmatrix} w_1(t) \\ w_2(t) \\ w_3(t) \end{bmatrix} \\
&+ \begin{bmatrix} 12 \\ -15 \\ -9 \end{bmatrix} \tilde{y}(t) + \begin{bmatrix} 0 \\ 24 \\ 0 \end{bmatrix} \tilde{y}(t-\tau) + \begin{bmatrix} 5 \\ 81 \\ -49 \end{bmatrix} y_1(t) \\
&+ \begin{bmatrix} -24 \\ 221 \\ -44 \end{bmatrix} y_1(t-\tau) + \begin{bmatrix} 0 \\ 200 \\ 0 \end{bmatrix} y_1(t-2\tau).
\end{aligned} \tag{5.27}
$$

Finally, to show how well $\hat{d}(t)$ tracks $d(t)$ using the above third-order observer, Fig. 5.3 depicts the trajectories of $d(t)$ and $\hat{d}(t)$, where asymptotic tracking is achieved. As expected, asymptotic tracking is achieved within one second which is significantly faster than the case of Option (i). For Option (ii), even though the order of the observer is higher than in Option (i), the other advantage is that its stability is independent of the time delay τ. Please refer to Numerical Example 5 (Chap. 13), Option (ii), for more details on the design of a third-order observer for this example.

5.4 Unknown Input Observer Design Subject to Constraint 2

Secondly, for the decomposition (5.12), from (5.3) and (5.7), we can obtain the following disturbance-free system with multiple time delays in both the state and output vectors

$$
\begin{aligned}
\dot{\bar{x}}_1(t) = {} & \bar{A}_{11}\bar{x}_1(t) + \tilde{A}_{d11}\bar{x}_1(t-\tau) + \tilde{A}_{dd11}\bar{x}_1(t-2\tau) \\
& + \bar{A}_{12}V_1 y(t) + \bar{A}_{d12}V_1 y(t-\tau),
\end{aligned} \tag{5.28}
$$

$$
\tilde{y}(t) = \tilde{C}_1\bar{x}_1(t) + \tilde{C}_{d1}\bar{x}_1(t-\tau) + \tilde{C}_{dd1}\bar{x}_1(t-2\tau), \tag{5.29}
$$

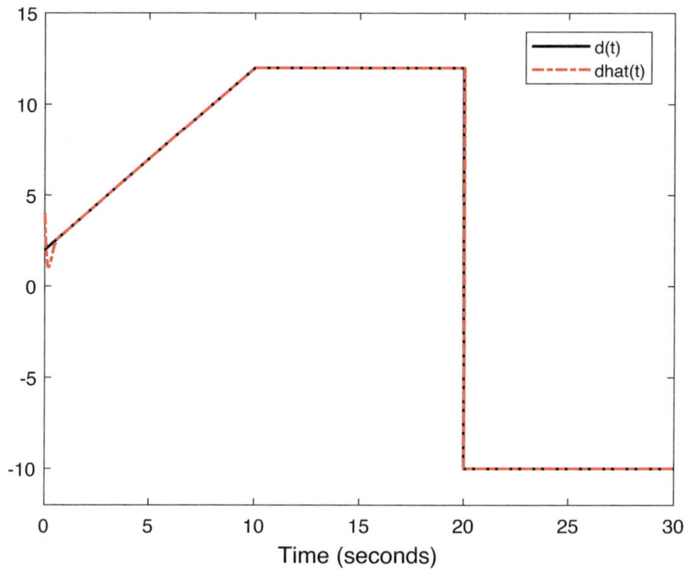

Fig. 5.3 Trajectories of $d(t)$ and $\hat{d}(t)$ (Numerical Example 2, third-order observer with no internal delay, Chap. 5)

where

$$\tilde{A}_{d11} = \bar{A}_{d11} - \bar{A}_{12}V_1C_dE, \tag{5.30}$$

$$\tilde{A}_{dd11} = -\bar{A}_{d12}V_1C_dE, \tag{5.31}$$

$$\tilde{C}_{dd1} = -V_2C_dDV_1C_dE. \tag{5.32}$$

Finally, by substituting (5.12) and (5.28) into (5.4), the unknown input, $d(t)$, is given as follows

$$\begin{aligned} d(t) &= V_1\dot{y}(t) - \bar{A}_{22}V_1y(t) - \tilde{A}_{12}y(t-\tau) - \tilde{A}_{d12}y(t-2\tau) \\ &\quad - \bar{A}_{21}\bar{x}_1(t) - \tilde{A}_{d21}\bar{x}_1(t-\tau) - \tilde{A}_{d22}\bar{x}_1(t-2\tau) - V_1C_dE\tilde{A}_{dd11}\bar{x}_1(t-3\tau), \end{aligned} \tag{5.33}$$

where

$$\tilde{A}_{12} = \bar{A}_{d22}V_1 + V_1C_dE\bar{A}_{12}V_1, \tag{5.34}$$

$$\tilde{A}_{d12} = V_1C_dE\bar{A}_{d12}V_1, \tag{5.35}$$

$$\tilde{A}_{d21} = \bar{A}_{d21} + V_1C_dE\bar{A}_{11} - \bar{A}_{22}V_1C_dE, \tag{5.36}$$

$$\tilde{A}_{d22} = V_1C_dE\tilde{A}_{d11} - \bar{A}_{d22}V_1C_dE. \tag{5.37}$$

Here, system (5.28)–(5.29) contains multiple time delays in both state and output vectors. Note that the unknown input, $d(t)$, as defined in (5.33) contains the first-derivative of one output measurement only but with multiple delays in the state vector $\bar{x}_1(t)$. As such, the design of an unknown input observer to estimate $d(t)$ is more complicated than those cases discussed previously. To estimate $d(t)$, Option 1 as discussed in Sect. 4.2 can be adopted, i.e., we design a state observer to estimate $\bar{x}_1(t)$. But as we will show later in this book, estimating $\bar{x}_1(t)$ is not often a good option (see, Example 1 in Chap. 15). In Chap. 12, we consider the design of functional observers to estimate a linear function of the state vector, $z(t) = F\bar{x}_1(t)$, of system (5.28)–(5.29).

On the other hand, by letting

$$\tilde{z}(t) = -\bar{A}_{21}\bar{x}_1(t) - \tilde{A}_{d21}\bar{x}_1(t-\tau) - \tilde{A}_{d22}\bar{x}_1(t-2\tau) - V_1 C_d E \tilde{A}_{dd11}\bar{x}_1(t-3\tau), \tag{5.38}$$

and provide that a functional observer, $\hat{\tilde{z}}(t)$, can be designed to asymptotically estimate $\tilde{z}(t)$ then the disturbance $d(t)$ can be estimated as follows

$$\hat{d}(t) = V_1\dot{y}(t) - \bar{A}_{22}V_1 y(t) - \tilde{A}_{12}y(t-\tau) - \tilde{A}_{d12}y(t-2\tau) + \hat{\tilde{z}}(t). \tag{5.39}$$

The problem of designing a generalized functional observer to estimate $\tilde{z}(t)$ as defined by (5.38) for system (5.28)–(5.29) is a new problem that has not yet been considered in the literature. This is the first time that the solution to this problem is needed and reported in this book. Indeed, this new problem arises from the unknown input estimation problem via our novel matrix decomposition technique. In Chap. 15, we discuss the design of various types of generalized observers to directly estimate $\tilde{z}(t)$ for system (5.28)–(5.29).

Let us now consider some numerical examples to illustrate the design of unknown input observers for the case where decomposition (5.12) is realized.

5.4.1 Numerical Example 3

In this example, we consider time-delay system (5.1)–(5.2), where $\tau = 1s$, matrices A, A_d, D, C and C_d are given below

$$A = \begin{bmatrix} -10 & 0 & 1 \\ -21 & -10 & 0 \\ -31 & 0 & -4 \end{bmatrix},\ A_d = \begin{bmatrix} -8 & 1 & 1 \\ -26 & -1 & 4 \\ -34 & 0 & -3 \end{bmatrix},\ D = \begin{bmatrix} 1 \\ 2 \\ 3 \end{bmatrix},\ C = \begin{bmatrix} 1 & 0 & 0 \\ 0 & 1 & 0 \end{bmatrix} \text{ and}$$
$$C_d = \begin{bmatrix} 2 & -1 & 0 \\ 1 & 1 & 1 \end{bmatrix}.$$

Our aim is to design an unknown observer to estimate the unknown input $d(t)$ by using the first-derivative of $y_1(t)$. With C_d given above, we notice that

$C_{d1} \neq 0$. We can check that condition (5.10) is satisfied, i.e., $C_{d1}D = 0$, and therefore decomposition (5.12) is realized. Accordingly, we obtain disturbance-free time-delay system (5.28)–(5.29), where $\bar{A}_{11} = \begin{bmatrix} -10 & -2 \\ 0 & -7 \end{bmatrix}$, $\tilde{A}_{d11} = \begin{bmatrix} -30 & 2 \\ -25 & -6 \end{bmatrix}$, $\tilde{A}_{dd11} = \begin{bmatrix} -10 & 0 \\ -34 & 0 \end{bmatrix}$, $\bar{A}_{12} = \begin{bmatrix} -27 \\ -22 \end{bmatrix}$, $\bar{A}_{d12} = \begin{bmatrix} -10 \\ -34 \end{bmatrix}$, $V_1 = \begin{bmatrix} 1 & 0 \end{bmatrix}$, $\tilde{C}_1 = \begin{bmatrix} 1 & 0 \end{bmatrix}$, $\tilde{C}_{d1} = \begin{bmatrix} 3 & 1 \end{bmatrix}$ and $\tilde{C}_{dd1} = \begin{bmatrix} 6 & 0 \end{bmatrix}$.

Notice that $\tilde{y}(t)$ as defined by (5.15) is related to $y(t)$ in (5.2) as $\tilde{y}(t) = -2y_1(t) + y_2(t) - 6y_1(t-\tau)$, which is the same as in Numerical Example 2.

From (5.33), the unknown input, $d(t)$, is given as follows

$$d(t) = \dot{y}_1(t) + 7y_1(t) - 24y_1(t-\tau) - 10y_1(t-2\tau) + \tilde{z}(t), \tag{5.40}$$

where $\tilde{z}(t)$ is as defined by (15.1), and matrices $F = \begin{bmatrix} 0 & -1 \end{bmatrix}$, $F_d = \begin{bmatrix} -4 & -3 \end{bmatrix}$, $F_{dd} = \begin{bmatrix} -27 & 2 \end{bmatrix}$ and $F_{3d} = \begin{bmatrix} -10 & 0 \end{bmatrix}$.

As can be seen from (5.40), an effective way to estimate $d(t)$ is to design a generalized observer to directly estimate $\tilde{z}(t)$. Please refer to Chap. 15 where we discuss the problem of designing such functional observers. Thus, if a generalized observer can be designed, then the unknown input is estimated as follows

$$\hat{d}(t) = \dot{y}_1(t) + 7y_1(t) - 24y_1(t-\tau) - 10y_1(t-2\tau) + \hat{\tilde{z}}(t). \tag{5.41}$$

In Numerical Example 1 (Chap. 15), we design a fifth-order observer with no internal delays to estimate $\tilde{z}(t)$. Interestingly, this is the first time that an observer of order much higher than n is designed. The designed observer is given by (15.56) and

$$\begin{aligned} \dot{w}(t) &= Nw(t) + G\tilde{y}(t) + G_d\tilde{y}(t-\tau) + G_{dd}\tilde{y}(t-2\tau) \\ &+ Jy(t) + J_d y(t-\tau) + J_{dd} y(t-2\tau) + J_{3d} y(t-3\tau), \end{aligned} \tag{5.42}$$

where N, G, G_d and G_{dd} are as given in Numerical Example 1 (Chap. 15), and J, J_d, J_{dd}, J_{3d} are obtained as follows $J = \begin{bmatrix} -78.5 & 0 \\ -27 & 0 \\ -22 & 0 \\ 0 & 0 \\ 0 & 0 \end{bmatrix}$, $J_d = \begin{bmatrix} -257.6111 & 0 \\ -10 & 0 \\ -34 & 0 \\ -27 & 0 \\ -22 & 0 \end{bmatrix}$,

$$J_{dd} = \begin{bmatrix} -202.5556 & 0 \\ 0 & 0 \\ 0 & 0 \\ -10 & 0 \\ -34 & 0 \end{bmatrix} \text{ and } J_{3d} = \begin{bmatrix} -128 & 0 \\ 0 & 0 \\ 0 & 0 \\ 0 & 0 \\ 0 & 0 \end{bmatrix}.$$

Finally, to show how well $\hat{d}(t)$ tracks $d(t)$ using the above fifth-order observer, Fig. 5.4 shows the trajectories of $d(t)$ and $\hat{d}(t)$ where asymptotic tracking is achieved.

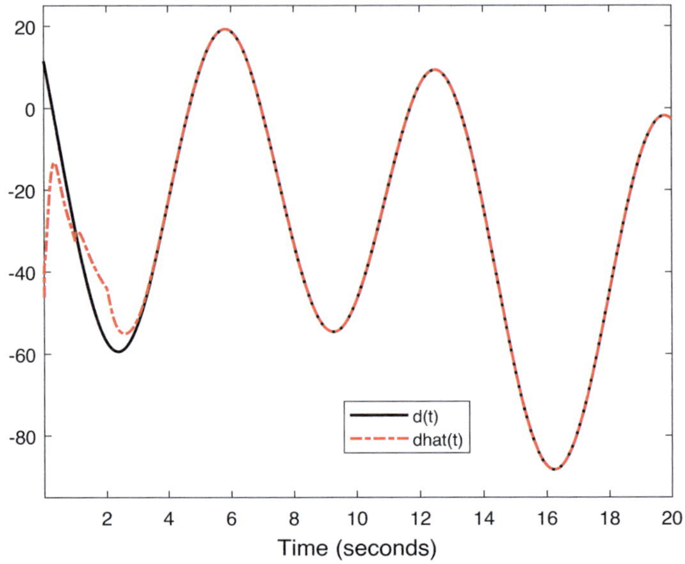

Fig. 5.4 Trajectories of $d(t)$ and $\hat{d}(t)$ (Numerical Example 3, fifth-order observer with no internal delays, Chap. 5)

5.4.2 *Numerical Example 4*

This example is given to show the design of an unknown input observer for the case, where matrix C associated with the output vector $y(t) \in \mathbb{R}^p$ as defined by (5.2) is not a full row matrix. Notice that to achieve the disturbance decoupling (see, (5.4)) based on the decomposition technique as presented in Sect. 2.2, it is necessary that $C_i D \neq 0$. The rest of the rows of C does not have to be non-zeros.

In this example, we reconsider Numerical Example 3, where we replace C with $C = \begin{bmatrix} 1\,0\,0 \\ 0\,0\,0 \end{bmatrix}$, i.e., C is now not a full row rank matrix. Matrices A, A_d, D and C_d are given in Numerical Example 3. Our aim is to design an unknown input observer to estimate $d(t)$, where $\dot{y}_1(t)$ is used in the observer.

In this example, we have $C_1 D = 1$, $C_{d1} D = 0$, and thus decomposition (5.12) is realized. Accordingly, we obtain disturbance-free time-delay system (5.28)–(5.29), where $\bar{A}_{11}$, $\tilde{A}_{d11}$, $\bar{A}_{dd11}$, $\bar{A}_{12}$, $\bar{A}_{d12}$, V_1 and $\tilde{C}_{dd1}$ are given in Numerical Example 3. Notice that the change in C results in the following new matrices: $V_2 = [0 \;\; 1]$, $\tilde{C}_1 = [0 \;\; 0]$ and $\tilde{C}_{d1} = [1 \;\; 1]$.

Observe that since $\tilde{C}_1 = [0 \;\; 0]$, $\tilde{y}(t)$ as defined by (5.29) does not contain the instantaneous vector $\bar{x}_1(t)$. Also, with the newly obtained V_2, $\tilde{y}(t)$ as defined by (5.15) is related to $y(t)$ in (5.2) as $\tilde{y}(t) = y_2(t) - 6y_1(t - \tau)$.

From (5.33), the unknown input $d(t)$ is obtained and it is the same as (5.40). Thus, the estimated unknown input, $\hat{d}(t)$, is obtained according to (5.41), where it

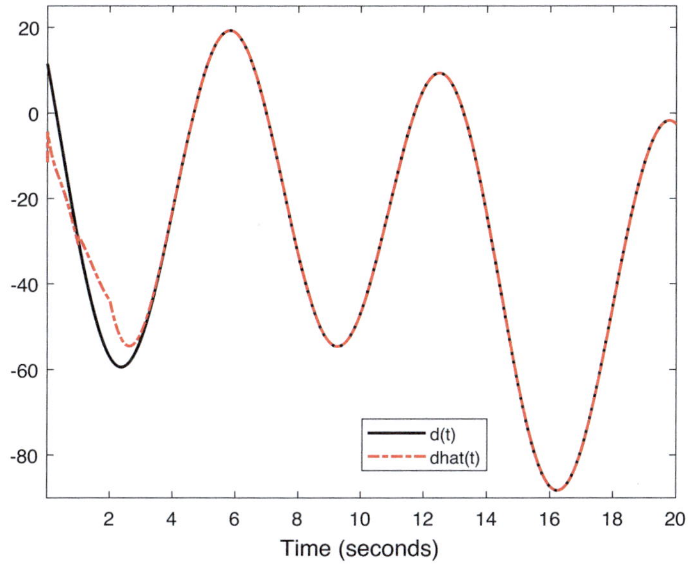

Fig. 5.5 Trajectories of $d(t)$ and $\hat{d}(t)$ (Numerical Example 4, fifth-order observer with no internal delays, Chap. 5)

is necessary to estimate $\tilde{z}(t)$, which is given by (15.42). Please refer to Numerical Example 2 (Chap. 15), where we design a fifth-order generalized observer with no internal delays to estimate $\tilde{z}(t)$.

From Chap. 15, the designed observer, $\hat{\tilde{z}}(t)$, is given by (15.59) and (5.42), where N, G, G_d and G_{dd} are as given in Numerical Example 2 (Chap. 15), and matrices J, J_d, J_{dd}, J_{3d} are obtained as follows $J = \begin{bmatrix} 22 & 0 \\ -27 & 0 \\ -22 & 0 \\ 0 & 0 \\ 0 & 0 \end{bmatrix}$, $J_d = \begin{bmatrix} -1.6111 & 0 \\ -10 & 0 \\ -34 & 0 \\ -27 & 0 \\ -22 & 0 \end{bmatrix}$, $J_{dd} = \begin{bmatrix} -135.8889 & 0 \\ 0 & 0 \\ 0 & 0 \\ -10 & 0 \\ -34 & 0 \end{bmatrix}$ and $J_{3d} = \begin{bmatrix} -128 & 0 \\ 0 & 0 \\ 0 & 0 \\ 0 & 0 \\ 0 & 0 \end{bmatrix}$.

Figure 5.5 shows the trajectories of $d(t)$ and $\hat{d}(t)$, where it is clear that asymptotic tracking is achieved.

5.5 Concluding Remark

In this chapter, we have discussed the design of unknown input observers for a class of time-delay systems (5.1)–(5.2), where a constant time delay, τ, occurs in the output vector, i.e., $y(t) = Cx(t) + C_d x(t - \tau)$. To solve this problem, we employ the matrix decomposition technique presented in Chap. 2, but now with a constraint imposed on matrix C_d, i.e., $V_1 C_d D = 0$. This constraint implies either $V_1 C_d = 0_{1\times n}$ or $C_{di} D = 0$. Consequently, we have discussed the design of unknown input observers with the above two structural constraints. With regard to the first constraint, i.e., $V_1 C_d = 0_{1\times n}$, the decomposition results in a disturbance-free time-delay system (5.13)–(5.14) with a time delay, τ, occurs in both state and output vectors. The unknown input is expressed in its entirety (4.5) which contains a generalized functional. Thus, the unknown input can be estimated by designing a generalized functional observer for time-delay system (5.13)–(5.14).

As for the second constraint, i.e., $C_{di} D = 0$, it covers a wider class of time-delay systems than the first constraint. On the other hand, the decomposition results in a more complex disturbance-free time-delay system (5.28)–(5.29) with multiple time delays in both state and output vectors. Additionally, the unknown input in its entirety now contains a generalized functional (5.38), which has more multiple state-delayed terms, which necessitates the design of generalized functional observers.

In conclusion, it is a novel idea to directly estimate $\bar{z}(t) = -\bar{A}_{21}\bar{x}_1(t) - \bar{A}_{d21}\bar{x}_1(t - \tau)$ and $\tilde{z}(t) = -\bar{A}_{21}\bar{x}_1(t) - \tilde{A}_{d21}\bar{x}_1(t - \tau) - \tilde{A}_{d22}\bar{x}_1(t - 2\tau) - V_1 C_d E \tilde{A}_{dd11} \bar{x}_1(t - 3\tau)$ of time-delay systems (5.13)–(5.14) and (5.28)–(5.29), respectively. This is the first time that the problem of estimating a generalized functional is reported. Since it was not given any attention before, its significant has not been fully realized and reported. The reader can refer to Chaps. 13 and 15, where we demonstrate the significance of the generalized functional estimation problem.

Reference

1. Nguyen, C.M., Tan, C.P., Trinh, H.: Sliding mode observer for estimating states and faults of linear time-delay systems with outputs subject to delays. Automatica **124**, 109,274 (2021)

Chapter 6
Unknown Input Observers Design Using Only Delayed Output Vector

6.1 Delayed Output Vector: $y(t) = C_d x(t - \tau)$

In this chapter, we consider a special case where the output vector $y(t)$ of time-delay system (5.1) and (5.2) is subject to the constraint $C = 0_{p\times n}$. Here, we investigate if it is possible to estimate the unknown input, $d(t)$, of the following time-delay system

$$\dot{x}(t) = Ax(t) + A_d x(t - \tau) + Dd(t), \ t \geq 0, \tag{6.1}$$

$$y(t) = C_d x(t - \tau). \tag{6.2}$$

Again, our aim is to use the first-derivative of one output measurement only, e.g., $y_i(t)$, together with a functional observer to asymptotically estimate the unknown input.

6.2 Main Results

To proceed, we first determine a state transformation, $x(t) = T\bar{x}(t)$, such that the unknown input, $d(t)$, can be expressed in its entirety. For this, we assume that there exists at least one output measurement, say, $y_i(t)$, such that $C_{di}D \neq 0$, and that its first-derivative will be used for estimating the unknown input. In the following, without loss of generality and for the sake of simplicity in the presentation, we let $y_i(t) = y_1(t)$, and matrix C_d takes the form $C_d = [I_p \ \ 0_{p\times(n-p)}]$.

Based on the matrix decomposition method presented in Sect. 2.2 (see Algorithm 1, Steps 1–4), we can select T and V as follows

H. Trinh et al., *Unknown Inputs Estimation in Linear Time-Delay Systems Using Generalized Functional Observers*, Studies in Systems, Decision and Control 648,
https://doi.org/10.1007/978-3-032-11870-7_6

$$T = [E \;\; D],\, E = \begin{bmatrix} 0_{1\times(n-1)} \\ I_{n-1} \end{bmatrix}, \tag{6.3}$$

$$V = [C_d D \;\; P],\, P = \begin{bmatrix} 0_{1\times(p-1)} \\ I_{p-1} \end{bmatrix}. \tag{6.4}$$

Then by applying the state transformation $x(t) = T\bar{x}(t)$ to (6.1) and (6.2), we obtain the following equations

$$\dot{\bar{x}}_1(t) = \bar{A}_{11}\bar{x}_1(t) + \bar{A}_{12}\bar{x}_2(t) + \bar{A}_{d11}\bar{x}_1(t-\tau) + \bar{A}_{d12}\bar{x}_2(t-\tau), \tag{6.5}$$

$$\dot{\bar{x}}_2(t) = \bar{A}_{21}\bar{x}_1(t) + \bar{A}_{22}\bar{x}_2(t) + \bar{A}_{d21}\bar{x}_1(t-\tau) + \bar{A}_{d22}\bar{x}_2(t-\tau) + d(t), \tag{6.6}$$

$$y(t) = [C_d E \;\; C_d D]\begin{bmatrix} \bar{x}_1(t-\tau) \\ \bar{x}_2(t-\tau) \end{bmatrix}. \tag{6.7}$$

Pre-multiplying both sides of (6.7) by $V^{-1} = \begin{bmatrix} V_1 \\ V_2 \end{bmatrix}$ and noting that $V_1 C_d E = 0_{1\times(n-1)}$, $V_1 C_d D = 1$ and $V_2 C_d D = 0_{(p-1)\times 1}$, we obtain the following two equations

$$V_1 y(t) = \bar{x}_2(t-\tau), \tag{6.8}$$

$$\bar{y}(t) = \tilde{C}_{d1}\bar{x}_1(t-\tau), \tag{6.9}$$

where $\bar{y}(t) = V_2 y(t)$ and $\tilde{C}_{d1} = V_2 C_d E$.

From (6.8) and (6.6), we have

$$\bar{x}_2(t) = V_1 y(t+\tau), \tag{6.10}$$

$$d(t) = V_1\dot{y}(t+\tau) - \bar{A}_{22}V_1 y(t+\tau) - \bar{A}_{d22}V_1 y(t) - \bar{A}_{21}\bar{x}_1(t) - \bar{A}_{d21}\bar{x}_1(t-\tau), \tag{6.11}$$

which is, however, a non-causal problem since it is not possible to obtain $y(t+\tau)$ at time t. Thus, for the case where $C = 0_{p\times n}$, and with our solution approach, it is not possible to estimate the instantaneous (without delays) unknown input $d(t)$. Thus, this problem remains an open problem for future research.

On the other hand, to avoid the non-causal problem, it might still be possible to estimate the unknown input but after a time delay, τ. Here, from (6.11), we can express $d(t-\tau)$ as

$$d(t-\tau) = V_1\dot{y}(t) - \bar{A}_{22}V_1 y(t) - \bar{A}_{d22}V_1 y(t-\tau) - \bar{A}_{21}\bar{x}_1(t-\tau) - \bar{A}_{d21}\bar{x}_1(t-2\tau). \tag{6.12}$$

Let us define $z_\tau(t)$ as follows

$$z_\tau(t) = -\bar{A}_{21}\bar{x}_1(t-\tau) - \bar{A}_{d21}\bar{x}_1(t-2\tau). \tag{6.13}$$

If we can design a functional observer, $\hat{z}_\tau(t)$, to estimate $z_\tau(t)$, then the delayed unknown input, $d(t-\tau)$, can be estimated, where

$$\hat{d}(t-\tau) = V_1\dot{y}(t) - \bar{A}_{22}V_1y(t) - \bar{A}_{d22}V_1y(t-\tau) + \hat{z}_\tau(t). \tag{6.14}$$

Now, to design an observer to estimate $z_\tau(t)$ as defined by (6.13), we examine (6.5) and consider the following two cases.

6.2.1 Case 1: $\bar{A}_{12} = 0_{(n-1)\times 1}$

Subject to the structural constraint $\bar{A}_{12} = 0_{(n-1)\times 1}$, and by substituting (6.8) into (6.5) we obtain the following disturbance-free time-delay system

$$\dot{\bar{x}}_1(t) = \bar{A}_{11}\bar{x}_1(t) + \bar{A}_{d11}\bar{x}_1(t-\tau) + \bar{A}_{d12}V_1y(t), \tag{6.15}$$

$$\bar{y}(t) = \tilde{C}_{d1}\bar{x}_1(t-\tau). \tag{6.16}$$

Therefore, we design an observer, $\hat{z}_\tau(t)$, to estimate $z_\tau(t)$ as defined by (6.13) for system (6.15) and (6.16). The reader can refer to Chap. 16, Sect. 16.1, for a design method. This is a special case of the general result of Sect. 16.1, where $C = 0_{p\times n}$ and $F = 0_{1\times n}$.

6.2.2 Case 2: $\bar{A}_{12} \neq 0_{(n-1)\times 1}$

When $\bar{A}_{12} \neq 0_{(n-1)\times 1}$, we design an observer to estimate $z_\tau(t)$ (where $z_\tau(t)$ is as defined by (6.13)) of the following time-delay system

$$\dot{\bar{x}}_1(t-\tau) = \bar{A}_{11}\bar{x}_1(t-\tau) + \bar{A}_{d11}\bar{x}_1(t-2\tau) + \bar{A}_{12}V_1y(t) + \bar{A}_{d12}V_1y(t-\tau), \tag{6.17}$$

$$\bar{y}(t) = \tilde{C}_{d1}\bar{x}_1(t-\tau). \tag{6.18}$$

Here, (6.17) is obtained by first delaying (6.5) by τ, and then replacing $\bar{A}_{12}\bar{x}_2(t-\tau)$ and $\bar{A}_{d12}\bar{x}_2(t-2\tau)$ with $\bar{A}_{12}V_1y(t)$ and $\bar{A}_{d12}V_1y(t-\tau)$, respectively.

To design an observer, $\hat{z}_\tau(t)$, for system (6.17) and (6.18), the reader can refer to Chap. 16, Sect. 16.1. This is a special case of the general result of Sect. 16.1, where $C_d = 0_{p\times n}$ and $F_{dd} = 0_{m\times n}$. Here, we note that Case 2 is applicable to a wider class of time-delay systems since it is not normally that sub-matrix $\bar{A}_{12} = 0_{(n-1)\times 1}$. In either case, the general result presented in Chap. 16, Sect. 16.1 covers both cases.

6.3 Numerical Examples

We consider two numerical examples to estimate delayed unknown input $d(t-\tau)$ as discussed in Case 1 and Case 2 above.

6.3.1 Numerical Example 1

This example is given to illustrate Case 1 in Sect. 6.2.1. Here, we consider a time-delay system (6.1) and (6.2) where $\tau = 1s$, matrices A, A_d, D and C_d are given as follows

$$A = \begin{bmatrix} 8 & -3 & -4 \\ 16 & -9 & -6 \\ 43 & -8 & -19 \end{bmatrix},\ A_d = \begin{bmatrix} 25 & -5 & -6 \\ 49 & -12 & -10 \\ 68 & -15 & -15 \end{bmatrix},\ D = \begin{bmatrix} 1 \\ 2 \\ 3 \end{bmatrix} \text{ and } C_d = \begin{bmatrix} 1 & 0 & 0 \\ 0 & 1 & 0 \end{bmatrix}.$$

As discussed in Sect. 6.2, when the output measurement vector takes the form $y(t) = C_d x(t-\tau)$ (i.e., $C = 0_{p\times n}$), we do not yet know how to estimate instantaneously the unknown input $d(t)$ for system (6.1) and (6.2). Instead, we carry out the design of an unknown input observer to estimate the delayed unknown input, $d(t-\tau)$.

Firstly, from (6.3) and (6.4), we obtain the following matrices $P = \begin{bmatrix} 0 \\ 1 \end{bmatrix}$, $E = \begin{bmatrix} 0 & 0 \\ 1 & 0 \\ 0 & 1 \end{bmatrix}$, $V = \begin{bmatrix} 1 & 0 \\ 2 & 1 \end{bmatrix}$, $V_1 = [1 \quad 0]$, $V_2 = [-2 \quad 1]$ and $T = \begin{bmatrix} 0 & 0 & 1 \\ 1 & 0 & 2 \\ 0 & 1 & 3 \end{bmatrix}$.

With T obtained above, we have

$$\bar{A} = T^{-1}AT = \begin{bmatrix} \bar{A}_{11} & \bar{A}_{12} \\ \bar{A}_{21} & \bar{A}_{22} \end{bmatrix} = \left[\begin{array}{cc|c} -3 & 2 & 0 \\ 1 & -7 & 0 \\ \hline -3 & -4 & -10 \end{array}\right],$$

$$\bar{A}_d = T^{-1}A_dT = \begin{bmatrix} \bar{A}_{d11} & \bar{A}_{d12} \\ \bar{A}_{d21} & \bar{A}_{d22} \end{bmatrix} = \left[\begin{array}{cc|c} -2 & 2 & 1 \\ 0 & 3 & 2 \\ \hline -5 & -6 & -3 \end{array}\right].$$

Here, sub-matrix $\bar{A}_{12}$ satisfies the structural constraint $\bar{A}_{12} = \begin{bmatrix} 0 \\ 0 \end{bmatrix}$, which belongs to Case 1 as discussed in Sect. 6.2.1. Accordingly, we obtain a disturbance-free time-delay system (6.15) and (6.16), where

$\bar{A}_{11} = \begin{bmatrix} -3 & 2 \\ 1 & -7 \end{bmatrix}$, $\bar{A}_{d11} = \begin{bmatrix} -2 & 2 \\ 0 & 3 \end{bmatrix}$, $\tilde{C}_{d1} = V_2 C_d E = \begin{bmatrix} 1 & 0 \end{bmatrix}$ and $\bar{A}_{d12} V_1 y(t) = \begin{bmatrix} 1 \\ 2 \end{bmatrix} y_1(t)$.

From (6.14), we can obtain $\hat{d}(t-1)$ by designing a functional observer $\hat{z}_\tau(t)$ to estimate $z_\tau(t)$, where

$$\begin{aligned} z_\tau(t) &= -\bar{A}_{21}\bar{x}_1(t-1) - \bar{A}_{d21}\bar{x}_1(t-2) \\ &= \begin{bmatrix} 3 & 4 \end{bmatrix}\bar{x}_1(t-1) + \begin{bmatrix} 5 & 6 \end{bmatrix}\bar{x}_1(t-2). \end{aligned} \tag{6.19}$$

The reader can refer to Chap. 16 (Numerical Example 5), where we design a first-order observer with no internal delay to estimate $\hat{z}_\tau(t)$. Thus, from (16.57) and (16.58), by replacing $y(t)$ with $\bar{y}(t)$ and $u(t)$ with $y_1(t)$, we obtain the following functional observer for system (6.15) and (6.16)

$$\hat{z}_\tau(t) = w(t) + 9\bar{y}(t) + 14\bar{y}(t-1), \tag{6.20}$$

$$\dot{w}(t) = -10w(t) - 38\bar{y}(t) - 45\bar{y}(t-1) + 18\bar{y}(t-2) + 2y_1(t-1) + 3y_1(t-2), \tag{6.21}$$

where

$$\bar{y}(t) = V_2 y(t) = [-2 \;\; 1]y(t) = -2y_1(t) + y_2(t). \tag{6.22}$$

Hence, by substituting (6.20) and (6.22) into (6.14), the delayed unknown input is obtained as follows

$$\hat{d}(t-1) = \dot{y}_1(t) - 8y_1(t) + 9y_2(t) - 25y_1(t-1) + 14y_2(t-1) + w(t), \tag{6.23}$$

$$\begin{aligned} \dot{w}(t) = &-10w(t) + 76y_1(t) - 38y_2(t) + 92y_1(t-1) - 45y_2(t-1) \\ &- 33y_1(t-2) + 18y_2(t-2). \end{aligned} \tag{6.24}$$

Figure 6.1 shows the trajectories of $d(t)$, $\hat{d}(t-1)$ and $d(t-1)$. Even though we cannot asymptotically estimate $d(t)$, but we can still asymptotically estimate the delayed unknown input, $d(t-1)$.

6.3.2 *Numerical Example 2*

This example is given to illustrate Case 2 in Sect. 6.2.2, where we design a delayed unknown input observer to estimate $d(t-\tau)$. We consider a time-delay system (6.1) and (6.2) where $\tau = 1s$, matrices A, A_d, D and C_d are given as follows

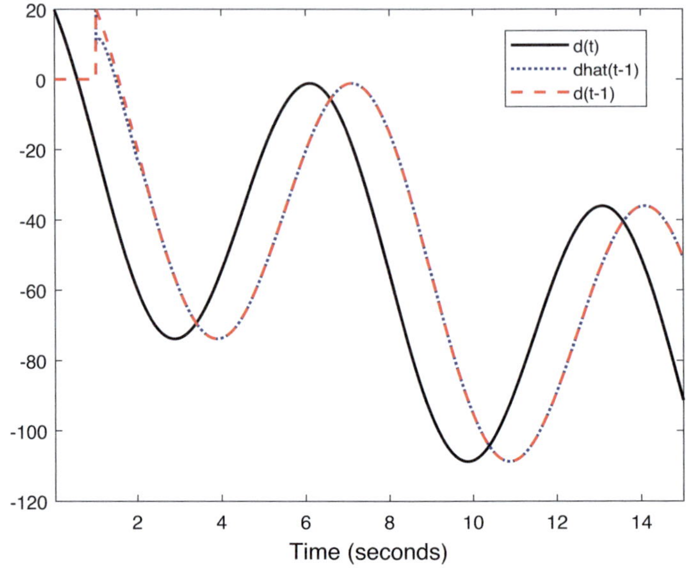

Fig. 6.1 Trajectories of $d(t)$, $\hat{d}(t-1)$ and $d(t-1)$ (Numerical Example 1, Chap. 6)

$$A = \begin{bmatrix} 12 & -3 & -4 \\ 19 & -9 & -6 \\ 47 & -8 & -16 \end{bmatrix}, A_d = \begin{bmatrix} 23 & -5 & -6 \\ 46 & -12 & -10 \\ 63 & -15 & -15 \end{bmatrix}, D = \begin{bmatrix} 1 \\ 2 \\ 3 \end{bmatrix} \text{ and } C_d = \begin{bmatrix} 1 & 0 & 0 \\ 0 & 1 & 0 \end{bmatrix}.$$

With T obtained in Numerical Example 1, we have

$$\bar{A} = T^{-1}AT = \begin{bmatrix} \bar{A}_{11} & \bar{A}_{12} \\ \bar{A}_{21} & \bar{A}_{22} \end{bmatrix} = \left[\begin{array}{cc|c} -3 & 2 & -5 \\ 1 & -4 & 1 \\ \hline -3 & -4 & -6 \end{array}\right],$$

$$\bar{A}_d = T^{-1}A_dT = \begin{bmatrix} \bar{A}_{d11} & \bar{A}_{d12} \\ \bar{A}_{d21} & \bar{A}_{d22} \end{bmatrix} = \left[\begin{array}{cc|c} -2 & 2 & 2 \\ 0 & 3 & 3 \\ \hline -5 & -6 & -5 \end{array}\right].$$

Here, we have sub-matrix $\bar{A}_{12} = \begin{bmatrix} -5 \\ 1 \end{bmatrix} \neq 0$, which belongs to Case 2 as discussed in Sect. 6.2.2. Thus, we need to design an observer to estimate $z_\tau(t)$ (see, (6.19)) for time-delay system (6.17) and (6.18). This task can be carried out by using the design method discussed in Sect. 16.1, where we let $C_d = 0_{p\times n}$ and $F_{dd} = 0_{m\times n}$. Accordingly, we design the following observer

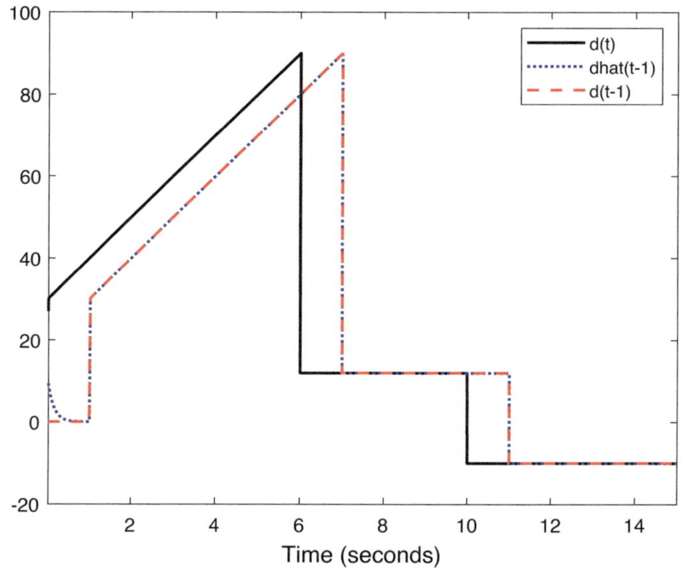

Fig. 6.2 Trajectories of $d(t)$, $\hat{d}(t-1)$ and $d(t-1)$ (Numerical Example 2, Chap. 6)

$$\hat{z}_\tau(t) = w(t) + 9\bar{y}(t) + 14\bar{y}(t-1), \tag{6.25}$$

$$\dot{w}(t) = -7w(t) - 20\bar{y}(t) - 18\bar{y}(t-1) + 18\bar{y}(t-2) + 34y_1(t) + 51y_1(t-1). \tag{6.26}$$

By substituting (6.25) and (6.22) into (6.14), we obtain the following delayed unknown input observer

$$\hat{d}(t-1) = \dot{y}_1(t) - 12y_1(t) + 9y_2(t) - 23y_1(t-1) + 14y_2(t-1) + w(t), \tag{6.27}$$

$$\dot{w}(t) = -7w(t) + 74y_1(t) - 20y_2(t) + 87y_1(t-1) - 18y_2(t-1) - 36y_1(t-2) + 18y_2(t-2). \tag{6.28}$$

Figure 6.2 shows the trajectories of $d(t)$, $\hat{d}(t-1)$ and $d(t-1)$. From the figure, it is clear that we can asymptotically estimate the delayed unknown input, $d(t-1)$.

6.4 Concluding Remark

In this Chapter, we have looked at the problem of designing an unknown input observer to asymptotically estimate an unknown input, $d(t)$, for a class of time-delay systems, where only delayed output vector is available, i.e.,

$$
\begin{aligned}
\dot{x}(t) &= Ax(t) + A_d x(t-\tau) + Dd(t),\ t \geq 0,\\
y(t) &= C_d x(t-\tau).
\end{aligned}
$$

However, we have shown that due to the non-causal problem (i.e., it is not possible to obtain $y(t+\tau)$ at time t), $d(t)$ can not be estimated using $y(t)$. However, we can estimate the unknown input but after a time delay, τ, i.e., $\hat{d}(t-\tau)$ asymptotically converges to $d(t-\tau)$ which is still a useful outcome. Accordingly, we have considered two cases for estimating $d(t-\tau)$. Both cases involved the design of functional observers for a class of time-delay systems, which will be covered extensively in Chap. 16.

Chapter 7
Unknown Input Observers Design: An Expanded Dimensional Approach

In Chap. 5, we discussed the design of unknown input observers for a class of time-delay systems (5.1) and (5.2), where a constant time delay, τ, occurs in the output vector. We used the matrix decomposition technique presented in Chap. 2 to obtain the unknown input in its entirety and a disturbance-free time-delay system. To achieve this outcome, we impose a structural constraint on matrix C_d, where $V_1 C_d D = 0$. Then the unknown input can be estimated by designing a generalized functional observer.

In this chapter, we investigate another way to achieve the decoupling of the unknown input by expanding the dimension of the time-delay system, where we augment $x(t)$ with a delayed state vector $\zeta(t) \equiv x(t - \tau)$.

7.1 Main Results

7.1.1 An Expanded Time-Delay System with Two Unknown Inputs

Let us consider time-delay system (5.1) and (5.2), i.e.,

$$\dot{x}(t) = Ax(t) + A_d x(t - \tau) + Dd(t), \tag{7.1}$$

$$y(t) = Cx(t) + C_d x(t - \tau). \tag{7.2}$$

Let us now introduce the following delayed state vector

$$\zeta(t) \equiv x(t - \tau). \tag{7.3}$$

H. Trinh et al., *Unknown Inputs Estimation in Linear Time-Delay Systems Using Generalized Functional Observers*, Studies in Systems, Decision and Control 648,
https://doi.org/10.1007/978-3-032-11870-7_7

From (7.1) and (7.3), we obtain the following equations

$$\dot{x}(t) = Ax(t) + A_d\zeta(t) + Dd(t), \tag{7.4}$$

$$\dot{\zeta}(t) = A\zeta(t) + A_d\zeta(t-\tau) + Dd(t-\tau). \tag{7.5}$$

Then by letting $x_a(t) = \begin{bmatrix} x(t) \\ \zeta(t) \end{bmatrix}$, $d_a(t) = \begin{bmatrix} d(t) \\ d(t-\tau) \end{bmatrix}$, $A_a = \begin{bmatrix} A & A_d \\ 0 & A \end{bmatrix}$, $A_{da} = \begin{bmatrix} 0 & 0 \\ 0 & A_d \end{bmatrix}$, $D_a = \begin{bmatrix} D_1 & D_2 \end{bmatrix} = \begin{bmatrix} D & 0 \\ 0 & D \end{bmatrix}$ and $C_a = \begin{bmatrix} C & C_d \end{bmatrix}$, we augment (7.4) and (7.5) and obtain the following expanded time-delay system

$$\dot{x}_a(t) = A_a x_a(t) + A_{da} x_a(t-\tau) + D_a d_a(t), \tag{7.6}$$

$$y(t) = C_a x_a(t). \tag{7.7}$$

Thus, using $\zeta(t)$ as an additional vector, system (7.6) and (7.7) now does not have a state-delayed term in the output vector. On the other hand, (7.6) and (7.7) is an expanded time-delay system with a higher dimension (i.e., $2n$) and there are now two unknown inputs in $d_a(t)$. To design an unknown input observer to estimate $d(t)$, we will consider two design approaches as presented in Chap. 3 for the decomposition of the expanded time-delay system (7.6) and (7.7). Note that Design Approach 2 (see, Sect. 7.1.4) does not need the structural constraint $V_1 C_d D = 0$ to be satisfied.

7.1.2 Design Approach 1

We apply Design Approach 1 in Sect. 3.2.1, Chap. 3, to achieve the decoupling of $d(t)$ from $d(t-\tau)$ for system (7.6) and (7.7). Let us assume that we use $\dot{y}_1(t)$ for the estimation of $d(t)$ and thus $C_1 D = \alpha_1 \neq 0$, where C_1 is the first row of matrix C in the canonical form.

Let us select T and V as follows

$$T = [E \;\; D_1],\; E = \begin{bmatrix} 0_{1\times(2n-1)} \\ I_{2n-1} \end{bmatrix}, \tag{7.8}$$

$$V = [C_a D_1 \;\; P],\; P = \begin{bmatrix} 0_{1\times(p-1)} \\ I_{p-1} \end{bmatrix}. \tag{7.9}$$

Then by applying the state transformation $x_a(t) = T\bar{x}_a(t)$ to (7.6) and (7.7), we obtain the following equations

$$\dot{\bar{x}}_{a1}(t) = \bar{A}_{11}\bar{x}_{a1}(t) + \bar{A}_{12}\bar{x}_{a2}(t) + \bar{A}_{d11}\bar{x}_{a1}(t-\tau) + \bar{D}_2 d(t-\tau), \tag{7.10}$$

$$\dot{\bar{x}}_{a2}(t) = \bar{A}_{21}\bar{x}_{a1}(t) + \bar{A}_{22}\bar{x}_{a2}(t) + d(t), \tag{7.11}$$

$$y(t) = [C_a E \;\; C_a D_1] \begin{bmatrix} \bar{x}_{a1}(t) \\ \bar{x}_{a2}(t) \end{bmatrix}. \tag{7.12}$$

Note that to obtain the above equations, we partition $\bar{x}_a(t)$ and $\bar{A} = T^{-1}A_aT$ as follows

$$\bar{x}_a(t) = \begin{bmatrix} \bar{x}_{a1}(t) \\ \bar{x}_{a2}(t) \end{bmatrix}, \ \bar{A} = T^{-1}A_aT = \begin{bmatrix} \bar{A}_{11} & \bar{A}_{12} \\ \bar{A}_{21} & \bar{A}_{22} \end{bmatrix}, \tag{7.13}$$

where $\bar{x}_{a1}(t) \in \mathbb{R}^{(2n-1)}$, $\bar{x}_{a2}(t) \in \mathbb{R}$, $\bar{A}_{11} \in \mathbb{R}^{(2n-1)\times(2n-1)}$, $\bar{A}_{12} \in \mathbb{R}^{(2n-1)}$, $\bar{A}_{21} \in \mathbb{R}^{1\times(2n-1)}$ and $\bar{A}_{22} \in \mathbb{R}$. Also, with $T = \begin{bmatrix} E & D_1 \end{bmatrix} = \begin{bmatrix} 0_{1\times(2n-1)} & \alpha_1 \\ I_{(2n-1)} & D_{12} \end{bmatrix}$, we have $T^{-1} = \begin{bmatrix} -\alpha_1^{-1}D_{12} & I_{(2n-1)} \\ \alpha_1^{-1} & 0_{1\times(2n-1)} \end{bmatrix}$. Since $A_{da} = \begin{bmatrix} 0 & 0 \\ 0 & A_d \end{bmatrix}$, we have

$$\bar{A}_d = T^{-1}A_{da}T = \begin{bmatrix} \bar{A}_{d11} & 0 \\ 0 & 0 \end{bmatrix}, \tag{7.14}$$

where $\bar{A}_{d11} = \begin{bmatrix} 0 & 0 \\ 0 & A_d \end{bmatrix} \in \mathbb{R}^{(2n-1)\times(2n-1)}$. From (7.10), we have $\bar{D}_2 = \begin{bmatrix} 0_{(n-1)} \\ D \end{bmatrix}$.

Pre-multiplying both sides of (7.12) by $V^{-1} = \begin{bmatrix} V_1 \\ V_2 \end{bmatrix}$ and noting that $V_1C_aD_1 = 1$ and $V_2C_aD_1 = 0_{(p-1)\times 1}$, we obtain the following two equations

$$V_1y(t) = V_1C_aE\bar{x}_{a1}(t) + \bar{x}_{a2}(t), \tag{7.15}$$

$$\bar{y}(t) = \tilde{C}_{d1}\bar{x}_{a1}(t), \tag{7.16}$$

where $\bar{y}(t) = V_2y(t)$ and $\tilde{C}_{d1} = V_2C_aE$.

From (7.15), we can express $\bar{x}_{a2}(t)$ as follows

$$\bar{x}_{a2}(t) = V_1y(t) - V_1C_aE\bar{x}_{a1}(t), \tag{7.17}$$

and hence

$$\dot{\bar{x}}_{a2}(t) = V_1\dot{y}(t) - V_1C_aE\dot{\bar{x}}_{a1}(t).$$

Note that (7.10) is not a disturbance-free time-delay system (i.e., we have the term $\bar{D}_2d(t-\tau)$), and thus from (7.11), $d(t)$ contains $d(t-\tau)$. We have

$$\begin{aligned} d(t) &= V_1\dot{y}(t) - (\bar{A}_{22}V_1 + V_1C_aE\bar{A}_{12}V_1)y(t) \\ &\quad - (\bar{A}_{21} + V_1C_aE\bar{A}_{11} - \bar{A}_{22}V_1C_aE - V_1C_aE\bar{A}_{12}V_1C_aE)\bar{x}_{a1}(t) \\ &\quad - V_1C_aE\bar{A}_{d11}\bar{x}_{a1}(t-\tau) - V_1C_aE\bar{D}_2d(t-\tau). \end{aligned} \tag{7.18}$$

To design an unknown input observer to asymptotically estimate $d(t)$, it is clear from (7.18) that we need to eliminate the term related to $d(t-\tau)$, and hence we impose the following constraint

$$V_1 C_a E \bar{D}_2 = 0. \tag{7.19}$$

Given that C is in the canonical form, E and V_1 are as defined by (7.8) and (2.4), respectively, it is clear that condition (7.19) holds if either (5.9) or (5.10) of Chap. 5 is satisfied.

Firstly, (5.9) is equivalent to

$$V_1 C_a E = 0_{1\times(2n-1)}. \tag{7.20}$$

Thus, subject to satisfying condition (7.20), we have $\bar{x}_{a2}(t) = V_1 y(t)$, and by substituting this term into (7.10) and (7.18), we have the following equations

$$\dot{\bar{x}}_{a1}(t) = \bar{A}_{11}\bar{x}_{a1}(t) + \bar{A}_{d11}\bar{x}_{a1}(t-\tau) + \bar{A}_{12}V_1 y(t) + \bar{D}_2 d(t-\tau), \tag{7.21}$$

$$d(t) = V_1\dot{y}(t) - \bar{A}_{22}V_1 y(t) - \bar{A}_{21}\bar{x}_{a1}(t). \tag{7.22}$$

Thus, from (7.22), $d(t)$ can be estimated by designing a functional observer, $\hat{z}(t)$, to asymptotically estimate the following function

$$z(t) = -\bar{A}_{21}\bar{x}_{a1}(t), \tag{7.23}$$

of the time-delay system as defined by (7.21) and (7.16), i.e.,

$$\begin{aligned}\dot{\bar{x}}_{a1}(t) &= \bar{A}_{11}\bar{x}_{a1}(t) + \bar{A}_{d11}\bar{x}_{a1}(t-\tau) + \bar{A}_{12}V_1 y(t) + \bar{D}_2 d(t-\tau),\\ \bar{y}(t) &= \tilde{C}_{d1}\bar{x}_{a1}(t).\end{aligned}$$

Compared with time-delay system (5.13) and (5.14) in Sect. 5.3, Chap. 5, the above time-delay system has a higher dimension (i.e., $(2n-1)$ instead of $(n-1)$) and contains an unknown input $d(t-\tau)$, but its output vector and $z(t)$ do not contain a state-delayed term.

Thus, provided that a functional observer $\hat{z}(t)$ can be designed to asymptotically estimate $z(t)$, the estimated unknown input is given as follows

$$\hat{d}(t) = V_1\dot{y}(t) - \bar{A}_{22}V_1 y(t) + \hat{z}(t). \tag{7.24}$$

Secondly, (5.10) is equivalent to (7.19). In this case, the unknown input $d(t)$ is presented as follows

$$\begin{aligned}d(t) &= V_1\dot{y}(t) - (\bar{A}_{22}V_1 + V_1 C_a E\bar{A}_{12}V_1)y(t)\\ &\quad - (\bar{A}_{21} + V_1 C_a E\bar{A}_{11} - \bar{A}_{22}V_1 C_a E - V_1 C_a E\bar{A}_{12}V_1 C_a E)\bar{x}_{a1}(t)\\ &\quad - V_1 C_a E\bar{A}_{d11}\bar{x}_{a1}(t-\tau).\end{aligned} \tag{7.25}$$

By substituting (7.17) into (7.10), we obtain the following time-delay system

$$\begin{aligned}\dot{\bar{x}}_{a1}(t) &= (\bar{A}_{11} - \bar{A}_{12}V_1C_aE)\bar{x}_{a1}(t) + \bar{A}_{d11}\bar{x}_{a1}(t-\tau) + \bar{A}_{12}V_1y(t)\\ &\quad + \bar{D}_2d(t-\tau), \qquad (7.26)\\ \bar{y}(t) &= \tilde{C}_{d1}\bar{x}_{a1}(t).\end{aligned}$$

Thus, by letting

$$\begin{aligned}z_d(t) &= -(\bar{A}_{21} + V_1C_aE\bar{A}_{11} - \bar{A}_{22}V_1C_aE - V_1C_aE\bar{A}_{12}V_1C_aE)\bar{x}_{a1}(t)\\ &\quad - V_1C_aE\bar{A}_{d11}\bar{x}_{a1}(t-\tau)\\ &\equiv F\bar{x}_{a1}(t) + F_d\bar{x}_{a1}(t-\tau), \qquad (7.27)\end{aligned}$$

we design a generalized observer, $\hat{z}_d(t)$, to asymptotically estimate $z_d(t)$ of the above time-delay system (7.26). If such an observer exists, then the unknown input can be estimated as follows

$$\hat{d}(t) = V_1\dot{y}(t) - (\bar{A}_{22}V_1 + V_1C_aE\bar{A}_{12}V_1)y(t) + \hat{z}_d(t). \qquad (7.28)$$

7.1.3 A Remark on the Expanded Dimensional Approach

The reader can refer to Chap. 18 for a discussion on the design of generalized functional observers for time-delay systems with unknown inputs. However, it should be noted that the Expanded Dimensional Approach leads to conservative observer design conditions. This drawback is caused by the high dimensionality of system (7.21), i.e., $(2n-1)$, but with a limited number of output measurement, i.e., $(p-1)$, together with the requirement that the designed observer be insensitive to the presence of the unknown input $d(t-\tau)$. In contrast to this, time-delay system (5.13) and (5.14) has a smaller dimension, $(n-1)$, but with the same number of output measurements and no unknown inputs.

To demonstrate the above point, let us relook at Numerical Example 1 as discussed in Sect. 5.3.1. In this example, it can be put into the expanded form (7.6) and (7.7). Since condition (5.9) is satisfied, we obtain time-delay system (7.21) and (7.16), where $\bar{A}_{11} = \begin{bmatrix} 0 & 0 & -10 & -1 & 2\\ 0 & 1 & -10 & 0 & 0\\ 0 & 0 & -10 & 0 & 0\\ 0 & 0 & -21 & 0 & 0\\ 0 & 0 & -31 & 0 & 1\end{bmatrix}$, $\bar{A}_{d11} = \begin{bmatrix} 0 & 0 & 0 & 0 & 0\\ 0 & 0 & 0 & 0 & 0\\ 0 & 0 & -8 & 0 & 1\\ 0 & 0 & -26 & -1 & 4\\ 0 & 0 & -34 & 0 & 3\end{bmatrix}$, $\bar{A}_{12}V_1y(t) = \begin{bmatrix}-1\\ 2\\ 0\\ 0\\ 0\end{bmatrix} y_1(t)$, $\bar{D}_2 = \begin{bmatrix}0 & 0 & 1 & 2 & 3\end{bmatrix}^T$ and $\tilde{C}_{d1} = \begin{bmatrix}1 & 0 & 1 & 1 & 1\end{bmatrix}$.

From (7.22), we have

$$d(t) = \dot{y}_1(t) + 10y_1(t) + \begin{bmatrix} 0 & 0 & 8 & 0 & -1 \end{bmatrix} \bar{x}_{a1}(t). \tag{7.29}$$

Thus, we proceed to design a functional observer to estimate the following function

$$z(t) = \begin{bmatrix} 0 & 0 & 8 & 0 & -1 \end{bmatrix} \bar{x}_{a1}(t),$$

for the above time-delay system. For this purpose, we test condition (18.50) for the case where $C_d = F_d = 0_{1\times 5}$. However, it is found that condition (18.50) is not satisfied, and thus we cannot design a first-order observer of the form (18.41) and (18.42) for time-delay system (7.21) and (7.16). On the other hand, we can design a first-order observer of the form (5.18) and (5.19) for time-delay system (5.13) and (5.14) to asymptotically estimate the unknown input $d(t)$. This analysis therefore highlights the attractiveness of the approach we used in Sect. 5.3 and the significance of those generalized observers reported in Chap. 13.

In the following, let us further investigate the Expanded Dimensional Approach to deal with the case, where the structural constraint (7.19) does not hold, i.e., $V_1 C_a E \bar{D}_2 \neq 0$.

7.1.4 Design Approach 2

Let us now consider the case where $V_1 C_a E \bar{D}_2 \neq 0$. This case is the same as $V_1 C_d D \neq 0$ in Chap. 5 that we could not solve.

Let us consider the expanded time-delay system (7.6) and (7.7), i.e.,

$$\begin{aligned} \dot{x}_a(t) &= A_a x_a(t) + A_{da} x_a(t-\tau) + D_a d_a(t), \\ y(t) &= C_a x_a(t). \end{aligned}$$

To overcome the requirement $V_1 C_a E \bar{D}_2 = 0$, it is now necessary to use the first-derivative of two output measurements. Therefore, we assume that $\text{rank}(C_a D_a) = \text{rank}(D_a) = 2$ and $p > 2$. This implies that the observer matching condition holds. Note that assumption $p > 2$ is restrictive as it requires more output measurements be available. However, this is a trade-off for not requiring $V_1 C_a E \bar{D}_2 = 0$. With the above assumptions satisfied, we can therefore pick two outputs, $y_1(t)$ and $y_2(t)$, and we rearrange the output vector $y(t)$ so that

$$y(t) = \begin{bmatrix} y_1(t) \\ y_2(t) \\ \hline y_r(t) \end{bmatrix} = \begin{bmatrix} C_{a1} \\ \hline C_{ar} \end{bmatrix} x_a(t), \tag{7.30}$$

where $y_r(t) \in \mathbb{R}^{(p-2)}$, $C_{a1} \in \mathbb{R}^{2\times 2n}$, $C_{ar} \in \mathbb{R}^{(p-2)\times 2n}$, and $\text{rank}(C_{a1} D_a) = 2$.

Thus, we will use first-derivative of both $y_1(t)$ and $y_2(t)$ for the estimation of $d_a(t)$. To achieve this, we let V be defined as follows

$$V = [C_a D_a \;\; P],\, P = \begin{bmatrix} 0_{2\times(p-2)} \\ I_{p-2} \end{bmatrix}. \tag{7.31}$$

With $\text{rank}(C_{a1}D_a) = 2$ and P defined above, we have $V^{-1} = \begin{bmatrix} V_1 \\ V_2 \end{bmatrix}$ where $V_1 \in \mathbb{R}^{2\times p}$ takes the following form

$$V_1 = [(C_{a1}D_a)^{-1} \;\; 0_{2\times(p-2)}]. \tag{7.32}$$

Let us consider the following transformation matrix

$$T = [E \;\; D_a], \tag{7.33}$$

where $E \in \mathbb{R}^{2n\times(2n-2)}$ is a rational basis of the null-space of $V_1 C_a$ (see, Sect. 2.1).

Then by applying the state transformation $x_a(t) = T\bar{x}_a(t)$ to (7.6) and (7.7), with T defined by (7.33), we obtain the following equations

$$\dot{\bar{x}}_{a1}(t) = \bar{A}_{11}\bar{x}_{a1}(t) + \bar{A}_{12}\bar{x}_{a2}(t) + \bar{A}_{d11}\bar{x}_{a1}(t-\tau) + \bar{A}_{d12}\bar{x}_{a2}(t-\tau), \tag{7.34}$$

$$\dot{\bar{x}}_{a2}(t) = \bar{A}_{21}\bar{x}_{a1}(t) + \bar{A}_{22}\bar{x}_{a2}(t) + \bar{A}_{d21}\bar{x}_{a1}(t-\tau) + \bar{A}_{d22}\bar{x}_{a2}(t-\tau) + d_a(t), \tag{7.35}$$

$$y(t) = [C_a E \;\; C_a D_a] \begin{bmatrix} \bar{x}_{a1}(t) \\ \bar{x}_{a2}(t) \end{bmatrix}, \tag{7.36}$$

where $\bar{x}_a(t)$, $\bar{A} = T^{-1}A_aT$ and $\bar{A}_d = T^{-1}A_{da}T$ are partitioned as follows

$$\bar{x}_a(t) = \begin{bmatrix} \bar{x}_{a1}(t) \\ \bar{x}_{a2}(t) \end{bmatrix},\; \bar{A} = \begin{bmatrix} \bar{A}_{11} & \bar{A}_{12} \\ \bar{A}_{21} & \bar{A}_{22} \end{bmatrix},\; \bar{A}_d = \begin{bmatrix} \bar{A}_{d11} & \bar{A}_{d12} \\ \bar{A}_{d21} & \bar{A}_{d22} \end{bmatrix}, \tag{7.37}$$

$\bar{x}_{a1}(t) \in \mathbb{R}^{(2n-2)}$, $\bar{x}_{a2}(t), \in \mathbb{R}^2$, $\bar{A}_{11}, \bar{A}_{d11} \in \mathbb{R}^{(2n-2)\times(2n-2)}$, $\bar{A}_{12}, \bar{A}_{d12} \in \mathbb{R}^{(2n-2)\times 2}$, $\bar{A}_{21}, \bar{A}_{d21} \in \mathbb{R}^{2\times(2n-2)}$ and $\bar{A}_{22}, \bar{A}_{d22} \in \mathbb{R}^{2\times 2}$.

Pre-multiplying both sides of (7.36) by $V^{-1} = \begin{bmatrix} V_1 \\ V_2 \end{bmatrix}$ where V is as defined by (7.31), we obtain the following two equations

$$V_1 y(t) = \bar{x}_{a2}(t), \tag{7.38}$$

$$\bar{y}(t) = \tilde{C}_1\bar{x}_{a1}(t), \tag{7.39}$$

where $\bar{y}(t) = V_2 y(t)$ and $\tilde{C}_1 = V_2 C_a E$.

Substituting (7.38) into (7.34) and together with (7.39), we obtain the following disturbance-free time-delay system

$$\dot{\bar{x}}_{a1}(t) = \bar{A}_{11}\bar{x}_{a1}(t) + \bar{A}_{d11}\bar{x}_{a1}(t-\tau) + \bar{A}_{12}V_1 y(t) + \bar{A}_{d12}V_1 y(t-\tau), \quad (7.40)$$

$$\bar{y}(t) = \tilde{C}_1\bar{x}_{a1}(t). \quad (7.41)$$

Finally, by substituting (7.38) into (7.35), $d_a(t)$ is obtained as follows

$$\begin{aligned} d_a(t) = {} & V_1\dot{y}(t) - \bar{A}_{22}V_1 y(t) - \bar{A}_{d22}V_1 y(t-\tau) \\ & - \bar{A}_{21}\bar{x}_{a1}(t) - \bar{A}_{d21}\bar{x}_{a1}(t-\tau). \end{aligned} \quad (7.42)$$

Let $(X)^1$ be defined as the first row of matrix X. From (7.42), $d(t)$ is obtained as follows

$$\begin{aligned} d(t) = {} & (V_1)^1\dot{y}(t) - (\bar{A}_{22}V_1)^1 y(t) - (\bar{A}_{d22}V_1)^1 y(t-\tau) \\ & - (\bar{A}_{21})^1\bar{x}_{a1}(t) - (\bar{A}_{d21})^1\bar{x}_{a1}(t-\tau). \end{aligned} \quad (7.43)$$

Thus, by letting

$$\begin{aligned} z_d(t) &= -(\bar{A}_{21})^1\bar{x}_{a1}(t) - (\bar{A}_{d21})^1\bar{x}_{a1}(t-\tau) \\ &\equiv F\bar{x}_{a1}(t) + F_d\bar{x}_{a1}(t-\tau), \end{aligned} \quad (7.44)$$

we design a generalized observer, $\hat{z}_d(t)$, to asymptotically estimate $z_d(t)$ of the above time-delay system (7.40) and (7.41). If such an observer exists (the reader can refer to Chap. 13, Sect. 13.1), then the unknown input can be estimated as follows

$$\hat{d}(t) = (V_1)^1\dot{y}(t) - (\bar{A}_{22}V_1)^1 y(t) - (\bar{A}_{d22}V_1)^1 y(t-\tau) + \hat{z}_d(t). \quad (7.45)$$

7.1.5 Numerical Example 1

In this numerical example, we consider time-delay system (5.1) and (5.2), where $A = \begin{bmatrix} -10 & 1 \\ 1 & -4 \end{bmatrix}$, $A_d = \begin{bmatrix} 1 & 1 \\ 0 & -3 \end{bmatrix}$, $D = \begin{bmatrix} 1 \\ 2 \end{bmatrix}$, $C = \begin{bmatrix} 1 & 0 \\ 0 & 1 \\ 1 & 1 \end{bmatrix}$ and $C_d = \begin{bmatrix} 1 & 0 \\ 1 & 1 \\ 0 & 1 \end{bmatrix}$.

Here, we have

$$\begin{aligned} y_1(t) &= x_1(t) + x_1(t-\tau), \\ y_2(t) &= x_2(t) + x_1(t-\tau) + x_2(t-\tau), \\ y_3(t) &= x_1(t) + x_2(t) + x_2(t-\tau). \end{aligned}$$

Solving for $x_1(t)$ and $x_2(t)$, we obtain

$$x_1(t) = 0.5y_1(t) - 0.5y_2(t) + 0.5y_3(t),$$

$$x_2(t) + x_2(t-\tau) = -0.5y_1(t) + 0.5y_2(t) + 0.5y_3(t).$$

It is clear from the above that even though there are three output measurements in this system, it is still not possible to express $x_2(t)$ in terms of the three outputs due to the presence of $x_2(t-\tau)$. Thus, an observer is needed.

Now, this system can be put into the expanded form (7.6) and (7.7). Note that Design Approach 1 discussed in Sect. 7.1.2 cannot be carried out since the requirement (7.19) is not met. Hence, we need to follow Design Approach 2 as outlined in Sect. 7.1.4.

From (7.31) and (7.33), we obtain the following matrices

$$V = \begin{bmatrix} 1 & 1 & 0 \\ 2 & 3 & 0 \\ 3 & 2 & 1 \end{bmatrix}, V_1 = \begin{bmatrix} 3 & -1 & 0 \\ -2 & 1 & 0 \end{bmatrix}, V_2 = \begin{bmatrix} -5 & 1 & 1 \end{bmatrix} \text{ and } T = \begin{bmatrix} -1 & 0 & 1 & 0 \\ -1 & -1 & 2 & 0 \\ 1 & 0 & 0 & 1 \\ 0 & 1 & 0 & 2 \end{bmatrix}.$$

Accordingly, we obtain a disturbance-free time-delay system (7.40) and (7.41), where $\bar{A}_{11} = \begin{bmatrix} -4 & 5 \\ 13 & 4 \end{bmatrix}$, $\bar{A}_{d11} = \begin{bmatrix} 2 & 5 \\ 2 & 5 \end{bmatrix}$, $\bar{A}_{12}V_1y(t) = \begin{bmatrix} -33 & 12 \\ -84 & 33 \end{bmatrix}\begin{bmatrix} y_1(t) \\ y_2(t) \end{bmatrix}$, $\bar{A}_{d12}V_1y(t-\tau) = \begin{bmatrix} -24 & 12 \\ -24 & 12 \end{bmatrix}\begin{bmatrix} y_1(t-\tau) \\ y_2(t-\tau) \end{bmatrix}$, $\bar{y}(t) = V_2y(t) = (-5y_1(t) + y_2(t) + y_3(t))$ and $\tilde{C}_1 = \begin{bmatrix} -2 & 0 \end{bmatrix}$.

From (7.43), the unknown input $d(t)$ is given below

$$d(t) = \begin{bmatrix} 3 & -1 \end{bmatrix}\begin{bmatrix} \dot{y}_1(t) \\ \dot{y}_2(t) \end{bmatrix} + \begin{bmatrix} 63 & -23 \end{bmatrix}\begin{bmatrix} y_1(t) \\ y_2(t) \end{bmatrix} + \begin{bmatrix} 24 & -12 \end{bmatrix}\begin{bmatrix} y_1(t-\tau) \\ y_2(t-\tau) \end{bmatrix} + z_d(t), \tag{7.46}$$

where

$$z_d(t) = F\bar{x}_{a1}(t) + F_d\bar{x}_{a1}(t-\tau) = \begin{bmatrix} -6 & -5 \end{bmatrix}\bar{x}_{a1}(t) + \begin{bmatrix} -2 & -5 \end{bmatrix}\bar{x}_{a1}(t-\tau).$$

Thus, to design an unknown input observer, $\hat{d}(t)$, to asymptotically estimate $d(t)$, we design a generalized observer, $\hat{z}_d(t)$, to asymptotically estimate $z_d(t)$. In this task, we refer the reader to Chap. 13, Sect. 13.1, where we consider an observer of the form (10.72) and (10.73), i.e.,

$$\begin{aligned} \hat{z}_d(t) &= w(t) + M\bar{y}(t) + M_d\bar{y}(t-\tau), \\ \dot{w}(t) &= Nw(t) + N_dw(t-\tau) + G\bar{y}(t) + G_d\bar{y}(t-\tau) + G_{dd}\bar{y}(t-2\tau) \\ &\quad + Jy(t) + J_dy(t-\tau) + J_{dd}y(t-2\tau). \end{aligned}$$

In this example, we find that condition (13.11) for the case where $N_d = 0$ is satisfied. Hence, we have designed a first-order observer where we select $N = -5$. The rest of the observer matrices are as given below

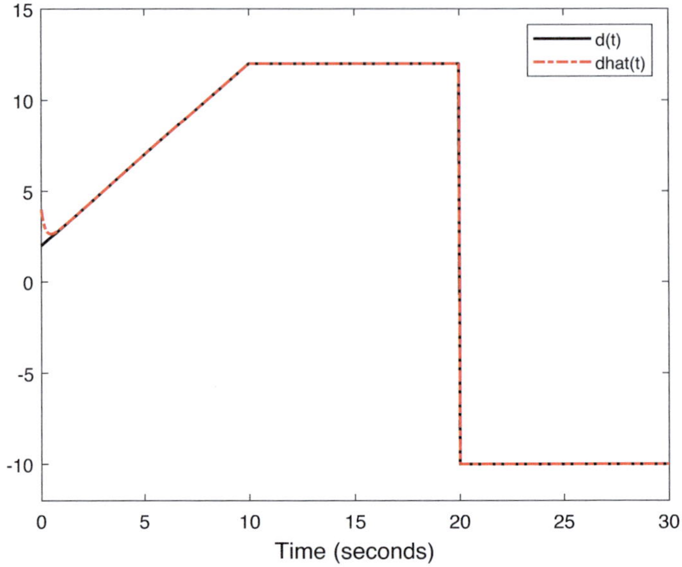

Fig. 7.1 Trajectories of $d(t)$ and $\hat{d}(t)$ (Numerical Example 1, Chap. 7)

$M = 7.5$, $M_d = 3.5$, $G = 28$, $G_d = 26$, $G_{dd} = 0$, $J = \begin{bmatrix} 123 & -57 & 0 \end{bmatrix}$, $J_d = \begin{bmatrix} 159 & -57 & 0 \end{bmatrix}$ and $J_{dd} = \begin{bmatrix} 0 & 0 & 0 \end{bmatrix}$.

Finally, noting that $\bar{y}(t) = V_2 y(t)$, $\hat{z}_d(t)$ is given as follows

$$\hat{z}_d(t) = w(t) + \begin{bmatrix} -37 & 7.5 & 7.5 \end{bmatrix} y(t) + \begin{bmatrix} -17.5 & 3.5 & 3.5 \end{bmatrix} y(t-\tau),$$
$$\dot{w}(t) = -5w(t) + \begin{bmatrix} -17 & -29 & 28 \end{bmatrix} y(t) + \begin{bmatrix} 29 & -31 & 26 \end{bmatrix} y(t-\tau).$$

When $\tau = 1s$, Fig. 7.1 shows the trajectories of $d(t)$ and $\hat{d}(t)$ where it is clear that asymptotic tracking is achieved.

7.2 Concluding Remark

In this chapter, we have looked at another design approach to design unknown input observers for a class of time-delay systems (5.1) and (5.2). In the Expanded Dimensional approach, an auxiliary state vector $\zeta(t) \equiv x(t-\tau)$ is augmented with the original state vector $x(t)$ to form an expanded time-delay system (7.6) and (7.7). In this expanded dimensional form, just like in Chap. 5, we have investigated two design approaches to design an unknown input observer for the system. The objective of Design Approach 1 is to use the first-derivative of only one output measurement. In this approach, we have arrived at the same structural constraint $V_1 C_d D = 0$ as in Chap. 5. However, compared with both cases (see, Sects. 5.3 and 5.4) in Chap. 5, the

Expanded Dimensional Approach results in a higher dimensional time-delay system (7.26), but it still contains an unknown input $d(t-\tau)$. Moreover, the design of generalized observers for time-delay systems with an unknown input is subject to more stringent observer existence conditions. Thus, this finding highlights the attractiveness of the approach we developed in Chap. 5.

On the other hand, as we have presented in Design Approach 2, the Expanded Dimensional Approach overcomes the requirement $V_1 C_d D = 0$. To achieve this, it is necessary to use the first-derivative of two output measurements and a higher number of output measurements be available, i.e., $p > 2$, and hence a trade-off is clearly required. Accordingly, we have obtained a disturbance-free time-delay system (7.40) and (7.41), and the unknown input can be asymptotically estimated by designing a generalized observer, which is the topic discussed extensively in Chap. 13.

We will continue our exploratory research in the next chapter. We will discuss the unknown input estimation problem for a class of time-delay systems with two non-commensurate delays in both the state and output vectors.

Chapter 8
Unknown Input Observers: Two Non-Commensurate Delays in State and Output Vectors

In this chapter, we discuss the unknown input estimation problem for a class of time-delay systems with two non-commensurate delays in both the state and output vectors.

8.1 System Description and Preliminaries

We consider the following time-delay system with an unknown input $d(t)$

$$\dot{x}(t) = Ax(t) + A_d x(t-\tau) + A_h x(t-h) + Dd(t),\ t \geq 0, \tag{8.1}$$

$$y(t) = Cx(t) + C_d x(t-\tau) + C_h x(t-h), \tag{8.2}$$

where $0 < \tau < h$, τ and h are known constant non-commensurate delays, and $A, A_d, A_h \in \mathbb{R}^{n\times n}$, $C, C_d, C_h \in \mathbb{R}^{p\times n}$ and $D \in \mathbb{R}^{n\times 1}$ are known constant matrices. As previously mentioned in Sect. 5.1, for the sake of simplicity in presentation, we assume that matrix C has full-row rank and of the form $C = [I_p \ \ 0_{p\times(n-p)}]$ (please see Numerical Example 2 where we discuss the case C is not a full rank matrix). Again, our aim is to use first-derivative of one output measurement only, e.g., $y_1(t)$ (note: here we require $C_1 D \neq 0$ where C_1 is the first row of C), together with a generalized functional observer to asymptotically estimate the unknown input $d(t)$. Due to the presence of non-commensurate delays in system (8.1) and (8.2), the design of an unknown input observer is more complicated than the one considered earlier for system (5.1) and (5.2). In the following, we provide some discussions on their designs as well as some open problems to be considered in the future.

We first determine a state transformation, $x(t) = T\bar{x}(t)$, such that the unknown input, $d(t)$, can be expressed in its entirety (see, for example, (4.5) or (5.33)). Based on the matrix decomposition technique presented in Sect. 2.2 (see Algorithm 1, Steps 1–4), we can select matrices T and V as follows

H. Trinh et al., *Unknown Inputs Estimation in Linear Time-Delay Systems Using Generalized Functional Observers*, Studies in Systems, Decision and Control 648,
https://doi.org/10.1007/978-3-032-11870-7_8

$$T = [E \ \ D],\, E = \begin{bmatrix} 0_{1\times(n-1)} \\ I_{n-1} \end{bmatrix}, \tag{8.3}$$

$$V = [CD \ \ P],\, P = \begin{bmatrix} 0_{1\times(p-1)} \\ I_{p-1} \end{bmatrix}. \tag{8.4}$$

Applying the state transformation $x(t) = T\bar{x}(t)$ to (8.1) and (8.2), we obtain the following equations

$$\begin{aligned} \dot{\bar{x}}_1(t) &= \bar{A}_{11}\bar{x}_1(t) + \bar{A}_{12}\bar{x}_2(t) + \bar{A}_{d11}\bar{x}_1(t-\tau) + \bar{A}_{d12}\bar{x}_2(t-\tau) \\ &\quad + \bar{A}_{h11}\bar{x}_1(t-h) + \bar{A}_{h12}\bar{x}_2(t-h), \end{aligned} \tag{8.5}$$

$$\begin{aligned} \dot{\bar{x}}_2(t) &= \bar{A}_{21}\bar{x}_1(t) + \bar{A}_{22}\bar{x}_2(t) + \bar{A}_{d21}\bar{x}_1(t-\tau) + \bar{A}_{d22}\bar{x}_2(t-\tau) \\ &\quad + \bar{A}_{h21}\bar{x}_1(t-h) + \bar{A}_{h22}\bar{x}_2(t-h) + d(t), \end{aligned} \tag{8.6}$$

$$\begin{aligned} y(t) &= [CE \ \ CD]\begin{bmatrix} \bar{x}_1(t) \\ \bar{x}_2(t) \end{bmatrix} + [C_dE \ \ C_dD]\begin{bmatrix} \bar{x}_1(t-\tau) \\ \bar{x}_2(t-\tau) \end{bmatrix} \\ &\quad + [C_hE \ \ C_hD]\begin{bmatrix} \bar{x}_1(t-h) \\ \bar{x}_2(t-h) \end{bmatrix}, \end{aligned} \tag{8.7}$$

where $\bar{A}_h = T^{-1}A_hT = \begin{bmatrix} \bar{A}_{h11} & \bar{A}_{h12} \\ \bar{A}_{h21} & \bar{A}_{h22} \end{bmatrix}$, $\bar{A}_{h11} \in \mathbb{R}^{(n-1)\times(n-1)}$, $\bar{A}_{h12} \in \mathbb{R}^{(n-1)\times 1}$, $\bar{A}_{h21} \in \mathbb{R}^{1\times(n-1)}$ and $\bar{A}_{h22} \in \mathbb{R}$.

Pre-multiplying both sides of (8.7) by $V^{-1} = \begin{bmatrix} V_1 \\ V_2 \end{bmatrix}$ and noting that $V_1CE = 0_{1\times(n-1)}$, $V_1CD = 1$, $V_2CD = 0_{(p-1)\times 1}$, we obtain the following two equations

$$\begin{aligned} V_1y(t) &= \bar{x}_2(t) + V_1C_dE\bar{x}_1(t-\tau) + V_1C_dD\bar{x}_2(t-\tau) \\ &\quad + V_1C_hE\bar{x}_1(t-h) + V_1C_hD\bar{x}_2(t-h), \end{aligned} \tag{8.8}$$

$$\begin{aligned} V_2y(t) &= V_2CE\bar{x}_1(t) + V_2C_dE\bar{x}_1(t-\tau) + V_2C_dD\bar{x}_2(t-\tau) \\ &\quad + V_2C_hE\bar{x}_1(t-h) + V_2C_hD\bar{x}_2(t-h). \end{aligned} \tag{8.9}$$

8.2 Structural Constraints on C_d and C_h

In a similar way as in Sect. 5.1, we impose two constraints on the output matrices C_d and C_h such that the following conditions hold

$$V_1C_dD = 0, \tag{8.10}$$

$$V_1C_hD = 0. \tag{8.11}$$

Substituting (8.10) and (8.11) into (8.8), we can now express $\bar{x}_2(t)$ as follows

$$\bar{x}_2(t) = V_1 y(t) - V_1 C_d E \bar{x}_1(t-\tau) - V_1 C_h E \bar{x}_1(t-h). \tag{8.12}$$

In the following, we discuss three cases arose from (8.12).

8.2.1 *Case 1:* $V_1 C_d E = V_1 C_h E = 0_{1\times(n-1)}$

In this case, we assume that matrices C_d and C_h satisfy the constraint $V_1 C_d E = V_1 C_h E = 0_{1\times(n-1)}$. Note that this case leads to a simpler disturbance-free time-delay system since (8.12) is reduced to (5.11), i.e., $\bar{x}_2(t) = V_1 y(t)$.

Hence, by substituting (5.11) into (8.5), (8.9) and (8.6), we obtain the following reduced-order disturbance-free time-delay system and the unknown input, $d(t)$,

$$\begin{aligned}\dot{\bar{x}}_1(t) &= \bar{A}_{11}\bar{x}_1(t) + \bar{A}_{d11}\bar{x}_1(t-\tau) + \bar{A}_{h11}\bar{x}_1(t-h) + \bar{A}_{12}V_1 y(t) \\ &\quad + \bar{A}_{d12}V_1 y(t-\tau) + \bar{A}_{h12}V_1 y(t-h), \end{aligned} \tag{8.13}$$

$$y_{\tau h}(t) = \tilde{C}_1\bar{x}_1(t) + \tilde{C}_{d1}\bar{x}_1(t-\tau) + \tilde{C}_{h1}\bar{x}_1(t-h), \tag{8.14}$$

$$d(t) = V_1\dot{y}(t) - \bar{A}_{22}V_1 y(t) - \bar{A}_{d22}V_1 y(t-\tau) - \bar{A}_{h22}V_1 y(t-h) + z_{\tau h}(t), \tag{8.15}$$

where

$$y_{\tau h}(t) = V_2 y(t) - V_2 C_d D V_1 y(t-\tau) - V_2 C_h D V_1 y(t-h), \tag{8.16}$$

$$\tilde{C}_{h1} = V_2 C_h E, \tag{8.17}$$

$$z_{\tau h}(t) = -\bar{A}_{21}\bar{x}_1(t) - \bar{A}_{d21}\bar{x}_1(t-\tau) - \bar{A}_{h21}\bar{x}_1(t-h). \tag{8.18}$$

Thus, from (8.15), the unknown input can be estimated by estimating the function $z_{\tau h}(t)$ as defined by (8.18) for the disturbance-free time-delay system (8.13) and (8.14), i.e.,

$$\hat{d}(t) = V_1\dot{y}(t) - \bar{A}_{22}V_1 y(t) - \bar{A}_{d22}V_1 y(t-\tau) - \bar{A}_{h22}V_1 y(t-h) + \hat{z}_{\tau h}(t). \tag{8.19}$$

Note that unlike the system described by (5.28) and (5.29), time-delay system (8.13) and (8.14) contains two different (non-commensurate) time delays, τ and h, in state and output vectors. So far, the problem of designing generalized observers to estimate $z_{\tau h}(t)$ has not yet been considered in the literature. By following similar steps as presented in Chap. 15, we can develop a method for designing generalized observers, $\hat{z}_{\tau h}(t)$, to estimate $z_{\tau h}(t)$ for systems with two non-commensurate time delays in both the state and output vectors. For completeness, the reader can refer to Chap. 17 for a design method.

8.2.2 Case 2: Either $V_1C_dE = 0_{1\times(n-1)}$ or $V_1C_hE = 0_{1\times(n-1)}$

In this case, either $V_1C_dE = 0_{1\times(n-1)}$ or $V_1C_hE = 0_{1\times(n-1)}$. The following discussion is for the case where $V_1C_dE \neq 0_{1\times(n-1)}$ and $V_1C_hE = 0_{1\times(n-1)}$. In this situation, from (8.12), we have

$$\bar{x}_2(t) = V_1y(t) - V_1C_dE\bar{x}_1(t-\tau). \tag{8.20}$$

Firstly, similar to Case 1 above, we substitute (8.20) into (8.5) and (8.9) to obtain the following disturbance-free time-delay system

$$\begin{aligned}\dot{\bar{x}}_1(t) &= \bar{A}_{11}\bar{x}_1(t) + \tilde{A}_{d11}\bar{x}_1(t-\tau) + \tilde{A}_{dd11}\bar{x}_1(t-2\tau) + \bar{A}_{h11}\bar{x}_1(t-h)\\ &\quad + \tilde{A}_{h12}\bar{x}_1(t-\tau-h) + \bar{A}_{12}V_1y(t) + \bar{A}_{d12}V_1y(t-\tau) + \bar{A}_{h12}V_1y(t-h),\end{aligned} \tag{8.21}$$

$$\begin{aligned}y_{\tau h}(t) &= \tilde{C}_1\bar{x}_1(t) + \tilde{C}_{d1}\bar{x}_1(t-\tau) + \tilde{C}_{dd1}\bar{x}_1(t-2\tau) + \tilde{C}_{h1}\bar{x}_1(t-h)\\ &\quad + \tilde{C}_{\tau h}\bar{x}_1(t-\tau-h),\end{aligned} \tag{8.22}$$

where

$$\tilde{A}_{h12} = -\bar{A}_{h12}V_1C_dE, \tag{8.23}$$

$$\tilde{C}_{\tau h} = -V_2C_hDV_1C_dE. \tag{8.24}$$

To recap, in (8.21) and (8.22), $\tilde{C}_1 = V_2CE$, $\tilde{A}_{d11}$, $\tilde{A}_{dd11}$, $\tilde{C}_{d1}$, $\tilde{C}_{dd1}$ and $\tilde{C}_{h1}$ are as defined by (5.30), (5.31), (5.16), (5.32) and (8.17), respectively. Note that the disturbance-free time-delay system (8.21) and (8.22) contains mixed time delays in both the state and output vectors.

Then, we substitute (8.20) into (8.6) and after some rearranging, we can express the unknown input, $d(t)$, as follows

$$\begin{aligned}d(t) &= V_1\dot{y}(t) - \bar{A}_{22}V_1y(t) - \tilde{A}_{12}y(t-\tau) - \tilde{A}_{d12}y(t-2\tau) - \bar{A}_{h22}V_1y(t-h)\\ &\quad - V_1C_dE\bar{A}_{h12}V_1y(t-\tau-h) + \bar{z}_{\tau h}(t),\end{aligned} \tag{8.25}$$

where

$$\begin{aligned}\bar{z}_{\tau h}(t) &= -\bar{A}_{21}\bar{x}_1(t) - \tilde{A}_{d21}\bar{x}_1(t-\tau) - \tilde{A}_{d22}\bar{x}_1(t-2\tau) - V_1C_dE\tilde{A}_{dd11}\bar{x}_1(t-3\tau)\\ &\quad - \bar{A}_{h21}\bar{x}_1(t-h) - \tilde{A}_{\tau h}\bar{x}_1(t-\tau-h) - V_1C_dE\tilde{A}_{h12}\bar{x}_1(t-2\tau-h),\end{aligned} \tag{8.26}$$

$$\tilde{A}_{\tau h} = V_1C_dE\bar{A}_{h11} - \bar{A}_{h22}V_1C_dE, \tag{8.27}$$

$\tilde{A}_{12}$, $\tilde{A}_{d12}$, $\tilde{A}_{d21}$ and $\tilde{A}_{d22}$ are as defined by (5.34), (5.35), (5.36) and (5.37), respectively.

Thus, from (8.25), the unknown input can be estimated by estimating the function $\bar{z}_{\tau h}(t)$ defined by (8.26) for the time-delay system (8.21) and (8.22). Again, note that the problem of estimating directly the function $\bar{z}_{\tau h}(t)$ for system (8.21) and (8.22) has not yet been considered in the literature. In Chap. 19, we provide a discussion on the design of generalized observers for a class of systems with mixed non-commensurate time delays in both the state and output vectors (i.e., system (8.21)-8.22)). However, this topic still remains an open research problem for future research.

8.2.3 Case 3: $V_1C_dE \neq 0_{1\times(n-1)}$ and $V_1C_hE \neq 0_{1\times(n-1)}$

In this case, we have both $V_1C_dE \neq 0_{1\times(n-1)}$ and $V_1C_hE \neq 0_{1\times(n-1)}$. In this situation, we have $\bar{x}_2(t)$ is as given by (8.12). By following similar logic as in Case 2 above, we substitute (8.12) into (8.5), (8.9) and (8.6). Then we can obtain a disturbance-free time-delay system and the unknown input expressed in its entirety. However, the derivation is quite a drawn-out process and the design of such generalized observers still remains an open research problem. We would like to leave this case for the reader for further investigation.

8.3 Numerical Examples

In this section, we provide two numerical examples to demonstrate the design of unknown inputs for system (8.1) and (8.2). Note that both examples are used to demonstrate Case 1 only. However, Cases 2 and 3 remain as problems for future research.

8.3.1 Numerical Example 1

This example is given to illustrate Case 1 as presented in Sect. 8.2.1. We consider a time-delay system (8.1) and (8.2) where $\tau = 0.8\,s$, $h = 1.1\,s$, matrices A, A_d, A_h, D, C, C_d and C_h are given as follows

$$A = \begin{bmatrix} 3 & -1 & -2 \\ 7 & -5 & -2 \\ 30 & -2 & -13 \end{bmatrix}, A_d = \begin{bmatrix} 8 & -3 & -4 \\ 21 & -10 & -7 \\ 27 & -9 & -13 \end{bmatrix}, A_h = \begin{bmatrix} 27 & -5 & -6 \\ 52 & -12 & -10 \\ 72 & -15 & -15 \end{bmatrix}, D = \begin{bmatrix} 1 \\ 2 \\ 3 \end{bmatrix},$$

$$C = \begin{bmatrix} 1 & 0 & 0 \\ 0 & 1 & 0 \end{bmatrix}, C_d = \begin{bmatrix} 0 & 0 & 0 \\ 1 & 0 & 0 \end{bmatrix} \text{ and } C_h = \begin{bmatrix} 0 & 0 & 0 \\ -3 & 0 & 1 \end{bmatrix}.$$

Our aim is to design an observer to estimate the unknown input $d(t)$. Firstly, based on the matrix decomposition technique presented in Sect. 2.2, we obtain the following matrices

$P = \begin{bmatrix} 0 \\ 1 \end{bmatrix}$, $E = \begin{bmatrix} 0 & 0 \\ 1 & 0 \\ 0 & 1 \end{bmatrix}$, $V = [CD \;\; P] = \begin{bmatrix} 1 & 0 \\ 2 & 1 \end{bmatrix}$, $V_1 = [1 \;\; 0]$, $V_2 = [-2 \;\; 1]$ and $T = [E \;\; D] = \begin{bmatrix} 0 & 0 & 1 \\ 1 & 0 & 2 \\ 0 & 1 & 3 \end{bmatrix}$.

With V_1, V_2 and E obtained above, we can check that conditions (8.10) and (8.11) are satisfied. Furthermore, we also have $V_1 C_d E = V_1 C_h E = [0 \;\; 0]$. Hence from (8.13) and (8.14), we obtain a disturbance-free time-delay system, where $\bar{A}_{11} = \begin{bmatrix} -3 & 2 \\ 1 & -7 \end{bmatrix}$, $\bar{A}_{d11} = \begin{bmatrix} -4 & 1 \\ 0 & -1 \end{bmatrix}$, $\bar{A}_{h11} = \begin{bmatrix} -2 & 2 \\ 0 & 3 \end{bmatrix}$, $\bar{A}_{12} = \begin{bmatrix} 1 \\ 2 \end{bmatrix}$, $\bar{A}_{d12} = \bar{A}_{h12} = \begin{bmatrix} 0 \\ 0 \end{bmatrix}$, $\tilde{C}_1 = [1 \;\; 0]$, $\tilde{C}_{d1} = [0 \;\; 0]$ and $\tilde{C}_{h1} = [0 \;\; 1]$. Also, from (8.16), $y_{\tau h}(t)$ is obtained as follows

$$y_{\tau h}(t) = -2y_1(t) + y_2(t) - y_1(t-\tau). \tag{8.28}$$

We also obtain $\bar{A}_{21} = -[1 \;\; 2]$, $\bar{A}_{d21} = -[3 \;\; 4]$, $\bar{A}_{h21} = -[5 \;\; 6]$, $\bar{A}_{22} = -5$, $\bar{A}_{d22} = -10$ and $\bar{A}_{h22} = -1$. From (8.15), the unknown input $d(t)$ is obtained as follows

$$d(t) = \dot{y}_1(t) + 5y_1(t) + 10y_1(t-\tau) + y_1(t-h) + z_{\tau h}(t), \; t \geq 0, \tag{8.29}$$

where

$$z_{\tau h}(t) = [1 \;\; 2]\bar{x}_1(t) + [3 \;\; 4]\bar{x}_1(t-\tau) + [5 \;\; 6]\bar{x}_1(t-h). \tag{8.30}$$

Notice from (8.29) that $d(t)$ contains first-derivative of only one output measurement, i.e. $\dot{y}_1(t)$. The unknown input can now be estimated by estimating the function $z_{\tau h}(t)$ as defined by (8.30) for the time-delay system (8.13) and (8.14). Hence, we have

$$\begin{aligned} \hat{d}(t) = {} & \dot{y}_1(t) + 5y_1(t) + 10y_1(t-\tau) + y_1(t-h) \\ & + \hat{z}_{\tau h}(t), \; t \geq 0. \end{aligned} \tag{8.31}$$

The reader can refer to Chap. 17 where we design a generalized observer, $\hat{z}_{\tau h}(t)$, to estimate $z_{\tau h}(t)$ for systems with two non-commensurate time delays in both state and output vectors (i.e., systems (8.13) and (8.14)). Also, the design of a first-order generalized observer to estimate $z_{\tau h}(t)$ as defined by (8.30) is given in Numerical Example 2 of Chap. 17. Hence, we have

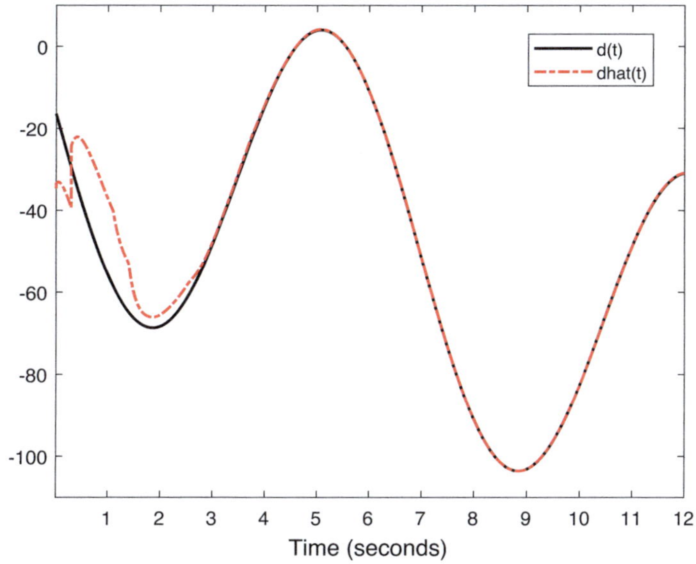

Fig. 8.1 Trajectories of $d(t)$ and $\hat{d}(t)$ (Numerical Example 1, Chap. 8)

$$\hat{z}_{\tau h}(t) = w(t) + y_{\tau h}(t) + 3y_{\tau h}(t-\tau) + 5y_{\tau h}(t-h), \tag{8.32}$$

$$\begin{aligned} \dot{w}(t) = & -7w(t) - w(t-\tau) + 2w(t-h) + 2y_{\tau h}(t) \\ & + 4y_{\tau h}(t-\tau) + 5y_{\tau h}(t-h) - 5y_{\tau h}(t-2h) \\ & - 3y_{\tau h}(t-\tau-h) + 4y_1(t) + 8y_1(t-\tau) \\ & + 10y_1(t-h) - 10y_1(t-2h) - 6y_1(t-\tau-h), \end{aligned} \tag{8.33}$$

where $y_{\tau h}(t)$ is as defined by (8.28). Figure 8.1 shows the trajectories of $d(t)$ and $\hat{d}(t)$ where it is clear that asymptotic tracking is achieved.

8.3.2 Numerical Example 2

For the sake of simplicity in presentation, in Sect. 8.1 we assumed that matrix C has full-row rank and of the form $C = [I_p \;\; 0_{p\times(n-p)}]$. Note that C does not need to be a full rank matrix, but instead we only require $CD \neq 0$. In this example, we consider the case where C is not full rank, and we provide more details on the state transformation to achieve the decomposition (8.13)–(8.15).

Let us now consider time-delay system (8.1) and (8.2) where $\tau = 0.8\,s$, $h = 1.1\,s$, matrices A, A_d, A_h, D, C, C_d and C_h are given as follows

$$A = \begin{bmatrix} -6 & -1 & 1 \\ 4 & -5 & -1 \\ -3 & -2 & -2 \end{bmatrix},\ A_d = \begin{bmatrix} -7 & 0 & -1 \\ -6 & -4 & -2 \\ -15 & 0 & -5 \end{bmatrix},\ A_h = \begin{bmatrix} 0 & 1 & -1 \\ 3 & -1 & -1 \\ 8 & 2 & -5 \end{bmatrix},\ D = \begin{bmatrix} 1 \\ 2 \\ 3 \end{bmatrix},$$
$$C = \begin{bmatrix} 1 & 0 & 0 \\ 0 & 0 & 0 \\ 0 & 0 & 0 \end{bmatrix},\ C_d = \begin{bmatrix} 0 & 0 & 0 \\ 0 & 0 & 1 \\ 1 & 0 & 1 \end{bmatrix} \text{ and } C_h = \begin{bmatrix} 0 & 0 & 0 \\ 0 & 1 & 1 \\ 1 & 0 & 1 \end{bmatrix}.$$

Our aim is to design an observer to estimate the unknown input $d(t)$. Firstly, we note that C is not of full rank. Since we have $CD = \begin{bmatrix} 1 \\ 0 \\ 0 \end{bmatrix}$, we shall undertake the state transformation so that the first-derivative of $y_1(t)$ will be used for the estimation of $d(t)$. We select P and E as follows

$$P = \begin{bmatrix} 0 & 0 \\ 1 & 0 \\ 0 & 1 \end{bmatrix},\ E = \begin{bmatrix} 0 & 0 \\ 1 & 0 \\ 0 & 1 \end{bmatrix}.$$

Hence, we obtain the state transformation matrix T as $T = [E\ \ D] = \begin{bmatrix} 0 & 0 & 1 \\ 1 & 0 & 2 \\ 0 & 1 & 3 \end{bmatrix}$, and from $V = [CD\ \ P] = I_3$, we have $V_1 = [1\ \ 0\ \ 0]$ and $V_2 = \begin{bmatrix} 0 & 1 & 0 \\ 0 & 0 & 1 \end{bmatrix}$.

With V_1, V_2 and E obtained above, we can verify that conditions (8.10) and (8.11) are satisfied. Furthermore, we also have $V_1 C_d E = V_1 C_h E = [0\ \ 0]$. By applying the state transformation $x(t) = T\bar{x}(t)$ to the time-delay system (8.1) and (8.2), we obtain disturbance-free time-delay system (8.13) and (8.14), where
$\bar{A}_{11} = \begin{bmatrix} -3 & -3 \\ 1 & -5 \end{bmatrix}$, $\bar{A}_{d11} = \begin{bmatrix} -4 & 0 \\ 0 & -2 \end{bmatrix}$, $\bar{A}_{h11} = \begin{bmatrix} -3 & 1 \\ -1 & -2 \end{bmatrix}$, $\bar{A}_{12} = \begin{bmatrix} 1 \\ 2 \end{bmatrix}$, $\bar{A}_{d12} =$
$\bar{A}_{h12} = \begin{bmatrix} 0 \\ 0 \end{bmatrix}$, $\tilde{C}_1 = 0_{2\times 2}$, $\tilde{C}_{d1} = \begin{bmatrix} 0 & 1 \\ 0 & 1 \end{bmatrix}$ and $\tilde{C}_{h1} = \begin{bmatrix} 1 & 1 \\ 0 & 1 \end{bmatrix}$.

From (8.16), $y_{\tau h}(t)$ is obtained as follows

$$y_{\tau h}(t) = \begin{bmatrix} y_2(t) - 3y_1(t-\tau) - 5y_1(t-h) \\ y_3(t) - 4y_1(t-\tau) - 4y_1(t-h) \end{bmatrix}. \tag{8.34}$$

We also obtain $\bar{A}_{21} = [-1\ \ 1]$, $\bar{A}_{d21} = [0\ \ -1]$, $\bar{A}_{h21} = [1\ \ -1]$, $\bar{A}_{22} = -5$, $\bar{A}_{d22} = -10$ and $\bar{A}_{h22} = -1$. From (8.15), the unknown input $d(t)$ is obtained as follows

$$d(t) = \dot{y}_1(t) + 5y_1(t) + 10y_1(t-\tau) + y_1(t-h) + z_{\tau h}(t),\ t \geq 0, \tag{8.35}$$

where

$$z_{\tau h}(t) = [1\ \ -1]\bar{x}_1(t) + [0\ \ 1]\bar{x}_1(t-\tau) + [-1\ \ 1]\bar{x}_1(t-h). \tag{8.36}$$

Therefore, the unknown input can now be estimated by estimating $z_{\tau h}(t)$ as defined by (8.36). Note that we have $\tilde{C}_1 = 0_{2\times 2}$. The reader can refer to Numerical Example 3 (Chap. 17), where we design a third-order generalized observer without internal

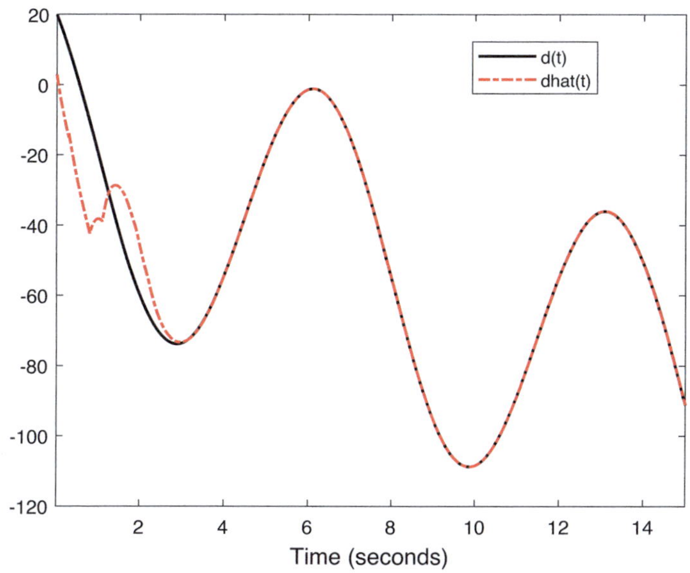

Fig. 8.2 Trajectories of $d(t)$ and $\hat{d}(t)$ (Numerical Example 2, Chap. 8)

delays to estimate $z_{\tau h}(t)$. Finally, Fig. 8.2 shows the trajectories of $d(t)$ and $\hat{d}(t)$, where asymptotic tracking is achieved.

8.4 Concluding Remark

In this chapter, we have looked at the problem of designing unknown input observers to estimate an unknown input, $d(t)$, for a class of time-delay systems with two non-commensurate delays in both the state and output vectors, i.e.,

$$\dot{x}(t) = Ax(t) + A_d x(t-\tau) + A_h x(t-h) + Dd(t), \ t \geq 0,$$
$$y(t) = Cx(t) + C_d x(t-\tau) + C_h x(t-h).$$

Due to the presence of two non-commensurate delays, τ and h, the design of an unknown input observer is complicated and there are problems that still need to be solved in the future. With the structural constraint $V_1 C_d D = 0$, $V_1 C_h D = 0$ and $V_1 C_d E = V_1 C_h E = 0_{1\times(n-1)}$, i.e., Case 1 in Sect. 8.2.1, we obtain a disturbance-free time-delay system (8.13) and (8.14), and express the unknown input $d(t)$ in its entirety according to (8.15). As such, the estimation of $d(t)$ rests with the successful design of an observer to estimate the required function $z_{\tau h}(t)$ as defined by (8.18) for the disturbance-free time-delay system (8.13) and (8.14). This is a new observer design problem that has not yet been considered before. In Chap. 17, we present some results on the design of such observers, and we provide three numerical examples, in which two of them have been used in this chapter.

On the other hand, with a weaker constraint imposed on matrix C_d or C_h (see, Case 2, where either $V_1 C_d E \neq 0_{1\times(n-1)}$ or $V_1 C_d E \neq 0_{1\times(n-1)}$), we obtain a disturbance-free time-delay system (8.21) and (8.22) and the unknown input $d(t)$ as given by (8.25). Thus, in Chap. 19, we provide a discussion on the design of generalized observers for a class of systems with mixed non-commensurate time delays in both the state and output vectors. However, in Cases 2 and 3 discussed in this chapter, the design of generalized observers for such systems remains an open research problem for future research.

Chapter 9
Multiple Unknown Inputs: Time Delays in State and Output Vectors

9.1 Background and Motivations

In Chaps. 4 and 5, we considered the problem of estimating an unknown input $d(t)$ for various classes of time-delay systems. Specifically, only a single unknown input was present and considered in those time-delay systems. On the other hand, as discussed in Chap. 3, we often face a situation where we need to estimate a particular unknown input in the presence of multiple unknown inputs. Thus, in this chapter, we discuss the problem of designing an unknown input observer to estimate a specific unknown input, say, $d_i(t)$, in the presence of multiple unknown inputs for a class of time-delay systems with both the state vector and output vector subject to a constant time delay. The reader can refer to Chap. 3 where we addressed the problem of estimating an unknown input in the presence of multiple unknown inputs for linear systems with no time delays. Thanks to our detailed discussion which includes two approaches, we can design an unknown input observer with the least number of differentiators used for the estimation of the unknown input.

In this chapter, let us reconsider time-delay system (5.1) and (5.2), i.e.,

$$\dot{x}(t) = Ax(t) + A_d x(t-\tau) + Dd(t),\ t \geq 0,$$
$$y(t) = Cx(t) + C_d x(t-\tau),$$

but instead of a single unknown input, $d(t)$, we now consider that $d(t)$ has two unknown inputs, i.e., $d(t) = \begin{bmatrix} d_1(t) \\ d_2(t) \end{bmatrix}$. Our objective is to design an unknown input observer to estimate, for example, $d_1(t)$, in the presence of the other unknown input $d_2(t)$. Here, for the sake of simplicity in presenting our result and discussion, we only consider two unknown inputs, $d_1(t)$ and $d_2(t)$, in the time-delay system (5.1) and (5.2). Note: the reader can refer to Sect. 3.2.4, Design Approach 2, where we discussed a more general case with q unknown inputs based on some coordinate changes that result in the least number of differentiators used for estimating $d_1(t)$.

H. Trinh et al., *Unknown Inputs Estimation in Linear Time-Delay Systems Using Generalized Functional Observers*, Studies in Systems, Decision and Control 648,
https://doi.org/10.1007/978-3-032-11870-7_9

9.2 Main Results

To facilitate the design of an unknown input observer to estimate $d_1(t)$ in the presence of $d_2(t)$ for system (5.1) and (5.2), we shall follow the two design approaches presented in Chap. 3.

9.2.1 *Design Approach 1*

Firstly, in this approach, we assume that there exists an output measurement, denoted by $y_i(t)$, where

$$y_i(t) = C_i x(t) + C_{di} x(t-\tau), \tag{9.1}$$

$C_i, C_{di} \in \mathbb{R}^{1\times n}$, such that $C_i D_1 = \alpha_i \neq 0$ and $C_i D_2 = 0$, where D is partitioned as $D = [D_1 \ D_2]$. If such an output $y_i(t)$ exists, we can use only one differentiator, i.e., $\dot{y}_i(t)$, for the purpose of estimating the unknown input $d_1(t)$. Furthermore, we can use matrices T and V_1, where $T = [E \ D_1]$, and V_1 is as defined by (2.4) to achieve the decoupling of $d_1(t)$ from the unknown input $d_2(t)$.

In the following, for ease of presentation, we consider the design of an unknown input observer, where we use the first-time derivative of the output $y_1(t)$ to estimate $d_1(t)$. For this, we consider time-delay system (5.1) and (5.2), where matrix D is now assumed to take the following form

$$D = [D_1 \ D_2] = \begin{bmatrix} \alpha_1 & 0 \\ D_{12} & D_{22} \end{bmatrix}, \tag{9.2}$$

where $\alpha_1 \neq 0$ and $D_{12}, D_{22} \in \mathbb{R}^{(n-1)}$. Thus, by choosing T as $T = [E \ D_1]$, where $E = \begin{bmatrix} 0_{1\times(n-1)} \\ I_{n-1} \end{bmatrix}$, and with D as defined by (9.2), we can easily verify the following

$$\begin{aligned} T^{-1}D &= \begin{bmatrix} -\alpha_1^{-1}D_{12} & I_{(n-1)} \\ \alpha_1^{-1} & 0_{1\times(n-1)} \end{bmatrix} \begin{bmatrix} \alpha_1 & 0 \\ D_{12} & D_{22} \end{bmatrix} \\ &= \begin{bmatrix} 0_{(n-1)} & D_{22} \\ 1 & 0 \end{bmatrix}. \end{aligned} \tag{9.3}$$

Thanks to (9.3), we can now apply the state transformation $x(t) = T\bar{x}(t)$ to (5.1) and (5.2) to decouple $d_1(t)$ from the unknown input $d_2(t)$ (see, (9.5)). Thus, we obtain the following equations

$$\dot{\bar{x}}_1(t) = \bar{A}_{11}\bar{x}_1(t) + \bar{A}_{12}\bar{x}_2(t) + \bar{A}_{d11}\bar{x}_1(t-\tau) + \bar{A}_{d12}\bar{x}_2(t-\tau) + D_{22}d_2(t), \tag{9.4}$$

$$\dot{\bar{x}}_2(t) = \bar{A}_{21}\bar{x}_1(t) + \bar{A}_{22}\bar{x}_2(t) + \bar{A}_{d21}\bar{x}_1(t-\tau) + \bar{A}_{d22}\bar{x}_2(t-\tau) + d_1(t), \tag{9.5}$$

$$y(t) = [CE \ CD_1]\begin{bmatrix} \bar{x}_1(t) \\ \bar{x}_2(t) \end{bmatrix} + [C_dE \ C_dD_1]\begin{bmatrix} \bar{x}_1(t-\tau) \\ \bar{x}_2(t-\tau) \end{bmatrix}. \tag{9.6}$$

Next, we consider matrix $V = [CD_1 \ P]$, where $P = \begin{bmatrix} 0_{1\times(p-1)} \\ I_{p-1} \end{bmatrix}$. Noting that with $V^{-1} = \begin{bmatrix} V_1 \\ V_2 \end{bmatrix}$, we have $V_1CE = 0_{1\times(n-1)}$, $V_1CD_1 = 1$ and $V_2CD_1 = 0_{(p-1)}$. By pre-multiplying both sides of (9.6) by $\begin{bmatrix} V_1 \\ V_2 \end{bmatrix}$, we obtain the following two equations

$$V_1y(t) = \bar{x}_2(t) + V_1C_dE\bar{x}_1(t-\tau) + V_1C_dD_1\bar{x}_2(t-\tau), \tag{9.7}$$
$$V_2y(t) = V_2CE\bar{x}_1(t) + V_2C_dE\bar{x}_1(t-\tau) + V_2C_dD_1\bar{x}_2(t-\tau). \tag{9.8}$$

In a similar way as in Sect. 5.1, to be able to express $\bar{x}_2(t)$ in terms of $y_1(t)$, we impose a constraint on the output matrix C_d such that

$$V_1C_d = 0_{1\times n}. \tag{9.9}$$

The reader can refer to Sect. 5.1 for a discussion on the practical implication of condition (9.9).

Thus, if condition (9.9) holds, we have $\bar{x}_2(t) = V_1y(t))$. By substituting this into (9.4) and (9.8), we obtain the following reduced-order time-delay with an unknown input $d_2(t)$

$$\begin{aligned} \dot{\bar{x}}_1(t) &= \bar{A}_{11}\bar{x}_1(t) + \bar{A}_{d11}\bar{x}_1(t-\tau) + D_{22}d_2(t) + \bar{A}_{12}V_1y(t) \\ &\quad + \bar{A}_{d12}V_1y(t-\tau), \ t \geq 0, \end{aligned} \tag{9.10}$$

$$\tilde{y}(t) = \tilde{C}_1\bar{x}_1(t) + \tilde{C}_{d1}\bar{x}_1(t-\tau), \tag{9.11}$$

where $\tilde{y}(t) = V_2y(t) - V_2C_dD_1V_1y(t-\tau)$, $\tilde{C}_1 = V_2CE$ and $\tilde{C}_{d1} = V_2C_dE$. Also, from (9.5), we can express the unknown input $d_1(t)$ as follows

$$d_1(t) = V_1\dot{y}(t) - \bar{A}_{22}V_1y(t) - \bar{A}_{d22}V_1y(t-\tau) - \bar{A}_{21}\bar{x}_1(t) - \bar{A}_{d21}\bar{x}_1(t-\tau), \ t \geq 0. \tag{9.12}$$

Accordingly, the unknown input $d_1(t)$ can be estimated by designing a generalized observer to estimate the function $\bar{z}(t) = -\bar{A}_{21}\bar{x}_1(t) - \bar{A}_{d21}\bar{x}_1(t-\tau)$ of time-delay system (9.10) and (9.11). Note that system (9.10) and (9.11) is not a disturbance-free system as it contains the unknown input $d_2(t)$. The reader can refer to Chap. 18 (Remark 3) for a design method.

Remark 1 Here, unlike the case $C_{di}D = 0$ as discussed in Sect. 5.2.2 in Chap. 5, this constraint does not lead to the complete decoupling of $d_1(t)$ from $d_2(t)$. This is explained below.

When $C_{d1}D_1 = 0$, then $V_1C_dD_1 = 0$, and from (9.7), we obtain Eq. (5.12), i.e., $\bar{x}_2(t) = V_1y(t) - V_1C_dE\bar{x}_1(t-\tau)$. From (9.4) and (9.8), we obtain the following time-delay system with an unknown input $d_2(t)$

$$\dot{\bar{x}}_1(t) = \bar{A}_{11}\bar{x}_1(t) + \tilde{A}_{d11}\bar{x}_1(t-\tau) + \tilde{A}_{dd11}\bar{x}_1(t-2\tau) + \bar{A}_{12}V_1y(t) + \bar{A}_{d12}V_1y(t-\tau) + D_{22}d_2(t), \tag{9.13}$$

$$\tilde{y}(t) = \tilde{C}_1\bar{x}_1(t) + \tilde{C}_{d1}\bar{x}_1(t-\tau) + \tilde{C}_{dd1}\bar{x}_1(t-2\tau), \tag{9.14}$$

where $\tilde{C}_{dd1} = -V_2C_dD_1V_1C_dE$, and $\tilde{A}_{d11}$ and $\tilde{A}_{dd11}$ are defined by (5.30) and (5.31), respectively.

Finally, by substituting $\bar{x}_2(t) = V_1y(t) - V_1C_dE\bar{x}_1(t-\tau)$ into (9.5) and using (9.13), the unknown input $d_1(t)$ is as given below

$$d_1(t) = \mathscr{R}_{(5.33)} - V_1C_dED_{22}d_2(t-\tau), \tag{9.15}$$

where for brevity $\mathscr{R}_{(5.33)}$ is used to denote all the terms on the right-hand side of Eq. (5.33).

It is clear from (9.15) that $d_1(t)$ includes $d_2(t-\tau)$. Thus, the constraint $C_{d1}D_1 = 0$ is not enough to decouple $d_1(t)$ completely from $d_2(t)$. Here, we need to further impose a constraint where

$$V_1C_dED_{22} = 0. \tag{9.16}$$

Note that (9.16) can be easily checked and if it holds, we can proceed to design a generalized observer to estimate the function $\tilde{z}(t)$ as defined by (5.38), i.e.,

$$\tilde{z}(t) = -\bar{A}_{21}\bar{x}_1(t) - \tilde{A}_{d21}\bar{x}_1(t-\tau) - \tilde{A}_{d22}\bar{x}_1(t-2\tau) - V_1C_dE\tilde{A}_{dd11}\bar{x}_1(t-3\tau),$$

for the unknown input time-delay system (9.13) and (9.14).

The problem of designing a generalized observer to estimate $\tilde{z}(t)$ as defined by (5.38) for the unknown input time-delay system (9.13) and (9.14) is a new problem that has not yet been reported in the literature. By extending the design method reported in Chap. 15 to incorporate unknown inputs, we can derive observer existence conditions and a procedure for designing an observer to estimate $\tilde{z}(t)$. For completeness, the reader can refer to Chap. 18 for an observer design procedure.

9.2.2 Design Approach 2

When it is not possible to find an output measurement $y_i(t), \forall i = 1, 2, \cdots p$, such that $C_i D_1 = \alpha_i \neq 0$ and $C_i D_2 = 0$, then we need to use the first-derivative of two output measurements. In such a case, the observer matching condition is needed, i.e., $\text{rank}(CD) = 2$ and we also require $p > 2$. This is similar to the situation as discussed in Remark 4 of Chap. 3. Subject to this condition, there always exist at least two output measurements, denoted by $y_o(t) = C_o x(t) + C_{do} x(t-\tau)$ where $C_o, C_{do} \in \mathbb{R}^{2\times n}$, such that $\text{rank}(C_o D) = 2$. Accordingly, we rearrange the output vector, $y(t)$, as follows

$$y(t) = \begin{bmatrix} y_o(t) \\ y_r(t) \end{bmatrix} = \begin{bmatrix} C_o \\ C_r \end{bmatrix} x(t) + \begin{bmatrix} C_{do} \\ C_{dr} \end{bmatrix} x(t-\tau) \triangleq Cx(t) + C_d x(t-\tau), \tag{9.17}$$

where $y_r(t) \in \mathbb{R}^{p-2}$ and $C_r, C_{dr} \in \mathbb{R}^{(p-2)\times n}$. By letting $P = \begin{bmatrix} 0_{2\times(p-2)} \\ I_{p-2} \end{bmatrix}$ and $V = \begin{bmatrix} C_o D & 0_{2\times(p-2)} \\ C_r D & I_{p-2} \end{bmatrix}$, we have

$$V^{-1} = \begin{bmatrix} V_1 \\ V_2 \end{bmatrix} = \begin{bmatrix} (C_o D)^{-1} & 0_{2\times(p-2)} \\ -C_r D (C_o D)^{-1} & I_{p-2} \end{bmatrix}. \tag{9.18}$$

From (9.18), we have $V_1 = \begin{bmatrix} (C_o D)^{-1} & 0_{2\times(p-2)} \end{bmatrix} \in \mathbb{R}^{2\times p}$ and $V_2 = \begin{bmatrix} -C_r D (C_o D)^{-1} & I_{p-2} \end{bmatrix} \in \mathbb{R}^{(p-2)\times p}$. Also, we select $T = \begin{bmatrix} E & D \end{bmatrix}$ where $E \in \mathbb{R}^{n\times(n-2)}$ is a rational basis of the null-space of $V_1 C$, i.e., we have $V_1 C E = 0_{2\times(n-2)}$.

With T given above, we apply the state transformation $x(t) = T\bar{x}(t)$ to time-delay system (5.1) and (5.2) to obtain the following equations

$$\dot{\bar{x}}_1(t) = \bar{A}_{11}\bar{x}_1(t) + \bar{A}_{12}\bar{x}_2(t) + \bar{A}_{d11}\bar{x}_1(t-\tau) + \bar{A}_{d12}\bar{x}_2(t-\tau), \tag{9.19}$$

$$\dot{\bar{x}}_2(t) = \bar{A}_{21}\bar{x}_1(t) + \bar{A}_{22}\bar{x}_2(t) + \bar{A}_{d21}\bar{x}_1(t-\tau) + \bar{A}_{d22}\bar{x}_2(t-\tau) + d(t), \tag{9.20}$$

$$y(t) = [CE \;\; CD]\begin{bmatrix} \bar{x}_1(t) \\ \bar{x}_2(t) \end{bmatrix} + [C_d E \;\; C_d D]\begin{bmatrix} \bar{x}_1(t-\tau) \\ \bar{x}_2(t-\tau) \end{bmatrix}, \tag{9.21}$$

where $\bar{x}_1(t) \in \mathbb{R}^{n-2}$ and $\bar{x}_2(t) \in \mathbb{R}^2$.

Next, we pre-multiply both sides of (9.21) by V^{-1} where V^{-1} is as defined by (9.18), and noting that $V_1 C E = 0_{2\times(n-2)}$, $V_1 C D = I_2$, $V_2 C D = 0_{(p-2)\times 2}$, we obtain the following two equations

$$V_1 y(t) = \bar{x}_2(t) + V_1 C_d E \bar{x}_1(t-\tau) + V_1 C_d D \bar{x}_2(t-\tau), \tag{9.22}$$

$$V_2 y(t) = V_2 C E \bar{x}_1(t) + V_2 C_d E \bar{x}_1(t-\tau) + V_2 C_d D \bar{x}_2(t-\tau). \tag{9.23}$$

As is the case discussed in Sect. 5.2 of Chap. 5, we need to impose a constraint on the output matrix C_d such that the following condition holds

$$V_1 C_d D = 0_{2\times 2}. \tag{9.24}$$

Note that (9.24) holds if either of the following conditions hold

$$V_1 C_d = 0_{2\times n}, \tag{9.25}$$
$$C_{do} D = 0_{2\times 2}. \tag{9.26}$$

With condition (9.24) satisfied (here, we do not go into either (9.25) or (9.26)), from (9.22), we have

$$\bar{x}_2(t) = V_1 y(t) - V_1 C_d E \bar{x}_1(t-\tau). \tag{9.27}$$

Substituting (9.27) into (9.19) and (9.23), we obtain a reduced-order disturbance-free time-delay system as described by (5.28) and (5.29). Here, we note that $\bar{x}_1(t) \in \mathbb{R}^{n-2}$.

Finally, by substituting (9.27) into (9.20), the unknown input $d(t)$ is obtained and given by (5.33). As $d(t)$ now contains two unknown inputs, i.e., $d_1(t)$ and $d_2(t)$, we can, if needed, estimate both by designing a generalized observer to estimate the following generalized vector

$$\tilde{z}(t) = -\bar{A}_{21}\bar{x}_1(t) - \tilde{A}_{d21}\bar{x}_1(t-\tau) - \tilde{A}_{d22}\bar{x}_1(t-2\tau) - V_1 C_d E \tilde{A}_{dd11}\bar{x}_1(t-3\tau). \tag{9.28}$$

On the other hand, if we are interested in estimating $d_1(t)$ only, then we design a generalized observer to estimate the following function

$$\begin{aligned}\tilde{z}_1(t) = &-(\bar{A}_{21})^1\bar{x}_1(t) - (\tilde{A}_{d21})^1\bar{x}_1(t-\tau) - (\tilde{A}_{d22})^1\bar{x}_1(t-2\tau)\\ &- (V_1 C_d E \tilde{A}_{dd11})^1\bar{x}_1(t-3\tau),\end{aligned} \tag{9.29}$$

where $(X)^1$ is used to denote the first row of matrix X.

Thus, if a generalized observer, $\hat{\tilde{z}}_1(t)$, can be designed to asymptotically estimate $\tilde{z}_1(t)$ then the disturbance $d_1(t)$ can be estimated as follows

$$\hat{d}_1(t) = (V_1)^1\dot{y}(t) - (\bar{A}_{22}V_1)^1 y(t) - (\tilde{A}_{12})^1 y(t-\tau) - (\tilde{A}_{d12})^1 y(t-2\tau) + \hat{\tilde{z}}_1(t). \tag{9.30}$$

The reader can refer to Chap. 15 where we discuss the design of various types of generalized observers to directly estimate $\tilde{z}_1(t)$.

Finally, note that when condition (9.25) holds, i.e., $V_1 C_d = 0_{2\times n}$, then we have from (9.27), $\bar{x}_2(t) = V_1 y(t)$, and hence a simpler time-delay system (5.13) and (5.14). Also, the unknown input $d(t)$ is as given by (4.5).

9.3 Numerical Examples

9.3.1 Numerical Example 1

In this example, we consider time-delay system (5.1) and (5.2) with two unknown inputs, $d_1(t)$ and $d_2(t)$, where $\tau = 1s$, matrices A, A_d, D, C and C_d are given below $A = \begin{bmatrix} -2 & 0 & -1 \\ 23 & -10 & -4 \\ 17 & 0 & -10 \end{bmatrix}$, $A_d = \begin{bmatrix} 1 & -1 & -1 \\ 3 & -5 & 0 \\ 27 & -6 & -9 \end{bmatrix}$, $D = [D_1\ D_2] = \begin{bmatrix} 1 & 0 \\ 2 & 1 \\ 3 & -1 \end{bmatrix}$, $C = \begin{bmatrix} 1 & 0 & 0 \\ 0 & 1 & 0 \\ 0 & 0 & 0 \end{bmatrix}$ and $C_d = \begin{bmatrix} 0 & 0 & 0 \\ 1 & 1 & 1 \\ 0 & 1 & 0 \end{bmatrix}$.

As there are two external unknown inputs in this system, our aim is to design an unknown observer to asymptotically estimate $d_1(t)$ in the presence of $d_2(t)$.

We first test to see if we can use the first-derive of only one output measurement (i.e., Design Approach 1 as discussed in Sect. 9.2.1). For this, we can check that output measurement $y_1(t)$ can be used for this purpose since we have $C_1 D_1 = 1 \neq 0$ and $C_1 D_2 = 0$.

Next, we select matrices V and T as follows

$$V = [CD_1\ P] = \begin{bmatrix} 1 & 0 & 0 \\ 2 & 1 & 0 \\ 0 & 0 & 1 \end{bmatrix}, T = [E\ D_1] = \begin{bmatrix} 0 & 0 & 1 \\ 1 & 0 & 2 \\ 0 & 1 & 3 \end{bmatrix}.$$

With V given above, we have $V_1 = \begin{bmatrix} 1 & 0 & 0 \end{bmatrix}$ and $V_2 = \begin{bmatrix} -2 & 1 & 0 \\ 0 & 0 & 1 \end{bmatrix}$. With C_d given above, condition (9.9) is satisfied, and hence we obtain a reduced-order time-delay (9.10) and (9.11), where
$\bar{A}_{11} = \begin{bmatrix} -10 & -2 \\ 0 & -7 \end{bmatrix}$, $\bar{A}_{d11} = \begin{bmatrix} -3 & 2 \\ -3 & -6 \end{bmatrix}$, $\bar{A}_{12} = \begin{bmatrix} 1 \\ 2 \end{bmatrix}$, $\bar{A}_{d12} = \begin{bmatrix} 1 \\ 0 \end{bmatrix}$, $D_{22} = \begin{bmatrix} 1 \\ -1 \end{bmatrix}$, $\tilde{C}_1 = \begin{bmatrix} 1 & 0 \\ 0 & 0 \end{bmatrix}$ and $\tilde{C}_{d1} = \begin{bmatrix} 1 & 1 \\ 1 & 0 \end{bmatrix}$. Note that as $\tilde{y}(t) = V_2 y(t) - V_2 C_d D_1 V_1 y(t-\tau)$, we thus have

$$\tilde{y}(t) = \begin{bmatrix} \tilde{y}_1(t) \\ \tilde{y}_2(t) \end{bmatrix} = \begin{bmatrix} -2y_1(t) + y_2(t) - 6y_1(t-\tau) \\ y_3(t) - 2y_1(t-\tau) \end{bmatrix}.$$

With the state transformation $x(t) = T\bar{x}(t)$, we obtain $\bar{A}_{22} = -5$, $\bar{A}_{d22} = -4$, $\bar{A}_{21} = [0\ -1]$ and $\bar{A}_{d21} = [-1\ -1]$. Hence from (9.12), the unknown input $d_1(t)$ is obtained as follows

$$d_1(t) = \dot{y}_1(t) + 5y_1(t) + 4y_1(t-\tau) + \begin{bmatrix} 0 & 1 \end{bmatrix}\bar{x}_1(t) + \begin{bmatrix} 1 & 1 \end{bmatrix}\bar{x}_1(t-\tau),\ t \geq 0. \tag{9.31}$$

Thus, from (9.31), we design a generalized observer to estimate the following function

$$\bar{z}(t) = \begin{bmatrix}0 & 1\end{bmatrix}\bar{x}_1(t) + \begin{bmatrix}1 & 1\end{bmatrix}\bar{x}_1(t-\tau) \tag{9.32}$$

for time-delay (9.10) and (9.11). Then $d_1(t)$ is estimated as follows

$$\hat{d}_1(t) = \dot{y}_1(t) + 5y_1(t) + 4y_1(t-\tau) + \hat{\bar{z}}(t). \tag{9.33}$$

The reader can refer to Numerical Example 1 of Chap. 18, where we design a first-order generalized observer to asymptotically estimate $\hat{z}_\tau(t)$ for time-delay system (18.29) and (18.30). Here, by replacing $Bu(t)$ with $[\bar{A}_{12}V_1 \;\; \bar{A}_{d12}V_1]\begin{bmatrix} y(t) \\ y(t-\tau)\end{bmatrix}$, we obtain the following generalized observer to estimate $\bar{z}(t)$

$$\hat{\bar{z}}(t) = w(t) + M\tilde{y}(t) + M_d\tilde{y}(t-\tau), \tag{9.34}$$

$$\begin{aligned}\dot{w}(t) &= Nw(t) + N_d w(t-\tau) + G\tilde{y}(t) + G_d\tilde{y}(t-\tau) + G_{dd}\tilde{y}(t-2\tau)\\ &\quad + 3y_1(t) + 7y_1(t-\tau) + 1.7668y_1(t-2\tau) - 0.0777y_1(t-3\tau),\end{aligned} \tag{9.35}$$

where M, M_d, N, N_d, G, G_d and G_{dd} are given in Numerical Example 1 of Chap. 18.

Figure 9.1 shows the trajectories of $d_1(t)$ and $\hat{d}_1(t)$ in the presence of an external unknown input $d_2(t)$. From the figure, it is clear that $\hat{d}_1(t)$ asymptotically converges to $d_1(t)$ regardless of the other unknown input, $d_2(t)$, affecting the system.

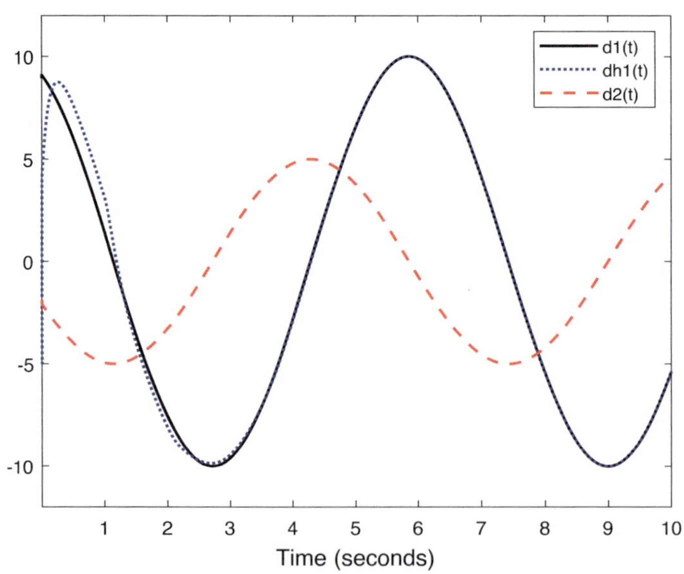

Fig. 9.1 Trajectories of $\hat{d}_1(t)$ and $d_1(t)$ in the presence of $d_2(t)$ (Numerical Example 1, Chap. 9)

9.3.2 Numerical Example 2

In this example, we consider time-delay system (5.1) and (5.2) with two unknown inputs, $d_1(t)$ and $d_2(t)$, where $\tau = 1s$, matrices A, A_d, D, C and C_d are given below $A = \begin{bmatrix} -6 & -1 & -1 \\ -3 & -12 & -5 \\ -2 & 3 & -4 \end{bmatrix}$, $A_d = \begin{bmatrix} -8 & -3 & -4 \\ 1 & -11 & -6 \\ -17 & 6 & -3 \end{bmatrix}$, $D = [D_1 \; D_2] = \begin{bmatrix} 1 & 0 \\ 2 & 1 \\ -3 & -1 \end{bmatrix}$, $C = \begin{bmatrix} 1 & 0 & 0 \\ 0 & 0 & 0 \end{bmatrix}$ and $C_d = \begin{bmatrix} 0 & 0 & 0 \\ 1 & 1 & 2 \end{bmatrix}$.

Our aim is to design an unknown observer to asymptotically estimate $d_1(t)$ in the presence of $d_2(t)$. In this example, we have $C_1 D_1 = 1 \neq 0$ and $C_1 D_2 = 0$. Thus, we select V and T as follows

$$V = [CD_1 \;\; P] = \begin{bmatrix} 1 & 0 \\ 0 & 1 \end{bmatrix}, T = [E \;\; D_1] = \begin{bmatrix} 0 & 0 & 1 \\ 1 & 0 & 2 \\ 0 & 1 & -3 \end{bmatrix}.$$

With V given above, we have $V_1 = \begin{bmatrix} 1 & 0 \end{bmatrix}$, $V_2 = \begin{bmatrix} 0 & 1 \end{bmatrix}$, condition (9.9) is satisfied, and we obtain a reduced-order time-delay (9.10) and (9.11) with an unknown input $d_2(t)$, where $\bar{A}_{11} = \begin{bmatrix} -10 & -3 \\ 0 & -7 \end{bmatrix}$, $\bar{A}_{d11} = \begin{bmatrix} -5 & 2 \\ -3 & -15 \end{bmatrix}$, $\bar{A}_{12} = \begin{bmatrix} -2 \\ 1 \end{bmatrix}$, $\bar{A}_{d12} = \begin{bmatrix} 1 \\ -2 \end{bmatrix}$, $D_{22} = \begin{bmatrix} 1 \\ -1 \end{bmatrix}$, $\tilde{C}_1 = \begin{bmatrix} 0 & 0 \end{bmatrix}$ and $\tilde{C}_{d1} = \begin{bmatrix} 1 & 2 \end{bmatrix}$.

Note that due to $\tilde{C}_1 = \begin{bmatrix} 0 & 0 \end{bmatrix}$, the output measurement $\tilde{y}(t)$ does not contain any instantaneous state variables of $\bar{x}_1(t)$, i.e., we have

$$\tilde{y}(t) = \bar{x}_{11}(t-\tau) + 2\bar{x}_{12}(t-\tau),$$

where $\bar{x}_1(t) = \begin{bmatrix} \bar{x}_{11}(t) \\ \bar{x}_{12}(t) \end{bmatrix}$. Note also that from $\tilde{y}(t) = V_2 y(t) - V_2 C_d D_1 V_1 y(t-\tau)$, we have

$$\tilde{y}(t) = y_2(t) + 3y_1(t-\tau).$$

Since in this example, we have $y_2(t) = x_1(t-\tau) + x_2(t-\tau) + 2x_3(t-\tau)$ and thus $\tilde{y}(t)$ does not contain any instantaneous state variables of $x(t)$. This is an intriguing case under consideration in this example.

Now, with the state transformation $x(t) = T\bar{x}(t)$, we obtain $\bar{A}_{22} = -5$, $\bar{A}_{d22} = -2$, $\bar{A}_{21} = [-1 \;\; -1]$ and $\bar{A}_{d21} = [-3 \;\; -4]$. Hence from (9.12), the unknown input $d_1(t)$ is obtained as follows

$$d_1(t) = \dot{y}_1(t) + 5y_1(t) + 2y_1(t-\tau) + \begin{bmatrix} 1 & 1 \end{bmatrix}\bar{x}_1(t) + \begin{bmatrix} 3 & 4 \end{bmatrix}\bar{x}_1(t-\tau), \; t \geq 0. \quad (9.36)$$

Thus, from (9.36), we design a generalized observer to estimate the following function

$$\bar{z}(t) = \begin{bmatrix} 1 & 1 \end{bmatrix}\bar{x}_1(t) + \begin{bmatrix} 3 & 4 \end{bmatrix}\bar{x}_1(t-\tau) \tag{9.37}$$

for time-delay (9.10) and (9.11). Then $d_1(t)$ is estimated as follows

$$\hat{d}_1(t) = \dot{y}_1(t) + 5y_1(t) + 2y_1(t-\tau) + \hat{\bar{z}}(t). \tag{9.38}$$

Note that since the output measurement $\tilde{y}(t)$ does not contain any instantaneous state variables of $\bar{x}_1(t)$ (i.e., $\tilde{C}_1 = \begin{bmatrix} 0 & 0 \end{bmatrix}$), the design of a generalized observer to asymptotically estimate $\bar{z}(t)$ as defined by (9.37) is subject to more stringent conditions (the reader can refer to Numerical Examples 2 and 3 of Chap. 18). From Numerical Example 3 of Chap. 18, we obtain the following generalized observer

$$\hat{\bar{z}}(t) = w(t) + \tilde{y}(t), \tag{9.39}$$

$$\begin{aligned} \dot{w}(t) = &-10w(t) - 3w(t-\tau) - 5\tilde{y}(t) - 10\tilde{y}(t-\tau) \\ &- y_1(t) - 3y_1(t-\tau) - 2y_1(t-2\tau). \end{aligned} \tag{9.40}$$

Figure 9.2 shows the trajectories of $d_1(t)$ and $\hat{d}_1(t)$ in the presence of an external unknown input $d_2(t)$. In the figure, it is clear that $\hat{d}_1(t)$ asymptotically converges to $d_1(t)$ regardless of the other unknown input, $d_2(t)$, affecting the system. Also, Fig. 9.3 shows the trajectories of $d_1(t)$ and $\hat{d}_1(t)$ in the presence of an external

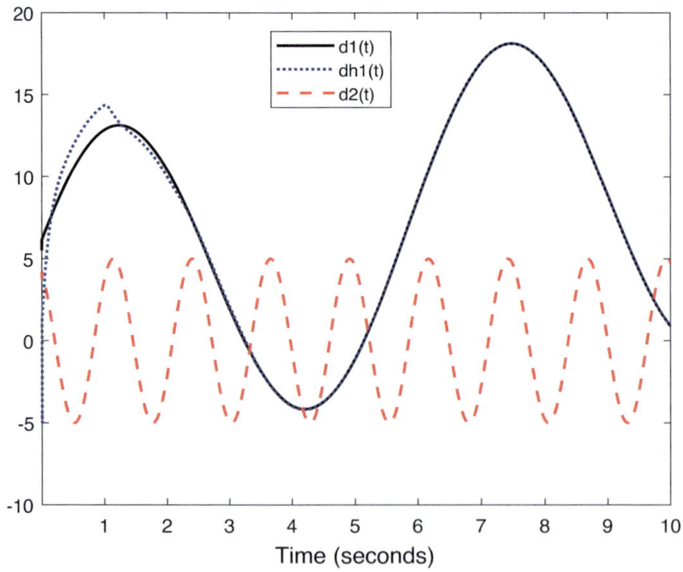

Fig. 9.2 Trajectories of $\hat{d}_1(t)$ and $d_1(t)$ in the presence of $d_2(t)$ (Numerical Example 2, Chap. 9)

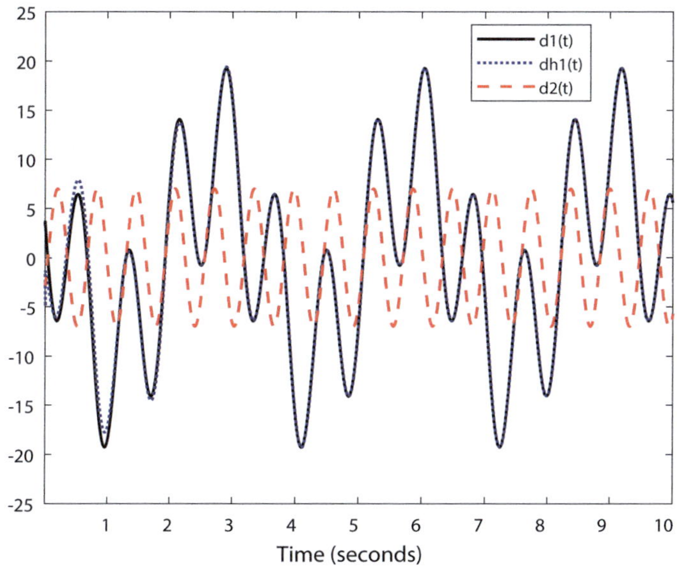

Fig. 9.3 Trajectories of $\hat{d}_1(t)$ and $d_1(t)$ in the presence of $d_2(t)$ (Numerical Example 2, Chap. 9)

unknown input $d_2(t)$. Here, both $d_1(t)$ and $d_2(t)$ are faster time-varying unknown inputs than the ones shown in Fig. 9.2. Again, from the figure, it is clear that $\hat{d}_1(t)$ asymptotically converges to $d_1(t)$ regardless of the other unknown input, $d_2(t)$, affecting the system.

9.3.3 Numerical Example 3

In this example, we consider time-delay system (5.1) and (5.2) with two unknown inputs, $d_1(t)$ and $d_2(t)$, where $\tau = 1s$, matrices A, A_d, D, C and C_d are given below

$$A = \begin{bmatrix} -6 & -1 & -1 \\ -15 & -12 & -9 \\ -42 & 3.5 & -17 \end{bmatrix}, A_d = \begin{bmatrix} -8 & -3 & -4 \\ -40 & -8.5 & -18 \\ 15.5 & 10 & 10.5 \end{bmatrix}, D = [D_1 \ D_2] = \begin{bmatrix} 1 & 0 \\ 2 & 1 \\ -3 & 0 \end{bmatrix},$$

$$C = \begin{bmatrix} 1 & 0 & 0 \\ 0 & 1 & 0 \\ 0 & 0 & 0 \end{bmatrix} \text{ and } C_d = \begin{bmatrix} 3 & 0 & 1 \\ 1 & 0 & 1 \\ 2.5 & 0.5 & 1 \end{bmatrix}.$$

Our aim is to design an unknown observer to asymptotically estimate $d_1(t)$ in the presence of $d_2(t)$. In this example, we have $C_1D_1 = 1 \neq 0$ and $C_1D_2 = 0$. Thus, we select V and T as follows

$$V = [CD_1 \ P] = \begin{bmatrix} 1\,0\,0 \\ 2\,1\,0 \\ 0\,0\,1 \end{bmatrix}, T = [E \ D_1] = \begin{bmatrix} 0\,0 & 1 \\ 1\,0 & 2 \\ 0\,1 & -3 \end{bmatrix}.$$

With V and T given above, we have $V_1 = \begin{bmatrix} 1 \ 0 \ 0 \end{bmatrix}$, $V_2 = \begin{bmatrix} -2\,1\,0 \\ 0\,0\,1 \end{bmatrix}$ and $D_{22} = \begin{bmatrix} 1 \\ 0 \end{bmatrix}$. Here, note that condition (9.9) is not satisfied since $V_1 C_d = \begin{bmatrix} 3 \ 0 \ 1 \end{bmatrix} \neq 0_{1\times 3}$. Thus, we cannot obtain a reduced-order time-delay system (9.10) and (9.11).

On the other hand, with C_d given above, we can check that $V_1 C_d D_1 = 0$ and thus as discussed in Remark 1, we further check condition (9.16) and find that it is satisfied, i.e., $V_1 C_d E D_{22} = 0$. As a result, we obtain a time-delay system (9.13) and (9.14) with an unknown input $d_2(t)$, where
$\bar{A}_{11} = \begin{bmatrix} -10 & -7 \\ 0.5 & -20 \end{bmatrix}$, $\tilde{A}_{d11} = \begin{bmatrix} -2.5 & -8 \\ 1 & -2.5 \end{bmatrix}$, $\tilde{A}_{dd11} = \begin{bmatrix} 0 & -1 \\ 0 & 2 \end{bmatrix}$, $\bar{A}_{12} = \begin{bmatrix} -2 \\ 1 \end{bmatrix}$, $\bar{A}_{d12} = \begin{bmatrix} 1 \\ -2 \end{bmatrix}$, $D_{22} = \begin{bmatrix} 1 \\ 0 \end{bmatrix}$, $\tilde{C}_1 = \begin{bmatrix} 1\,0 \\ 0\,0 \end{bmatrix}$, $\tilde{C}_{d1} = \begin{bmatrix} 0 & -1 \\ 0.5 & 1 \end{bmatrix}$ and $\tilde{C}_{dd1} = \begin{bmatrix} 0 & 2 \\ 0 & -0.5 \end{bmatrix}$.

With the state transformation $x(t) = T\bar{x}(t)$, we obtain $\bar{A}_{22} = -5$, $\bar{A}_{d22} = -2$, $\bar{A}_{21} = [-1 \ -1]$ and $\bar{A}_{d21} = [-3 \ -4]$. From (5.34) and (5.35), we obtain $\tilde{A}_{12} = [-1 \ 0 \ 0]$ and $\tilde{A}_{d12} = [-2 \ 0 \ 0]$.

Thus, with $V_1 C_d E D_{22} = 0$, from (9.15), the unknown input $d_1(t)$ is obtained as follows

$$d_1(t) = \mathscr{R}_{(5.33)} = \dot{y}_1(t) + 5y_1(t) + y_1(t-\tau) + 2y_1(t-2\tau) + \tilde{z}(t), \tag{9.41}$$

where $\tilde{z}(t)$ is as defined by (5.38), i.e.,

$$\begin{aligned} \tilde{z}(t) = &\begin{bmatrix} 1 \ 1 \end{bmatrix} \bar{x}_1(t) + \begin{bmatrix} 2.5 \ 19 \end{bmatrix} \bar{x}_1(t-\tau) + \begin{bmatrix} -1 \ 0.5 \end{bmatrix} \bar{x}_1(t-2\tau) \\ &+ \begin{bmatrix} 0 \ -2 \end{bmatrix} \bar{x}_1(t-3\tau). \end{aligned} \tag{9.42}$$

Note that from $\tilde{y}(t) = V_2 y(t) - V_2 C_d D_1 V_1 y(t-\tau)$, we have

$$\tilde{y}(t) = \begin{bmatrix} \tilde{y}_1(t) \\ \tilde{y}_2(t) \end{bmatrix} = \begin{bmatrix} -2y_1(t) + y_2(t) + 2y_1(t-\tau) \\ y_3(t) - 0.5y_1(t-\tau) \end{bmatrix}.$$

Thus, to estimate the unknown input $d_1(t)$ we design a generalized observer to estimate $\tilde{z}(t)$ as defined by (9.42). The reader can refer to Numerical Example 4 of Chap. 18. Accordingly, by replacing $Bu(t)$ with $[\tilde{A}_{12} V_1 \ \tilde{A}_{d12} V_1] \begin{bmatrix} y(t) \\ y(t-\tau) \end{bmatrix}$, we obtain the following generalized observer to estimate $\tilde{z}(t)$

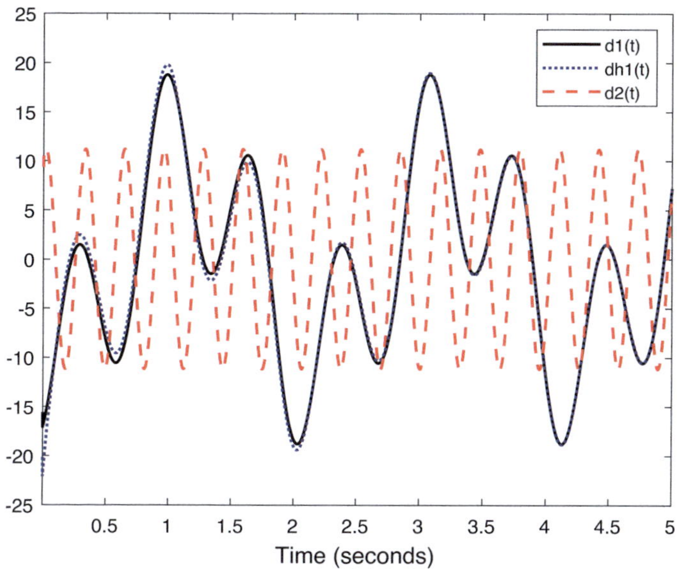

Fig. 9.4 Trajectories of $\hat{d}_1(t)$ and $d_1(t)$ in the presence of $d_2(t)$ (Numerical Example 3, Chap. 9)

$$\hat{\bar{z}}(t) = w(t) + \begin{bmatrix} 1 & 1.2945 \end{bmatrix} \tilde{y}(t) + \begin{bmatrix} 1.8528 & -2 \end{bmatrix} \tilde{y}(t-1), \tag{9.43}$$

$$\begin{aligned} \dot{w}(t) = & -20w(t) - 4w(t-1) + 2w(t-2) + \begin{bmatrix} 0.5 & 2 \end{bmatrix} \tilde{y}(t) + \begin{bmatrix} 9.3528 & 37.4110 \end{bmatrix} \\ & \tilde{y}(t-1) + \begin{bmatrix} 1.5 & 6 \end{bmatrix} \tilde{y}(t-2) + \begin{bmatrix} -3.3528 & -13.4110 \end{bmatrix} \tilde{y}(t-3) + y_1(t) + 16.7055 \\ & y_1(t-1) - 34.411 y_1(t-2) - 12.7055 y_1(t-3) + 13.411 y_1(t-4). \end{aligned} \tag{9.44}$$

Finally, with $\tau = 1s$ and from (9.41), $\hat{d}_1(t)$ is obtained as follows

$$\hat{d}_1(t) = \dot{y}_1(t) + 5y_1(t) + y_1(t-1) + 2y_1(t-2) + \hat{\bar{z}}(t). \tag{9.45}$$

Figure 9.4 shows the trajectories of $d_1(t)$ and $\hat{d}_1(t)$ in the presence of an external unknown input $d_2(t)$. In the figure, it is clear that $\hat{d}_1(t)$ asymptotically converges to $d_1(t)$ regardless of the other unknown input, $d_2(t)$, affecting the system.

9.3.4 Numerical Example 4

In this example, we consider time-delay system (5.1) and (5.2) with two unknown inputs, $d_1(t)$ and $d_2(t)$, where $\tau = 1s$, matrices A, A_d, D, C and C_d are given below

$$A = \begin{bmatrix} -6 & -1 & -0.8 \\ -15 & -12 & -9 \\ -42 & 3.5 & -17 \end{bmatrix},\ A_d = \begin{bmatrix} -8 & -3 & -4 \\ -40 & -8.5 & -18 \\ 15.5 & 10 & 10.5 \end{bmatrix},\ D = [D_1\ D_2] = \begin{bmatrix} 1 & 2 \\ 3 & 4 \\ 5 & 6 \end{bmatrix},$$
$$C = \begin{bmatrix} 1 & 0 & 0 \\ 0 & 1 & 0 \\ 0 & 0 & 0 \end{bmatrix} \text{ and } C_d = \begin{bmatrix} 0 & 0 & 0 \\ 0 & 0 & 0 \\ 1 & 2 & 3 \end{bmatrix}.$$

Again, our aim is to design an unknown observer to asymptotically estimate $d_1(t)$ in the presence of $d_2(t)$. Note that in this example, we cannot find an output measurement, $y_i(t)$, such that $C_i D_1 \neq 0$ and $C_i D_2 = 0$, since $CD = \begin{bmatrix} 1 & 2 \\ 3 & 4 \\ 0 & 0 \end{bmatrix}$. On the other hand, the observer matching condition is satisfied, i.e., $\text{rank}(CD) = 2$, and we also have $p > 2$. Thus, as discussed in Sect. 9.2.2, we use the first-derivative of two output measurements, $y_1(t)$ and $y_2(t)$, to estimate $d_1(t)$ (or, if needed, both $d_1(t)$ and $d_2(t)$).

Based on the development in Sect. 9.2.2, we select P and V as follows

$$P = \begin{bmatrix} 0 \\ 0 \\ 1 \end{bmatrix},\quad V = [CD\ P] = \begin{bmatrix} 1 & 2 & 0 \\ 3 & 4 & 0 \\ 0 & 0 & 1 \end{bmatrix}.$$

With V given above, we have $V_1 = \begin{bmatrix} -2 & 1 & 0 \\ 1.5 & -0.5 & 0 \end{bmatrix}$, $V_2 = \begin{bmatrix} 0 & 0 & 1 \end{bmatrix}$ and $E = \begin{bmatrix} 0 \\ 0 \\ 1 \end{bmatrix}$ (note that E is a rational basis for the null-space of $V_1 C$, i.e., $V_1 C E = 0_{2\times 1}$).

Hence, T is obtained as follows

$$T = [E\ D] = \begin{bmatrix} 0 & 1 & 2 \\ 0 & 3 & 4 \\ 1 & 5 & 6 \end{bmatrix}.$$

With C_d given above, condition (9.24) is satisfied (note that (9.25) is also satisfied and hence we obtain a less complex reduced-order time-delay system). From (5.28) and (5.29), we obtain the following reduced-order disturbance-free time-delay system

$$\dot{\bar{x}}_1(t) = 0.2\bar{x}_1(t) + 42.5\bar{x}_1(t-\tau) + \begin{bmatrix} -18.2 & 26.9 & 0 \end{bmatrix} y(t) + \begin{bmatrix} 45 & 109 & 0 \end{bmatrix} y(t-\tau), \tag{9.46}$$

$$\tilde{y}(t) = 3\bar{x}_1(t-\tau), \tag{9.47}$$

From $\tilde{y}(t) = V_2 y(t) - V_2 C_d D V_1 y(t-\tau)$, we have

$$\tilde{y}(t) = y_3(t) + 2y_1(t-\tau) - 8y_2(t-\tau).$$

From (5.33), the unknown input vector, $d(t)$, is obtained as follows

$$\begin{bmatrix} d_1(t) \\ d_2(t) \end{bmatrix} = \begin{bmatrix} -2 & 1 \\ 1.5 & -0.5 \end{bmatrix} \begin{bmatrix} \dot{y}_1(t) \\ \dot{y}_2(t) \end{bmatrix} + \begin{bmatrix} -4.4 & 24.8 \\ 4.8 & -11.1 \end{bmatrix} \begin{bmatrix} y_1(t) \\ y_2(t) \end{bmatrix} + \begin{bmatrix} 14 & 22.5 \\ -5 & -5.75 \end{bmatrix} \begin{bmatrix} y_1(t-\tau) \\ y_2(t-\tau) \end{bmatrix} + \tilde{z}(t), \tag{9.48}$$

where

$$\tilde{z}(t) = \begin{bmatrix} \tilde{z}_1(t) \\ \tilde{z}_2(t) \end{bmatrix} = \begin{bmatrix} 7.4 \\ -3.3 \end{bmatrix} \bar{x}_1(t) + \begin{bmatrix} 10 \\ -3 \end{bmatrix} \bar{x}_1(t-\tau). \tag{9.49}$$

Thus, to estimate the unknown input $d_1(t)$ we design a generalized observer to estimate $\tilde{z}_1(t)$, where

$$\tilde{z}_1(t) = 7.4\bar{x}_1(t) + 10\bar{x}_1(t-\tau). \tag{9.50}$$

From (9.48), $\hat{d}_1(t)$ is obtained as follows

$$\hat{d}_1(t) = -2\dot{y}_1(t) + \dot{y}_2(t) - 4.4y_1(t) + 24.8y_2(t) + 14y_1(t-\tau) + 22.5y_2(t-\tau) + \hat{\tilde{z}}_1(t). \tag{9.51}$$

Let us now turn our attention to time-delay system (9.46) and (9.47). From (9.47), we have $\bar{x}_1(t-\tau) = \frac{1}{3}\tilde{y}(t)$, and thus we only know the delayed state variable $\bar{x}_1(t-\tau)$. However, from (9.50), knowledge of $\bar{x}_1(t-\tau)$ is not enough to realize $\tilde{z}_1(t)$ since instantaneous information of $\bar{x}_1(t)$ is needed. Thus, we need to design a generalized observer to estimate $\tilde{z}_1(t)$ using the delayed output measurement $\tilde{y}(t)$. The reader can refer to Chap. 18, where we discuss the design of generalized observers including the case where $C = 0_{p\times n}$.

Let us now employ a general observer of the form (18.41) and (18.42) to estimate $\tilde{z}_1(t)$. Thus, we apply the result presented in Remark 3, Chap. 18 (see, Eqs. (18.29) and (18.31)), by letting $C = 0$ and $D = 0$. Note that since $D = 0$, condition (18.43) is always satisfied. We then solve for the unknown matrix X as defined by (15.32). Here, since $D = 0$, (18.47) and (18.50) are replaced with the following equation and condition, respectively,

$$X\bar{\Theta}_\tau = \bar{\Upsilon}_\tau, \tag{9.52}$$

$$\text{rank}\begin{bmatrix} \bar{\Theta}_\tau \\ \bar{\Upsilon}_\tau \end{bmatrix} = \text{rank}\begin{bmatrix} \bar{\Theta}_\tau \end{bmatrix}, \tag{9.53}$$

where

$$\bar{\Theta}_\tau = \begin{bmatrix} F & F_d & 0 & 0 & 0 \\ 0 & F & F_d & 0 & 0 \\ 0 & 0 & F & F_d & 0 \\ 0 & C_d & 0 & 0 & 0 \\ 0 & 0 & C_d & 0 & 0 \\ 0 & 0 & 0 & C_d & 0 \\ 0 & 0 & 0 & 0 & C_d \\ 0 & C_\tau & C_d A_d & 0 & 0 \\ 0 & 0 & C_\tau & C_d A_d & 0 \end{bmatrix}, \tag{9.54}$$

$$\bar{\Upsilon}_\tau = \begin{bmatrix} FA & FA_d + F_d A & F_d A_d & 0 & 0 \end{bmatrix}. \tag{9.55}$$

Note that for this example, an observer with no internal delays, i.e., $N_d = N_{dd} = 0$, does not exist, since we have $N = 0.2$ as the fixed-pole, resulting in an unstable error dynamic equation. On the other hand, by employing an observer with an internal delay, i.e., $N_d w(t-\tau)$, we obtain a stable error dynamic equation. This is done as follows.

We first remove the third row from $\bar{\Theta}_\tau$ and then we test condition (9.53). It is found that the condition is satisfied. By solving Eq. (9.52) (i.e., $X = \bar{\Upsilon}_\tau \bar{\Theta}_\tau^+ + Z(I - \bar{\Theta}_\tau \bar{\Theta}_\tau^+)$), we obtain the following error dynamic equation

$$\dot{e}(t) = 0.2e(t) + (36.4401 + ZN_{d2})e(t-1), \tag{9.56}$$

where $N_{d2}^T = \begin{bmatrix} 0 & 0.1426 & -0.3495 & 0.0014 & 0 & 0 & -0.0112 & 0 \end{bmatrix}$. Thus, we now need to find a Z such that the error dynamic equation (9.56) is stable. This can be easily done since $N_{d2} \neq 0$. Here, we make $(36.4401 + ZN_{d2}) = -0.45$, which results in a stable error dynamic equation $\dot{e}(t) = 0.2e(t) - 0.45e(t-1)$ (there are two dominant eigenvalues at $-0.799 \pm j0.0546$). We obtain the following observer

$$\hat{\tilde{z}}_1(t) = w(t) + 3.3784\tilde{y}(t), \tag{9.57}$$

$$\begin{aligned} \dot{w}(t) &= 0.2w(t) - 0.45w(t-1) + 105.9433\tilde{y}(t) - 1.9362\tilde{y}(t-1) + 0.002\tilde{y}(t-2) \\ &\quad - 134.68y_1(t) + 199.06y_2(t) + 335.4614y_1(t-1) + 802.9621y_2(t-1) \\ &\quad - 6.0883y_1(t-2) - 14.7374y_2(t-2) + 0.0062y_1(t-3) + 0.015y_2(t-3). \end{aligned} \tag{9.58}$$

Finally, with $\tau = 1s$, Fig. 9.5 shows the trajectories of $d_1(t)$ and $\hat{d}_1(t)$ in the presence of an external unknown input $d_2(t)$. In the figure, it is clear that $\hat{d}_1(t)$ asymptotically converges to $d_1(t)$ regardless of the other unknown input, $d_2(t)$, affecting the system.

As a matter of interest, we can also design an unknown input observer to estimate $d_2(t)$. This is done by designing an observer to asymptotically estimate $\tilde{z}_2(t)$, where

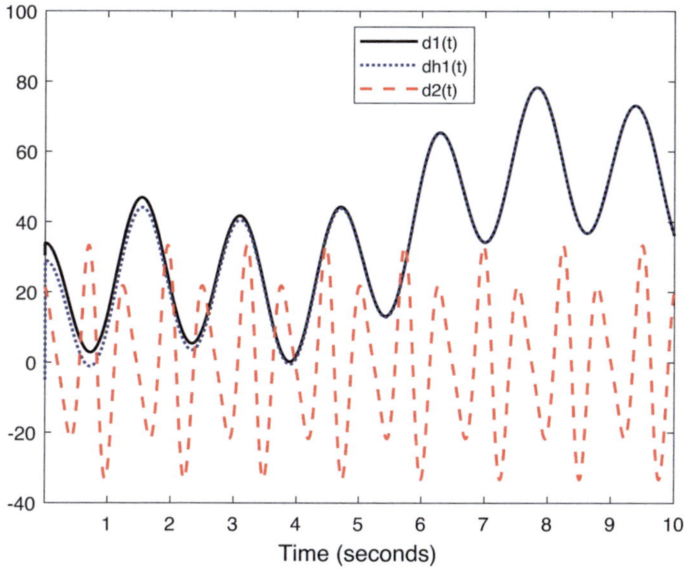

Fig. 9.5 Trajectories of $\hat{d}_1(t)$ and $d_1(t)$ in the presence of $d_2(t)$ (Numerical Example 4, Chap. 9)

$$\tilde{z}_2(t) = -3.3\bar{x}_1(t) - 3\bar{x}_1(t-1). \tag{9.59}$$

From (9.48), $\hat{d}_2(t)$ is obtained as follows

$$\begin{aligned}\hat{d}_2(t) = {} & 1.5\dot{y}_1(t) - 0.5\dot{y}_2(t) + 4.8y_1(t) - 11.1y_2(t) - 5y_1(t-1) - 5.75y_2(t-1) \\ & + \hat{\tilde{z}}_2(t).\end{aligned} \tag{9.60}$$

Here, the design of $\hat{\tilde{z}}_2(t)$ is omitted, and we would like to leave it to the reader to complete the observer design. Finally, Fig. 9.6 shows the trajectories of $d_2(t)$ and $\hat{d}_2(t)$ in the presence of an external unknown input $d_1(t)$. From the figure, it is clear that $\hat{d}_2(t)$ asymptotically converges to $d_2(t)$ regardless of the other unknown input, $d_1(t)$, affecting the system.

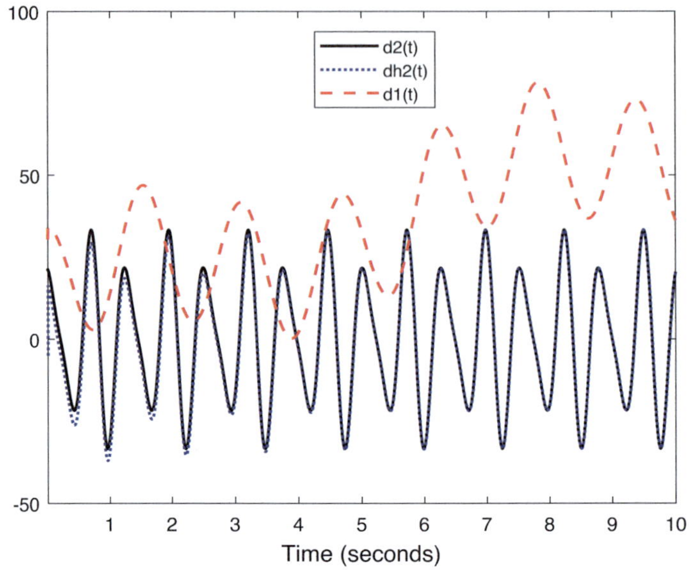

Fig. 9.6 Trajectories of $\hat{d}_2(t)$ and $d_2(t)$ in the presence of $d_1(t)$ (Numerical Example 4, Chap. 9)

9.4 Concluding Remark

In this chapter, we have considered the problem of designing an observer to asymptotically estimate an unknown input in the presence of another unknown input for a class of time-delay systems, with both state and output vectors subject to a constant time delay. This chapter extends the results in Chap. 5 to deal with two unknown inputs. Further work can be directed to deal with multiple unknown inputs using the least number of differentiators to estimate a selected unknown input.

Chapter 10
Design of Functional Observers for Time-Delay Systems

10.1 Preliminaries

In this chapter, we first present some well-known results for designing an m-th order functional observer to estimate m functions of the state vector of time-delay system as described by (4.3) and (4.4), i.e.,

$$\begin{aligned}\dot{\bar{x}}_1(t) &= \bar{A}_{11}\bar{x}_1(t) + \bar{A}_{d11}\bar{x}_1(t-\tau) + \bar{A}_{12}V_1y(t) + \bar{A}_{d12}V_1y(t-\tau),\ t \geq 0,\\ \bar{y}(t) &= \tilde{C}_1\bar{x}_1(t).\end{aligned}$$

Then based on these results, we provide a procedure for designing low-order scalar functional observers, which will be used to estimate the unknown input $d(t)$ as discussed in Chap. 4.

10.1.1 Observers Without Internal Delays

To estimate $z(t) = F\bar{x}_1(t)$, where $F \in \mathbb{R}^{m\times(n-1)}$ and $\operatorname{rank}\begin{bmatrix}\tilde{C}_1\\ F\end{bmatrix} = (p-1+m)$, $m \leq (n-p)$, we consider the following minimum-order functional observer

$$\hat{z}(t) = w(t) + M\bar{y}(t), \tag{10.1}$$

$$\dot{w}(t) = Nw(t) + G\bar{y}(t) + G_d\bar{y}(t-\tau) + Jy(t) + J_dy(t-\tau), \tag{10.2}$$

where $w(t) \in \mathbb{R}^m$ with the initial condition $w(\theta) = \rho(\theta),\ \forall\theta \in [-\tau, 0]$. Taking the derivative of both sides of (4.17) and substituting (10.2) and (4.3), we have

$$\begin{aligned}\dot{\epsilon}(t) &= \dot{w}(t) - L\dot{\bar{x}}_1(t)\\ &= N\epsilon(t) + (NL + G\tilde{C}_1 - L\bar{A}_{11})\bar{x}_1(t)\end{aligned}$$

H. Trinh et al., *Unknown Inputs Estimation in Linear Time-Delay Systems Using Generalized Functional Observers*, Studies in Systems, Decision and Control 648,
https://doi.org/10.1007/978-3-032-11870-7_10

$$+ (G_d \tilde{C}_1 - L\bar{A}_{d11})\bar{x}_1(t-\tau) + (J - L\bar{A}_{12}V_1)y(t)$$
$$+ (J_d - L\bar{A}_{d12}V_1)y(t-\tau). \tag{10.3}$$

From (10.3), it is clear that if conditions (4.18) and (4.20), and (4.22) and (4.23) are satisfied then the following error equation is obtained

$$\dot{\epsilon}(t) = N\epsilon(t). \tag{10.4}$$

Thus with a stable matrix N, the error vector $\epsilon(t)$ converges asymptotically to zero. Now, from (4.16), the error vector $e(t)$ can be expressed as

$$\begin{aligned} e(t) &= \hat{z}(t) - z(t) = w(t) + M\bar{y}(t) - F\bar{x}_1(t) \\ &= \epsilon(t) + (L + M\tilde{C}_1 - F)\bar{x}_1(t). \end{aligned} \tag{10.5}$$

From (10.5), it is clear that if the following condition is satisfied

$$L + M\tilde{C}_1 - F = 0, \tag{10.6}$$

then $e(t) = \epsilon(t)$, and thus it asymptotically converges to zero. Note that the above condition is for the case of multiple functions, i.e., $F \in \mathbb{R}^{m\times(n-1)}$, and K is selected as $K = I_m$ (cf. condition (4.21) for the case of one scalar function to be estimated).

10.1.2 Observers Design: Minimum-Order

To achieve the design of a minimum-order functional observer (10.1) and (10.2), we need to solve for the unknown matrices N, L, M, G and G_d such that the following matrix equations are satisfied

$$NL + G\tilde{C}_1 - L\bar{A}_{11} = 0, \ \ N \text{ is Hurwitz}, \tag{10.7}$$
$$G_d\tilde{C}_1 - L\bar{A}_{d11} = 0, \tag{10.8}$$
$$L + M\tilde{C}_1 - F = 0. \tag{10.9}$$

Note that the necessary and sufficient condition for the solvability of matrix equations (10.7)–(10.9) can be derived as presented in [14] (see, Chap. 5). Here, for the sake of completeness and to provide more insights into the design of a minimum-order scalar functional observer to be used for the estimation of the unknown input $d(t)$, we discuss the solvability of (10.7)–(10.9) and present a solution approach to give us the smallest order m.

Let F, L, $\bar{A}_{11}$ and $\bar{A}_{d11}$ be partitioned as
$F = \begin{bmatrix} F_1 & F_2 \end{bmatrix}$, $L = \begin{bmatrix} L_1 & L_2 \end{bmatrix}$, $\bar{A}_{11} = \begin{bmatrix} \bar{A}_{11}^{11} & \bar{A}_{11}^{12} \\ \bar{A}_{11}^{21} & \bar{A}_{11}^{22} \end{bmatrix}$, $\bar{A}_{d11} = \begin{bmatrix} \bar{A}_{d11}^{11} & \bar{A}_{d11}^{12} \\ \bar{A}_{d11}^{21} & \bar{A}_{d11}^{22} \end{bmatrix}$, where

$F_1, L_1 \in \mathbb{R}^{m\times(p-1)}$ $F_2, L_2 \in \mathbb{R}^{m\times(n-p)}$, $\bar{A}_{11}^{11}, \bar{A}_{d11}^{11} \in \mathbb{R}^{(p-1)\times(p-1)}$, $\bar{A}_{11}^{12}, \bar{A}_{d11}^{12} \in \mathbb{R}^{(p-1)\times(n-p)}$, $\bar{A}_{11}^{21}, \bar{A}_{d11}^{21} \in \mathbb{R}^{(n-p)\times(p-1)}$ and $\bar{A}_{11}^{22}, \bar{A}_{d11}^{22} \in \mathbb{R}^{(n-p)\times(n-p)}$.

To avoid any trivial cases, matrix F_2 is assumed to have full-row rank, i.e., $\operatorname{rank}(F_2) = m$. This is ensured with rank $\begin{bmatrix} \tilde{C}_1 \\ F \end{bmatrix} = (p-1+m)$ as stated earlier.

Now, (10.9) can be decoupled into two equations, where

$$M = F_1 - L_1, \tag{10.10}$$

$$L_2 = F_2. \tag{10.11}$$

Similarly, (10.7) and (10.8) can be expressed as follows

$$G = L_1\bar{A}_{11}^{11} + F_2\bar{A}_{11}^{21} - NL_1, \tag{10.12}$$

$$NF_2 - L_1\bar{A}_{11}^{12} - F_2\bar{A}_{11}^{22} = 0, \ \ N \text{ is Hurwitz}, \tag{10.13}$$

$$G_d = L_1\bar{A}_{d11}^{11} + F_2\bar{A}_{d11}^{21}, \tag{10.14}$$

$$L_1\bar{A}_{d11}^{12} = -F_2\bar{A}_{d11}^{22}. \tag{10.15}$$

Examination of the above equations reveals that we only need to solve Eqs. (10.13) and (10.15) for the unknown matrices N and L_1. Then M, G and G_d are obtained from (10.10), (10.12) and (10.14), respectively. Also, once L_1 is found, L is obtained as $L = \begin{bmatrix} L_1 & F_2 \end{bmatrix}$, and thus $J = L\bar{A}_{12}V_1$ and $J_d = L\bar{A}_{d12}V_1$.

In the following, we look at the solvability of (10.13) and (10.15) and derive conditions for the existence of L_1 and a stable matrix N. First, the necessary and sufficient condition for L_1 to satisfy (10.15) is

$$\operatorname{rank}\begin{bmatrix} \bar{A}_{d11}^{12} \\ F_2\bar{A}_{d11}^{22} \end{bmatrix} = \operatorname{rank}\big[\bar{A}_{d11}^{12}\big]. \tag{10.16}$$

Upon the satisfaction of (10.16), we can proceed to solve both (10.13) and (10.15) for N and L_1. For this, denoting $F_2^{\perp} \in \mathbb{R}^{(n-p)\times(n-p-m)}$ as an orthogonal basis of the null-space of F_2, i.e., $F_2F_2^{\perp} = 0$, and letting $F_2^{+} \in \mathbb{R}^{(n-p)\times m}$ be the Moore-Penrose inverse of F_2, i.e., $F_2F_2^{+} = I_m$. Thus $\begin{bmatrix} F_2^{+} & F_2^{\perp} \end{bmatrix} \in \mathbb{R}^{(n-p)\times(n-p)}$ is an invertible matrix, and by post-multiplying (10.13) by $\begin{bmatrix} F_2^{+} & F_2^{\perp} \end{bmatrix}$ we get

$$N = L_1\bar{A}_{11}^{12}F_2^{+} + F_2\bar{A}_{11}^{22}F_2^{+}, \tag{10.17}$$

$$L_1\bar{A}_{11}^{12}F_2^{\perp} = -F_2\bar{A}_{11}^{22}F_2^{\perp}. \tag{10.18}$$

Now, to solve for L_1 that can simultaneously satisfy (10.18) and (10.15), we combine both equations as follows

$$L_1A_c = F_c \tag{10.19}$$

where $A_c \in \mathbb{R}^{(p-1)\times(2n-2p-m)}$ and $F_c \in \mathbb{R}^{m\times(2n-2p-m)}$ are defined below

$$A_c = \left[\bar{A}_{11}^{12} F_2^{\perp} \quad \bar{A}_{d11}^{12} \right], \tag{10.20}$$

$$F_c = -\left[F_2 \bar{A}_{11}^{22} F_2^{\perp} \quad F_2 \bar{A}_{d11}^{22} \right]. \tag{10.21}$$

From (10.19), a solution for L_1 exists if and only if the following condition is satisfied

$$\text{rank} \begin{bmatrix} F_c \\ A_c \end{bmatrix} = \text{rank}\left[A_c \right]. \tag{10.22}$$

Note that condition (10.22) is the necessary and sufficient condition for the existence of a matrix L_1 that can satisfy both (10.18) and (10.15). If (10.22) holds, then a solution for L_1 is obtained, where

$$L_1 = F_c A_c^+ + Z(I_{p-1} - A_c A_c^+), \tag{10.23}$$

A_c^+ is the Moore-Penrose inverse of A_c and $Z \in \mathbb{R}^{m \times (p-1)}$ is an arbitrary matrix (see, [14]). As Z is an arbitrary matrix, it can be fully utilized in determining N. Thus, substituting (10.23) into (10.17), we obtain

$$N = N_1 - Z N_2 \tag{10.24}$$

where

$$N_1 = (F_c A_c^+ \bar{A}_{11}^{12} + F_2 \bar{A}_{11}^{22}) F_2^+, \tag{10.25}$$

$$N_2 = (A_c A_c^+ - I_{p-1}) \bar{A}_{11}^{12} F_2^+. \tag{10.26}$$

If $N_2 \neq 0$ and the pair (N_1, N_2) is observable, then we can always find a Z to place all the poles of N at any location in the complex s-plane. Hence a stable matrix N with any prescribed eigenvalues can be found. On the other hand, if the pair (N_1, N_2) is detectable then a stable matrix N is obtained but its poles cannot be arbitrarily assigned.

Therefore, to design a minimum-order functional observer to estimate a given function $z(t) = F\bar{x}_1(t)$, we can first proceed to check whether condition (10.16) is satisfied or not. If so, we then check condition (10.22). If it is satisfied and if the pair (N_1, N_2) is detectable/observable, a stable matrix N can be obtained, and the rest of the observer parameters can be derived.

10.1.3 Observers Design: Higher-Order

In Chap. 4, we discussed Design Option 2 where a scalar functional observer is used to estimate the unknown input $d(t)$ instead of Design Option 1, which uses a reduced-order state observer of order $(n - p)$. It is clear from (10.16) that the existence condition for a scalar functional observer is less restrictive than a reduced-order state observer. Since for $F_2 = I_{n-p}$ (i.e., the case of a reduced-order state observer), condition (10.16) is

$$\text{rank}\begin{bmatrix} \bar{A}_{d11}^{12} \\ \bar{A}_{d11}^{22} \end{bmatrix} = \text{rank}\big[\bar{A}_{d11}^{12}\big]. \tag{10.27}$$

In Scenario 2 of Chap. 4, our objective is to design a scalar functional observer to estimate a function $z(t) = F\bar{x}_1(t)$, where $F \in \mathbb{R}^{1\times(n-1)}$ is a known matrix. Based on the above derived results, we can now design a scalar functional observer of order m, where m is as small as possible. For this, first we show that condition (10.22) is equivalent to the following condition

$$\text{rank}\begin{bmatrix} F_2\bar{A}_{11}^{22} & F_2\bar{A}_{d11}^{22} \\ \bar{A}_{11}^{12} & \bar{A}_{d11}^{12} \\ F_2 & 0 \end{bmatrix} = \text{rank}\begin{bmatrix} \bar{A}_{11}^{12} & \bar{A}_{d11}^{12} \\ F_2 & 0 \end{bmatrix}. \tag{10.28}$$

By using the property that $\text{rank}(XY) = \text{rank}(X)$, where Y is square matrix of full rank, the right-hand side of (10.28) is equivalent to

$$\begin{aligned} &\text{rank}\begin{bmatrix} \bar{A}_{11}^{12} & \bar{A}_{d11}^{12} \\ F_2 & 0 \end{bmatrix}\begin{bmatrix} F_2^{+} & F_2^{\perp} & 0 \\ 0 & 0 & I_{n-p} \end{bmatrix} \\ &= m + \text{rank}(A_c), \end{aligned} \tag{10.29}$$

and the left-hand side of (10.28) is the same as

$$\begin{aligned} &\text{rank}\begin{bmatrix} F_2\bar{A}_{11}^{22} & F_2\bar{A}_{d11}^{22} \\ \bar{A}_{11}^{12} & \bar{A}_{d11}^{12} \\ F_2 & 0 \end{bmatrix}\begin{bmatrix} F_2^{+} & F_2^{\perp} & 0 \\ 0 & 0 & I_{n-p} \end{bmatrix} \\ &= m + \text{rank}\begin{bmatrix} F_c \\ A_c \end{bmatrix}. \end{aligned} \tag{10.30}$$

From (10.29) and (10.30), we can conclude that (10.28) is equivalent to (10.22). While they are equivalent, (10.28) is preferred over (10.22) since it can be checked without the need to calculate F_2^{+} and $F_2^{\perp}$.

When condition (10.16) is satisfied, but condition (10.28) is not satisfied, a first-order functional observer does not exist. On the other hand, when both conditions (10.16) and (10.28) are satisfied but the pair (N_1, N_2) is not observable/detectable, a first-order functional observer also does not exist. In such cases, we either stop or have an option to increase the order of the observer and carry out the following iterative steps, starting with $m = 2$. Here, we discuss two cases. The first case is where $\bar{A}_{d11}^{12}$ is a row matrix with $\text{rank}(\bar{A}_{d11}^{12}) \leq (p-1)$. The second case is where $\bar{A}_{d11}^{12}$ is a full-column matrix, i.e., $\text{rank}(\bar{A}_{d11}^{12}) = (n-p)$, and as such, condition (10.16) is always satisfied.

Case 1: $\text{rank}(\bar{A}_{d11}^{12}) \leq (p-1)$. In this case, condition (10.16) is not always satisfied for any arbitrary extra rows in F_2. There is a finite number of extra rows that can be added to F_2 to ensure that condition (10.16) holds.

First, we consider the case where $m = 1$, $F_2 \in \mathbb{R}^{1\times(n-p)}$ and both conditions (10.16) and (10.28) are satisfied but the pair (N_1, N_2) is not observable nor detectable, i.e., a stable matrix N cannot be found. In this situation, the following analysis shows that there is no need to look for a second-order functional observer (or high-order functional observers) since there will not be any. For this, we recall a definition of the detectability of the pair (N_1, N_2) as follows

$$\text{rank}\begin{bmatrix} s - N_1 \\ N_2 \end{bmatrix} = 1, \ \forall s \in \mathbb{C}, \Re(s) \geq 0. \tag{10.31}$$

Now, we can show that (10.31) is equivalent to the following

$$\begin{aligned} &\text{rank}\begin{bmatrix} (sF_2 - F_2\bar{A}_{11}^{22}) & -F_2\bar{A}_{d11}^{22} \\ \bar{A}_{11}^{12} & \bar{A}_{d11}^{12} \end{bmatrix} \\ &= \text{rank}\begin{bmatrix} \bar{A}_{12}^{12} & \bar{A}_{d11}^{12} \\ F_2 & 0 \end{bmatrix}, \forall s \in \mathbb{C}, \Re(s) \geq 0. \end{aligned} \tag{10.32}$$

The proof can be constructed by using the property that $\text{rank}(ZXY) = \text{rank}(X)$, where Z is a full-column rank matrix and Y is an invertible matrix. Thus, the left-hand side of (10.32) is the same as

$$\begin{aligned} &\text{rank}\begin{bmatrix} (sF_2 - F_2\bar{A}_{11}^{22}) & -F_2\bar{A}_{d11}^{22} \\ \bar{A}_{11}^{12} & \bar{A}_{d11}^{12} \end{bmatrix}\begin{bmatrix} F_2^+ & F_2^\perp & 0 \\ 0 & 0 & I_{n-p} \end{bmatrix} \\ &= \text{rank}\begin{bmatrix} (s - F_2\bar{A}_{11}^{22}F_2^+) & F_c \\ \bar{A}_{11}^{12}F_2^+ & A_c \end{bmatrix}. \end{aligned} \tag{10.33}$$

The right-hand side of (10.33) is the same as

$$\begin{aligned} &\text{rank}\begin{bmatrix} 1 & -F_cA_c^+ \\ 0 & (A_cA_c^+ - I_{p-1}) \\ 0 & A_cA_c^+ \end{bmatrix}\begin{bmatrix} (s - F_2\bar{A}_{11}^{22}F_2^+) & F_c \\ \bar{A}_{11}^{12}F_2^+ & A_c \end{bmatrix} \\ &= \text{rank}\begin{bmatrix} s - N_1 \\ N_2 \end{bmatrix} + \text{rank}(A_c), \forall s \in \mathbb{C}, \Re(s) \geq 0. \end{aligned} \tag{10.34}$$

On the other hand, for $m = 1$ and making use of (10.29), the right-hand side of (10.32) is $(1 + \text{rank}(A_c))$. Thus, the equivalency between (10.31) and (10.32) is proven.

Condition (10.32) is useful for deciding whether there is a need to increase the order of the observer or not. When a stable N is not found, this means that condition (10.32) is not satisfied, and hence the left-hand side of (10.32) is less than the right-hand side of (10.32) by one due to an unstable pole occurred at $\Re(s) \geq 0$. It is thus clear that if any additional row vector independent of F_2 is added to F_2, it can only increase the rank of both sides of (10.32) by one. As a result, there is still an imbalance between both sides of (10.32), i.e., the left-hand side is still less than the right-hand side by one. Thus, we will not be able to find a stable second-order functional observer. With this logic and together with the fact that there is a finite number of extra rows that can be added to F_2 to ensure condition (10.16) holds (see Remark 10.1 below), we therefore stop and do not search for any higher order functional observer. An alternative type of functional observers should be looked at (see, Sect. 10.2 for a method of designing functional observers with internal delay, $N_d w(t-\tau)$, which can result in less restrictive conditions).

Next, we consider the case where $m = 1$, $F_2 \in \mathbb{R}^{1\times(n-p)}$ with condition (10.16) satisfied but (10.28) not satisfied. In this case, we have an option to increase the order of the observer using an augmentation approach [4] in which we estimate additional functions. (The reader can refer to [4] for a comprehensive treatment on this approach, and on functional observability and the design of minimum order linear functional observers. In [15], the design of reduced-order linear functional state observers for linear systems with unknown inputs was also carried out based on the augmentation approach). This is discussed in the steps below.

Step 1: In this step, we increase the order of the observer to $m = 2$, and we also increase the number of functions to be estimated from one to two functions. For this, $z_{ex}(t) = F_{ex}\bar{x}_1(t)$ is an additional function to be estimated, and denote $z_a(t) \in \mathbb{R}^2$ as

$$z_a(t) = \begin{bmatrix} z(t) \\ z_{ex}(t) \end{bmatrix} = \begin{bmatrix} F \\ F_{ex} \end{bmatrix} \bar{x}_1(t). \tag{10.35}$$

Without loss of generality, let F_{ex} take the form

$$F_{ex} = \begin{bmatrix} 0 \ \bar{F}_{ex} \end{bmatrix}, \tag{10.36}$$

where $\bar{F}_{ex} \in \mathbb{R}^{1\times(n-p)} \neq 0$ is a row vector to be found such that rank $\begin{bmatrix} F_2 \\ \bar{F}_{ex} \end{bmatrix} = 2$, and condition (10.16) is satisfied, i.e.,

$$\text{rank} \begin{bmatrix} \bar{A}_{d11}^{12} \\ F_2 \bar{A}_{d11}^{22} \\ \bar{F}_{ex} \bar{A}_{d11}^{22} \end{bmatrix} = \text{rank}\begin{bmatrix} \bar{A}_{d11}^{12} \end{bmatrix}. \tag{10.37}$$

Remark 10.1 We note here that since $\text{rank}(\bar{A}_{d11}^{12}) \le (p-1)$, there exists only a finite number of $\bar{F}_{ex}$ such that condition (10.37) is satisfied. Any additional row, $\bar{F}_{ex}$, that can be added has to satisfy either: (i) $\bar{F}_{ex}\bar{A}_{d11}^{22}$ is linearly dependent on $\bar{A}_{d11}^{12}$; or (ii) $\bar{F}_{ex}\bar{A}_{d11}^{22} = 0$.

As for condition (i), $\bar{F}_{ex}$ can be found by checking if any row of $\bar{A}_{d11}^{22}$ (or any linear combinations of the rows of $\bar{A}_{d11}^{22}$) is linearly dependent on $\bar{A}_{d11}^{12}$.

For condition (ii), $\bar{F}_{ex} \neq 0$ exists if and only if $\text{rank}(\bar{A}_{d11}^{22}) = s$, where $s < (n-p)$. Denote $U \in \mathbb{R}^{(n-p-s)\times(n-p)}$ as an orthogonal basis of the left null-space of $\bar{A}_{d11}^{22}$

(i.e., $U\bar{A}_{d11}^{22} = 0$). Thus for any $\bar{F}_{ex} \subset U$, we have $\bar{F}_{ex}\bar{A}_{d11}^{22} = 0$, and there can only be a maximum of $(n - p - s)$ independent rows of $\bar{F}_{ex}$.

Step 2: Select a row $\bar{F}_{ex}$ found in Step 1 and check condition (10.28), i.e.,

$$\text{rank}\begin{bmatrix} F_2\bar{A}_{11}^{22} & F_2\bar{A}_{d11}^{22} \\ \bar{\boldsymbol{F}}_{\boldsymbol{ex}}\bar{A}_{11}^{22} & \bar{\boldsymbol{F}}_{\boldsymbol{ex}}\bar{A}_{d11}^{22} \\ \bar{A}_{11}^{12} & \bar{A}_{d11}^{12} \\ F_2 & 0 \\ \bar{\boldsymbol{F}}_{\boldsymbol{ex}} & 0 \end{bmatrix} = \text{rank}\begin{bmatrix} \bar{A}_{11}^{12} & \bar{A}_{d11}^{12} \\ F_2 & 0 \\ \bar{\boldsymbol{F}}_{\boldsymbol{ex}} & 0 \end{bmatrix}. \tag{10.38}$$

When condition (10.38) is satisfied, we next check whether the pair (N_1, N_2) is observable (or detectable) or not. If so, we obtain a stable second-order functional observer. If not, we select another row of $\bar{F}_{ex}$ from the set of $\bar{F}_{ex}$ found in Step 1 and repeat the above process.

When a second-order functional observer cannot be found after performing the above iterative process, we then eliminate all those rows of $\bar{F}_{ex}$ that were found to satisfy condition (10.38) but not result in the detectable/observable of the pair (N_1, N_2). We go to Step 3.

Step 3: If there are still two or more than two rows of $\bar{F}_{ex}$ left in the set of $\bar{F}_{ex}$, we then look for a third-order functional observer, i.e., $m = 3$. For this, we select any two rows of the remained set of $\bar{F}_{ex}$ and repeat the same process as in Step 2. We then stop if either a third-order functional observer is found or fewer than three rows of $\bar{F}_{ex}$ are left. Otherwise, Step 3 is repeated for $m = 4$. Since there is only a finite number of rows of $\bar{F}_{ex}$ that can satisfy condition (10.16), the search for a low-order scalar functional observer is not an exhaustive search.

Case 2: We consider the case where $\bar{A}_{d11}^{12}$ is a column matrix. Denoting by $\bar{A}_{d11}^{2} \in \mathbb{R}^{(n-1)\times(n-p)}$ as the following matrix

$$\bar{A}_{d11}^{2} = \begin{bmatrix} \bar{A}_{d11}^{12} \\ \bar{A}_{d11}^{22} \end{bmatrix}. \tag{10.39}$$

First, when $\bar{A}_{d11}^{2}$ is not a full-column rank matrix, i.e., $\text{rank}(\bar{A}_{d11}^{2}) = l < (n - p)$, we perform the following matrix decomposition by using singular value decomposition (SVD):

$$\bar{A}_{d11}^{2} = WS, \tag{10.40}$$

where $W \in \mathbb{R}^{(n-1)\times l}$ and $S \in \mathbb{R}^{l\times(n-p)}$ are, respectively, full-column rank and full-row rank matrices. Furthermore, let W be partitioned as follows

$$W = \begin{bmatrix} W_1 \\ W_2 \end{bmatrix}, \tag{10.41}$$

where $W_1 \in \mathbb{R}^{(p-1)\times l}$ and $W_2 \in \mathbb{R}^{(n-p)\times l}$. Now, Eq. (10.15) can be expressed as

$$\begin{bmatrix} L_1 & F_2 \end{bmatrix}\begin{bmatrix} \bar{A}_{d11}^{12} \\ \bar{A}_{d11}^{22} \end{bmatrix} = \begin{bmatrix} L_1 & F_2 \end{bmatrix}\begin{bmatrix} W_1 \\ W_2 \end{bmatrix} S = 0. \tag{10.42}$$

Since S is a full-row rank matrix, from (10.42), we have $\begin{bmatrix} L_1 & F_2 \end{bmatrix}\begin{bmatrix} W_1 \\ W_2 \end{bmatrix} = 0$, i.e.,

$$L_1 W_1 = -F_2 W_2, \tag{10.43}$$

and condition (10.16) is the same as

$$\text{rank}\begin{bmatrix} W_1 \\ F_2 W_2 \end{bmatrix} = \text{rank}\begin{bmatrix} W_1 \end{bmatrix}. \tag{10.44}$$

Now, we consider the case where matrix W_1 has full-column rank, i.e., $\text{rank}(W_1) = l$. Note that a necessary condition for this is $p \geq (l + 1)$. When W_1 has full-column rank, condition (10.44) is always satisfied. For any $F_2 \in \mathbb{R}^{m \times (n-p)}$, we next look at conditions (10.38) and (10.32), which are,

$$\text{rank}\begin{bmatrix} F_2 \bar{A}_{11}^{22} & F_2 W_2 \\ \bar{A}_{11}^{12} & W_1 \\ F_2 & 0 \end{bmatrix} = \text{rank}\begin{bmatrix} \bar{A}_{11}^{12} & W_1 \\ F_2 & 0 \end{bmatrix}, \tag{10.45}$$

and

$$\begin{aligned} &\text{rank}\begin{bmatrix} (sF_2 - F_2 \bar{A}_{11}^{22}) & -F_2 W_2 \\ \bar{A}_{11}^{12} & W_1 \end{bmatrix} \\ &= \text{rank}\begin{bmatrix} \bar{A}_{12}^{12} & W_1 \\ F_2 & 0 \end{bmatrix}, \forall s \in \mathbb{C}, \Re(s) \geq 0. \end{aligned} \tag{10.46}$$

Thus, to design a low-order scalar functional observer, we can now search for additional row vectors that can be added to F_2 to satisfy both conditions (10.45) and (10.46) (for more details, the reader can refer to Chap. 5 in [14]).

Finally, when $F_2 = I_{n-p}$ (i.e., the case for a minimum-order state observer to estimate $\bar{x}_1(t)$), condition (10.45) is satisfied and condition (10.46) is equivalent to the following

$$\begin{aligned} &\text{rank}\begin{bmatrix} sI_{n-p} - \bar{A}_{11}^{22} & -W_2 \\ \bar{A}_{11}^{12} & W_1 \end{bmatrix} \\ &= (n - p + l), \ \forall s \in \mathbb{C}, \Re(s) \geq 0. \end{aligned} \tag{10.47}$$

This is the same as the non-minimum phase of the triplet $(\bar{A}_{11}, W, \tilde{C}_1)$. Note that this requirement is not always satisfied, and thus other types of observers should be considered.

When an observer of the form (10.1) and (10.2) does not exist due to their restrictive existence conditions, we consider other Luenberger-type observers ([2, 11, 12]) that can further relax the conservatism of their existence conditions.

10.2 Observers with an Internal Delay

Here, we add an internal delay term, $N_d(t-\tau)$, to observer (10.1) and (10.2), i.e.,

$$\hat{z}(t) = w(t) + M\bar{y}(t), \tag{10.48}$$

$$\begin{aligned}\dot{w}(t) = Nw(t) + N_d w(t-\tau) + G\bar{y}(t) + G_d\bar{y}(t-\tau) \\ + Jy(t) + J_d y(t-\tau)\end{aligned} \tag{10.49}$$

where $N_d \in \mathbb{R}^{m\times m}$. The above observer has a more complex structure than the one described by (10.1) and (10.2) since an internal delay unit is used to generate a delayed state observer vector $w(t-\tau)$ in (10.49).

With the additional term, $N_d w(t-\tau)$, condition (10.8) is now changed to

$$N_d L + G_d\tilde{C}_1 - L\bar{A}_{d11} = 0. \tag{10.50}$$

Also, instead of requiring matrix N to be Hurtwitz, it is now required that the following time-delay system be asymptotically stable

$$\dot{\epsilon}(t) = N\epsilon(t) + N_d\epsilon(t-\tau) \tag{10.51}$$

with any initial condition $\epsilon(\theta) = \xi(\theta),\ \forall\theta \in [-\tau, 0]$.

Equation (10.50) can now be expressed as follows

$$G_d = L_1\bar{A}_{d11}^{11} + F_2\bar{A}_{d11}^{21} - N_d L_1, \tag{10.52}$$

$$N_d F_2 - L_1\bar{A}_{d11}^{12} - F_2\bar{A}_{d11}^{22} = 0. \tag{10.53}$$

Thus in contrast to the case without the internal delay, we do not require the condition (10.16) to be satisfied. By following similar lines as in (10.17) and (10.18), we can express (10.53) as

$$N_d = L_1\bar{A}_{d11}^{12}F_2^{+} + F_2\bar{A}_{d11}^{22}F_2^{+}, \tag{10.54}$$

$$L_1\bar{A}_{d11}^{12}F_2^{\perp} = -F_2\bar{A}_{d11}^{22}F_2^{\perp}. \tag{10.55}$$

We now combine (10.18) and (10.55) as follows

$$L_1\bar{A}_c = \bar{F}_c \tag{10.56}$$

where $\bar{A}_c \in \mathbb{R}^{(p-1)\times(2n-2p-2m)}$ and $\bar{F}_c \in \mathbb{R}^{m\times(2n-2p-2m)}$ are defined below

$$\bar{A}_c = \left[\bar{A}_{11}^{12} F_2^{\perp} \quad \bar{A}_{d11}^{12} F_2^{\perp}\right], \tag{10.57}$$

$$\bar{F}_c = -\left[F_2 \bar{A}_{11}^{22} F_2^{\perp} \quad F_2 \bar{A}_{d11}^{22} F_2^{\perp}\right]. \tag{10.58}$$

From (10.56), a solution for L_1 exists if and only if the following condition is satisfied

$$\text{rank} \begin{bmatrix} \bar{F}_c \\ \bar{A}_c \end{bmatrix} = \text{rank}\left[\bar{A}_c\right], \tag{10.59}$$

or equivalently

$$\text{rank} \begin{bmatrix} F_2 \bar{A}_{11}^{22} & F_2 \bar{A}_{d11}^{22} \\ \bar{A}_{11}^{12} & \bar{A}_{d11}^{12} \\ F_2 & 0 \\ 0 & F_2 \end{bmatrix} = \text{rank} \begin{bmatrix} \bar{A}_{11}^{12} & \bar{A}_{d11}^{12} \\ F_2 & 0 \\ 0 & F_2 \end{bmatrix}. \tag{10.60}$$

Observe that condition (10.60) is less conservative than condition (10.28) which is as expected since observer (10.48) and (10.49) is more complex than observer (10.1) and (10.2) as it contains an internal delay term $N_d w(t - \tau)$.

If (10.60) holds, L_1 is obtained as follows

$$L_1 = \bar{F}_c \bar{A}_c^{+} + \bar{Z}(I_{p-1} - \bar{A}_c \bar{A}_c^{+}), \tag{10.61}$$

where $\bar{A}_c^{+}$ is the Moore-Penrose inverse of $\bar{A}_c$, and $\bar{Z} \in \mathbb{R}^{m \times (p-1)}$ will be found later in order to stabilize the time-delay system (10.51). Thus, substituting (10.61) into (10.17) and (10.54), we obtain

$$N = \bar{N}_1 - \bar{Z}\bar{N}_2, \tag{10.62}$$

$$N_d = \bar{N}_{d1} - \bar{Z}\bar{N}_{d2}, \tag{10.63}$$

where

$$\bar{N}_1 = (\bar{F}_c \bar{A}_c^{+} \bar{A}_{11}^{12} + F_2 \bar{A}_{11}^{22}) F_2^{+}, \tag{10.64}$$

$$\bar{N}_2 = (\bar{A}_c \bar{A}_c^{+} - I_{p-1}) \bar{A}_{11}^{12} F_2^{+}, \tag{10.65}$$

$$\bar{N}_{d1} = (\bar{F}_c \bar{A}_c^{+} \bar{A}_{d11}^{12} + F_2 \bar{A}_{d11}^{22}) F_2^{+}, \tag{10.66}$$

$$\bar{N}_{d2} = (\bar{A}_c \bar{A}_c^{+} - I_{p-1}) \bar{A}_{d11}^{12} F_2^{+}. \tag{10.67}$$

Substituting (10.62) and (10.63) into (10.51), we obtain the following time-delay system

$$\dot{\epsilon}(t) = (\bar{N}_1 - \bar{Z}\bar{N}_2)\epsilon(t) + (\bar{N}_{d1} - \bar{Z}\bar{N}_{d2})\epsilon(t - \tau). \tag{10.68}$$

Since $\bar{N}_1$, $\bar{N}_2$, $\bar{N}_{d1}$ and $\bar{N}_{d2}$ are known matrices (see, (10.64)–(10.67)), we can now find $\bar{Z}$ so that system (10.68) is asymptotically stable (normally, exponential stability with a prescribed convergence rate is preferred) for any given value of τ and with any initial condition $\epsilon(\theta)$, $\forall\theta \in [-\tau, 0]$. A possible solution to the stabilization problem of the time-delay system (10.68), i.e., finding $\bar{Z}$, is based on the Lyapunov-Krasovskii theory, in which we can derive either delay-independent or delay-dependent sufficient stabilizability conditions in the form of LMIs. This approach is popular since LMIs can be cast into a convex optimization problem [1]. Provided that the considered LMI is feasible, $\bar{Z}$ can be effectively computed. The reader can refer to, for example, [3, 7, 8, 17], where stability and stabilizability conditions for (10.68) can be derived based on the Lyapunov-Krasovskii theory. For completeness, Chap. 20 provides some stability and stabilizability conditions for system (10.68).

Based on the above development, to design a low-order scalar functional observer to estimate a given function $z(t) = F\bar{x}_1(t)$, we first check whether condition (10.60) is satisfied or not. If it is satisfied, we obtain $\bar{N}_1$, $\bar{N}_2$, $\bar{N}_{d1}$ and $\bar{N}_{d2}$, and then find a matrix $\bar{Z}$ to ensure stability of the time-delay system (10.68). If either (10.60) is not satisfied or a $\bar{Z}$ is not found, we then follow similar logic in Step 1 as discussed above by adding a row vector to F_2 and repeat the above process. Note that when the order of the observer reaches $(n - p)$, i.e., $\text{rank}(F_2) = (n - p)$, condition (10.60) is satisfied since the matrix on the right-hand side of (10.60) is a full-column rank matrix.

On the other hand, when we need to design a reduced-order state observer to estimate $(n - p)$ unmeasurable state variables of the state vector $\bar{x}_1(t)$ of system (4.3) and (4.4), we can select $F_2 = I_{n-p}$. In such a case, from (10.13) and (10.53), we have

$$N = \bar{A}_{11}^{22} + L_1\bar{A}_{11}^{12}, \tag{10.69}$$
$$N_d = \bar{A}_{d11}^{22} + L_1\bar{A}_{d11}^{12}, \tag{10.70}$$

and the time-delay system (10.51) becomes

$$\dot{\epsilon}(t) = (\bar{A}_{11}^{22} + L_1\bar{A}_{11}^{12})\epsilon(t) + (\bar{A}_{d11}^{22} + L_1\bar{A}_{d11}^{12})\epsilon(t - \tau). \tag{10.71}$$

Hence the design of a reduced-order state observer amounts to finding a matrix L_1 such that the above time-delay is asymptotically stable.

10.3 Observers with an Internal Delay and Multiple Delayed Output Measurements

We consider another Luenberger-type observer with multiple delayed output measurements. In [11], it was reported that by employing a general Luenberger-type observer that has multiple delayed output and input measurements, a less conservative existence condition than the one reported in [2] can be obtained.

To estimate $z(t) = F\bar{x}_1(t)$, where $F \in \mathbb{R}^{m\times(n-1)}$, we consider the following observer

$$\hat{z}(t) = w(t) + M\bar{y}(t) + M_d\bar{y}(t-\tau), \tag{10.72}$$

$$\begin{aligned}\dot{w}(t) &= Nw(t) + N_d w(t-\tau) + G\bar{y}(t) + G_d\bar{y}(t-\tau) + G_{dd}\bar{y}(t-2\tau)\\ &\quad + Jy(t) + J_d y(t-\tau) + J_{dd}y(t-2\tau),\end{aligned} \tag{10.73}$$

with the initial condition $w(\theta) = \rho(\theta)$, $\forall\theta \in [0, \tau]$, $w(t) \in \mathbb{R}^m$ and $\hat{z}(t) \in \mathbb{R}^m$ is the estimate of $z(t)$. The above observer is an extended observer to the one considered in (10.48) and (10.49), where three extra delayed terms are introduced, namely, $M_d\bar{y}(t-\tau)$, $G_{dd}y(t-2\tau)$ and $J_{dd}y(t-2\tau)$.

In the following, for completeness, we derive their existence conditions and show that they are less conservative than those of observer (10.48) and (10.49). Consider the error vector (4.16) and from (10.72), we have

$$w(t) = e(t) + F\bar{x}_1(t) - M\bar{y}(t) - M_d\bar{y}(t-\tau). \tag{10.74}$$

The derivative of $e(t)$ is obtained as follows

$$\begin{aligned}\dot{e}(t) &= \dot{\hat{z}}(t) - \dot{z}(t)\\ &= \dot{w}(t) + M\dot{\bar{y}}(t) + M_d\dot{\bar{y}}(t-\tau) - F\dot{\bar{x}}_1(t).\end{aligned} \tag{10.75}$$

Substitute (10.73), (10.74), (4.3) and (4.4) into (10.75), and after some rearranging, we obtain the following error dynamic equation

$$\begin{aligned}\dot{e}(t) &= Ne(t) + N_d e(t-\tau) + (M\tilde{C}_1\bar{A}_{12}V_1 - F\bar{A}_{12}V_1 + J)y(t)\\ &\quad + (M\tilde{C}_1\bar{A}_{d12}V_1 + M_d\tilde{C}_1\bar{A}_{12}V_1 - F\bar{A}_{d12}V_1 + J_d)y(t-\tau)\\ &\quad + (J_{dd} + M_d\tilde{C}_1\bar{A}_{d12}V_1)y(t-2\tau)\\ &\quad + \big(NF + (G - NM)\tilde{C}_1 + M\tilde{C}_1\bar{A}_{11} - F\bar{A}_{11}\big)\bar{x}_1(t)\\ &\quad + \big(N_d F + (G_d - NM_d - N_d M)\tilde{C}_1 + M\tilde{C}_1\bar{A}_{d11} + M_d\tilde{C}_1\bar{A}_{11} - F\bar{A}_{d11}\big)\bar{x}_1(t-\tau)\\ &\quad + \big(M_d\tilde{C}_1\bar{A}_{d11} + (G_{dd} - N_d M_d)\tilde{C}_1\big)\bar{x}_1(t-2\tau).\end{aligned} \tag{10.76}$$

From (10.76), observer (10.72) and (10.73) exists such that $\hat{z}(t)$ converges asymptotically to $z(t)$ for any initial condition $e(\theta)$, $\forall\theta \in [0, \tau]$, if the following conditions hold

$$\dot{e}(t) = Ne(t) + N_d e(t-\tau) \text{ is stable}, \tag{10.77}$$

$$J = F\bar{A}_{12}V_1 - M\tilde{C}_1\bar{A}_{12}V_1, \tag{10.78}$$

$$J_d = F\bar{A}_{d12}V_1 - M\tilde{C}_1\bar{A}_{d12}V_1 - M_d\tilde{C}_1\bar{A}_{12}V_1, \tag{10.79}$$

$$J_{dd} = -M_d\tilde{C}_1\bar{A}_{d12}V_1, \tag{10.80}$$

$$NF + (G - NM)\tilde{C}_1 + M\tilde{C}_1\bar{A}_{11} - F\bar{A}_{11} = 0, \tag{10.81}$$

$$N_d F + (G_d - NM_d - N_d M)\tilde{C}_1 + M\tilde{C}_1\bar{A}_{d11} + M_d\tilde{C}_1\bar{A}_{11} - F\bar{A}_{d11} = 0, \tag{10.82}$$

$$M_d\tilde{C}_1\bar{A}_{d11} + (G_{dd} - N_d M_d)\tilde{C}_1 = 0. \tag{10.83}$$

The key to satisfying conditions (10.77)–(10.83) is the solvability of the three matrix equations (10.81)–(10.83). For this, we define $\tilde{T}$, $\tilde{W}$ and $\tilde{U}$ as

$$\tilde{T} = G - NM, \tag{10.84}$$

$$\tilde{W} = G_d - NM_d - N_d M, \tag{10.85}$$

$$\tilde{U} = G_{dd} - N_d M_d. \tag{10.86}$$

Then Eqs. (10.81)–(10.83) are the same as

$$NF + \tilde{T}\tilde{C}_1 + M\tilde{C}_1\bar{A}_{11} = F\bar{A}_{11}, \tag{10.87}$$

$$N_d F + \tilde{W}\tilde{C}_1 + M\tilde{C}_1\bar{A}_{d11} + M_d\tilde{C}_1\bar{A}_{11} = F\bar{A}_{d11}, \tag{10.88}$$

$$M_d\tilde{C}_1\bar{A}_{d11} + \tilde{U}\tilde{C}_1 = 0. \tag{10.89}$$

In (10.87)–(10.89), there are seven unknown matrices, namely, $\tilde{T}$, $\tilde{W}$, $\tilde{U}$, N, N_d, M and M_d to be solved for, while also ensuring the asymptotic stability of (10.77).

We now look at the solution to (10.87)–(10.89). With $\tilde{C}_1 = [I_{(p-1)} \;\; 0_{(p-1)\times(n-p)}]$, and matrices F, $\bar{A}_{11}$ and $\bar{A}_{d11}$ partitioned as per Sect. 10.1.2, we can decompose (10.87)–(10.89), respectively, into the following six simpler equations

$$\tilde{T} = F_1\bar{A}_{11}^{11} + F_2\bar{A}_{11}^{21} - NF_1 - M\bar{A}_{11}^{11}, \tag{10.90}$$

$$NF_2 + M\bar{A}_{11}^{12} = F_1\bar{A}_{11}^{12} + F_2\bar{A}_{11}^{22}, \tag{10.91}$$

$$\tilde{W} = F_1\bar{A}_{d11}^{11} + F_2\bar{A}_{d11}^{21} - N_d F_1 - M\bar{A}_{d11}^{11} - M_d\bar{A}_{11}^{11}, \tag{10.92}$$

$$N_d F_2 + M\bar{A}_{d11}^{12} + M_d\bar{A}_{11}^{12} = F_1\bar{A}_{d11}^{12} + F_2\bar{A}_{d11}^{22}, \tag{10.93}$$

$$\tilde{U} = -M_d\bar{A}_{d11}^{11}, \tag{10.94}$$

$$M_d\bar{A}_{d11}^{12} = 0. \tag{10.95}$$

We can see the advantage of the above decomposition since the number of unknown matrices has reduced from seven down to four, namely, N, N_d, M and M_d. Next, we consider the solution to the three matrix equations (10.91), (10.93) and (10.95). For this, we can further decompose (10.91) and (10.93) by post-multiplying them by $\left[F_2^{+} \;\; F_2^{\perp}\right]$ to get

$$N = (F_1\bar{A}_{11}^{12} + F_2\bar{A}_{11}^{22})F_2^{+} - \left[M_d \; M\right]\begin{bmatrix} 0_{(p-1)\times m} \\ \bar{A}_{11}^{12}F_2^{+} \end{bmatrix}, \tag{10.96}$$

$$N_d = (F_1 \bar{A}_{d11}^{12} + F_2 \bar{A}_{d11}^{22}) F_2^{+} - \begin{bmatrix} M_d & M \end{bmatrix} \begin{bmatrix} \bar{A}_{11}^{12} F_2^{+} \\ \bar{A}_{d11}^{12} F_2^{+} \end{bmatrix}, \tag{10.97}$$

$$M \bar{A}_{11}^{12} F_2^{\perp} = (F_1 \bar{A}_{11}^{12} + F_2 \bar{A}_{11}^{22}) F_2^{\perp}, \tag{10.98}$$

$$M_d \bar{A}_{11}^{12} F_2^{\perp} + M \bar{A}_{d11}^{12} F_2^{\perp} = (F_1 \bar{A}_{d11}^{12} + F_2 \bar{A}_{d11}^{22}) F_2^{\perp}. \tag{10.99}$$

We can express (10.95), (10.98) and (10.99) as follows

$$\begin{bmatrix} M_d & M \end{bmatrix} \tilde{A}_c = \tilde{F}_c, \tag{10.100}$$

where $\tilde{A}_c \in \mathbb{R}^{(2p-2)\times(3n-3p-2m)}$ and $\tilde{F}_c \in \mathbb{R}^{m\times(3n-3p-2m)}$ are defined below

$$\tilde{A}_c = \begin{bmatrix} \bar{A}_{d11}^{12} & 0 & \bar{A}_{11}^{12} F_2^{\perp} \\ 0 & \bar{A}_{11}^{12} F_2^{\perp} & \bar{A}_{d11}^{12} F_2^{\perp} \end{bmatrix}, \tag{10.101}$$

$$\tilde{F}_c = \begin{bmatrix} 0 & (F_1 \bar{A}_{11}^{12} + F_2 \bar{A}_{11}^{22}) F_2^{\perp} & (F_1 \bar{A}_{d11}^{12} + F_2 \bar{A}_{d11}^{22}) F_2^{\perp} \end{bmatrix}. \tag{10.102}$$

From (10.100), we can solve for $\begin{bmatrix} M_d & M \end{bmatrix}$ if and only if the following condition is satisfied

$$\operatorname{rank} \begin{bmatrix} \tilde{F}_c \\ \tilde{A}_c \end{bmatrix} = \operatorname{rank} \begin{bmatrix} \tilde{A}_c \end{bmatrix}. \tag{10.103}$$

We can also show that (10.103) is equivalent to the following condition

$$\begin{aligned} &\operatorname{rank} \begin{bmatrix} 0 & F_1 \bar{A}_{11}^{12} + F_2 \bar{A}_{11}^{22} & F_1 \bar{A}_{d11}^{12} + F_2 \bar{A}_{d11}^{22} \\ \bar{A}_{d11}^{12} & 0 & \bar{A}_{11}^{12} \\ 0 & \bar{A}_{11}^{12} & \bar{A}_{d11}^{12} \\ 0 & F_2 & 0 \\ 0 & 0 & F_2 \end{bmatrix} \\ &= \operatorname{rank} \begin{bmatrix} \bar{A}_{d11}^{12} & 0 & \bar{A}_{11}^{12} \\ 0 & \bar{A}_{11}^{12} & \bar{A}_{d11}^{12} \\ 0 & F_2 & 0 \\ 0 & 0 & F_2 \end{bmatrix}. \end{aligned} \tag{10.104}$$

Upon the satisfaction of (10.104), $\begin{bmatrix} M_d & M \end{bmatrix}$ is obtained as

$$\begin{bmatrix} M_d & M \end{bmatrix} = \tilde{F}_c \tilde{A}_c^{+} + \tilde{Z}(I_{2p-2} - \tilde{A}_c \tilde{A}_c^{+}), \tag{10.105}$$

$\tilde{A}_c^{+}$ is the Moore-Penrose inverse of $\tilde{A}_c$ and $\tilde{Z} \in \mathbb{R}^{m\times(2p-2)}$. Thus, substituting (10.105) into (10.96) and (10.97), we obtain

$$N = \tilde{N}_1 - \tilde{Z}\tilde{N}_2, \tag{10.106}$$

$$N_d = \tilde{N}_{d1} - \tilde{Z}\tilde{N}_{d2}, \tag{10.107}$$

where

$$\tilde{N}_1 = (F_1 \bar{A}_{11}^{12} + F_2 \bar{A}_{11}^{22}) F_2^+ - \tilde{F}_c \tilde{A}_c^+ \begin{bmatrix} 0_{(p-1)\times m} \\ \bar{A}_{11}^{12} F_2^+ \end{bmatrix}, \tag{10.108}$$

$$\tilde{N}_2 = (I_{2p-2} - \tilde{A}_c \tilde{A}_c^+) \begin{bmatrix} 0_{(p-1)\times m} \\ \bar{A}_{11}^{12} F_2^+ \end{bmatrix}, \tag{10.109}$$

$$\tilde{N}_{d1} = (F_1 \bar{A}_{d11}^{12} + F_2 \bar{A}_{d11}^{22}) F_2^+ - \tilde{F}_c \tilde{A}_c^+ \begin{bmatrix} \bar{A}_{11}^{12} F_2^+ \\ \bar{A}_{d11}^{12} F_2^+ \end{bmatrix}, \tag{10.110}$$

$$\tilde{N}_{d2} = (I_{2p-2} - \tilde{A}_c \tilde{A}_c^+) \begin{bmatrix} \bar{A}_{11}^{12} F_2^+ \\ \bar{A}_{d11}^{12} F_2^+ \end{bmatrix}. \tag{10.111}$$

Thus, the remaining task is to find a matrix $\tilde{Z}$ such that the following time-delay system is asymptotically stable

$$\dot{e}(t) = (\tilde{N}_1 - \tilde{Z}\tilde{N}_2) e(t) + (\tilde{N}_{d1} - \tilde{Z}\tilde{N}_{d2}) e(t - \tau). \tag{10.112}$$

Once $\tilde{Z}$ is found, $\begin{bmatrix} M_d & M \end{bmatrix}$, N and N_d are as given in (10.105), (10.106), and (10.107), respectively. Matrices $\tilde{T}$, $\tilde{W}$ and $\tilde{U}$ are obtained from (10.90), (10.92) and (10.94), respectively. Thus, from (10.84)–(10.86), we obtain $G = \tilde{T} + NM$, $G_d = \tilde{W} + NM_d + N_d M$ and $G_{dd} = \tilde{U} + N_d M_d$. Finally, we obtain matrices J, J_d and J_{dd} from (10.78)–(10.80), respectively.

Remark 10.2 It is expected that observer (10.72) and (10.73) offers a more relaxed existence condition than observer (10.48) and (10.49), however, at the expense of a more complex structure. Here, condition (10.104) is less conservative than condition (10.60). In other words, if condition (10.60) holds, then condition (10.104) also holds. But, the converse is not true. We can show this by a numerical example, where

$$\bar{A}_{11} = \begin{bmatrix} -1 & 2 & 1 \\ 1 & 0 & 2 \\ 1 & 3 & -2 \end{bmatrix}, \bar{A}_{d11} = \begin{bmatrix} -1 & 0 & 0 \\ 2 & 1 & 0 \\ 1 & 2 & -1 \end{bmatrix}, \bar{C}_1 = \begin{bmatrix} 1 & 0 & 0 \end{bmatrix} \text{ and } F = \begin{bmatrix} 0 & 0 & 1 \end{bmatrix}.$$

Here, we can confirm that in this example, condition (10.104) holds, but condition (10.60) does not hold.

The above discussion shows that it is possible to further relax existence conditions of linear functional observers by including extra terms with multiple delayed output measurements in their structure. However, it should be noted here that when $F_2 = I_{n-p}$, both conditions (10.104) and (10.60) are satisfied.

The reader can refer to Chap. 16 (Numerical Example 1), where we address the problem of designing a more general observer than (10.72) and (10.73) for a wider class of time-delay systems. We will show that we can obtain a less conservative existence condition than (10.104).

10.4 Observers Design Based on Finite Spectrum Assignment

As an alternative to those observer design methods discussed so far, it is possible to design finite-dimensional observers (i.e. delay-free Luenberger observers) based on the finite spectrum assignment theory, and based on the output injection approach.

Finite-dimensional observers were constructed [10] by first computing a set of marginally unstable eigenvalues, S_p, of a time-delay system where S_p is defined as follows

$$S_p = \{s : \det(I_n - A - A_d e^{-\tau s}) = 0,\ \mathrm{Re}(s) \geq -\alpha\} \tag{10.113}$$

and $\alpha > 0$ is a specified number providing the desired stability margin for the observer (the set of eigenvalues S_p can be obtained by using the Matlab package developed by Wu and Michiels [16]). Then by computing the left and right eigenvectors associated with $S_p = \{s_1, s_2, \cdots, s_p\}$, the triple matrix (F_c, G_c, H_c) can be constructed where $F_c = \mathrm{diag}(s_1, s_2, \cdots, s_p)$ (see, [10]). Then, which a finite-dimensional state observer with the order equal to the number of eigenvalues in the set S_p (i.e., p) can be designed. This approach is referred to as the finite spectrum assignment approach, which is essentially finding an observer gain L such that $(F_c - LCH_c)$ is Hurwitzian with a stability margin α.

Another alternative design method is based on finite-dimensional approximation of a given time-delay system [5]. Here, a finite-dimensional system of order $(p + n)$ is derived based on the approximation theory for time-delay systems [6]. Then, a delay-free Luenberger-type observer of order $(p + n)$ is designed for the time-delay system. It should be noted here that the order of the designed observers based on the above methods can be, in some cases, very high. Thus, if a faster rate of observer error decaying to zero is required (i.e., a larger α), a higher order observer is often needed. Also, as pointed out in [13], the designed method of [10] results in steady-state errors when a step change in the control input is applied. An alternative structure was proposed in [13] to rectify this problem.

10.5 Observers Design Based on the Output Injection Approach

This section discusses a state observer design method for time-delay systems based on a different approach to those methods discussed earlier in this chapter. Hou, Zítek and Patton in [9] proposed a design method based on an output injection approach. In this approach, the authors proposed a novel coordinate change such that in the new coordinates, all time-delay terms in the system description are associated with the output and input only. Thus, the design of a state observer becomes a trivial task and the structure of the observer is a Luenberger-type observer with multiple delayed output and input measurements.

For the sake of completeness, the following discussion outlines how such a coordinate change can be obtained [9]. Denoted by $d = \{d_i\}$ the time-delay operators, in which for a specific i the operator d_i is defined by $d_i x(t) = x(t - \tau_i)$ with $\tau_i > 0$ being the associated time-delay constant. Thus, for a time-delay system with multiple delays in the state, input and output vectors, it can be conveniently represented by the following compact form

$$\dot{x}(t) = A(d)x(t) + B(d)u(t), \tag{10.114}$$

$$y(t) = C(d)x(t), \tag{10.115}$$

where $A(d)$, $B(d)$ and $C(d)$ are polynomial matrices in the time-delay operators $d = \{d_i\}$. For example, the following time-delay system

$$\dot{x}(t) = \begin{bmatrix} 0 & 1 \\ -1 & 0 \end{bmatrix} x(t) + \begin{bmatrix} -1 & 1 \\ 0 & 1 \end{bmatrix} x(t - \tau_1) + \begin{bmatrix} 0 & 0 \\ 1 & 1 \end{bmatrix} x(t - \tau_2),$$
$$y(t) = \begin{bmatrix} 1 & 1 \end{bmatrix} x(t) + \begin{bmatrix} 0 & 1 \end{bmatrix} x(t - \tau_1),$$

can be written in the form (10.114) and (10.115), where $C(d) = [1 \quad 1 + d_1]$ and $A(d) = \begin{bmatrix} -d_1 & 1 + d_1 \\ -1 + d_2 & d_1 + d_2 \end{bmatrix}$.

The key in the design of state observers for system (10.114) and (10.115) is to obtain a generalized coordinate $\xi(t) = T(d)x(t)$, where $\xi(t) \in \mathbb{R}^{n_z}$, $n_z \geq n$, and $T(d)$ is a polynomial matrix in d, such that system (10.114) and (10.115) is transformed into the following ξ-coordinates

$$\dot{\xi}(t) = \bar{A}\xi(t) + \bar{F}(d)y(t) + \bar{B}(d)u(t), \tag{10.116}$$

$$y(t) = \bar{C}\xi(t), \tag{10.117}$$

where the matrix pair $(\bar{A}, \bar{C})$ is observable, $\bar{F}(d)y(t)$ is the term of output injection, and $\bar{F}(d)$ and $\bar{B}(d)$ are polynomial matrices in d. Note that $n_z = n$ for a single-output case, and $n_z \geq n$ holds generally in the multiple-output case [9].

Thanks to the coordinate change, as can be seen in the ξ-coordinates, all the delay terms are associated with the output and input vectors only. Therefore, a reduced-order Luenberger-type observer (with no internal delay) which includes multiple delayed output and input measurements for (10.116) and (10.117) can be designed to estimate $\xi(t)$. Provided that a polynomial left inverse of $T(d)$, denoted as $T^L(d)$, exists such that $T^L(d)T(d) = I_n$, then $\hat{x}(t)$ can be constructed from $\hat{\xi}(t)$ where $\hat{x}(t) = T^L(d)\hat{\xi}(t)$. Note that a polynomial matrix $M(s) \in \mathbb{R}^{n \times m}[s]$ is said to be column unimodular if $\text{rank}(M(s)) = m$ for all s in $\mathbb{C}$, and if $n = m$, $M(s)$ is said to be unimodular. There exists a left inverse $M^L(s)$ for $M(s)$ if and only if $M(s)$ is column unimodular.

As detailed in [9], a generalized coordinate change $\xi(t) = T(d)x(t)$, with $T(s) \in \mathbb{R}^{n_z \times n}[s]$ being column unimodular, exists for the transformation of the system (10.114) and (10.115) into (10.116) and (10.117) if and only if the following observability matrix, $Q_k(s)$, is column unimodular (or unimodular for the case of a single output)

$$Q_k(s) = \begin{bmatrix} C(s) \\ C(s)A(s) \\ \vdots \\ C(s)A^{k-1}(s) \end{bmatrix}, \tag{10.118}$$

with $k \leq n$ being the smallest integer such that $\text{rank}(Q_k(s)) = \text{rank}(Q_n(s))$ for all s in $\mathbb{C}$.

Upon $Q_k(s)$ being column unimodular, $T(d) = \text{col.}[T_1(d)\ T_2(d)\ \cdots\ T_k(d)]$ can be constructed as follows [9]

$$\begin{aligned} &T_1(d) = C(d), \\ &T_{i+1}(d) = T_i(d)A(d) - \bar{F}_i(d)C(d), i = 1, \ldots, k-1, \end{aligned} \tag{10.119}$$

where $\bar{F}_i(d), i = 1, ..., k$ are obtained as

$$[\bar{F}_k(d)\ \bar{F}_{k-1}(d)\ \cdots\ \bar{F}_1(d)] = C(d)A^k(d)Q_k^L(d). \tag{10.120}$$

Thus, the construction of the coordinate change $T(d)$ requires the existence of a polynomial left inverse of $Q_k(d)$, which can be achieved if and only if $Q_k(s)$ is column unimodular.

In the following, a numerical example is given to show how $T(d)$ is obtained. Consider the below linear time-delay system

$$\dot{x}(t) = Ax(t) + A_d x(t-\tau) + Bu(t),\ t \geq 0, \tag{10.121}$$

$$y(t) = Cx(t), \tag{10.122}$$

where $A = \begin{bmatrix} -1 & 1 \\ -1 & 0 \end{bmatrix}$, $A_d = \begin{bmatrix} 1 & 0 \\ 0 & -1 \end{bmatrix}$, $B = \begin{bmatrix} 1 \\ 2 \end{bmatrix}$ and $C = \begin{bmatrix} 1 & 0 \end{bmatrix}$. First, this system is put into the form (10.114) and (10.115), where $C(d) = C$, $B(d) = B$ and $A(d) = \begin{bmatrix} d-1 & 1 \\ -1 & -d \end{bmatrix}$.

Now, we have $CA(d) = \begin{bmatrix} d-1 & 1 \end{bmatrix}$. Hence, $Q_2(d) = \begin{bmatrix} C \\ CA(d) \end{bmatrix} = \begin{bmatrix} 1 & 0 \\ d-1 & 1 \end{bmatrix}$. For an arbitrary $s \in \mathbb{C}$, and consider the matrix $Q_2(s) = \begin{bmatrix} 1 & 0 \\ s-1 & 1 \end{bmatrix}$ which is unimodular since its determinant is non-zero, i.e., $|Q_2(s)| = 1 \neq 0$. Since $Q_2(s)$ is unimodular, one has $Q_2^{-1}(d) = Q_2^L(d) = \begin{bmatrix} 1 & 0 \\ 1-d & 1 \end{bmatrix}$. Now, from (10.120), we have

$$[\bar{F}_2(d)\ \bar{F}_1(d)] = C(d)A^2(d)Q_2^{-1}(d) = \left[d^2 - d - 1\ -1\right],$$

and hence the desired coordinate change is

$$T(d) = \begin{bmatrix} C \\ CA(d) - \bar{F}_1(d)C \end{bmatrix} = \begin{bmatrix} 1 & 0 \\ d & 1 \end{bmatrix}.$$

Thus in this example, a coordinate change where

$$\xi_1(t) = x_1(t),\ \xi_2(t) = x_1(t-\tau) + x_2(t), \tag{10.123}$$

transforms the system into the following ξ-coordinates

$$\dot{\xi}(t) = \begin{bmatrix} 0 & 1 \\ 0 & 0 \end{bmatrix}\xi(t) + \begin{bmatrix} -1 \\ d^2 - d - 1 \end{bmatrix}y(t) + \begin{bmatrix} 1 \\ d+2 \end{bmatrix}u(t), \tag{10.124}$$

$$y(t) = \begin{bmatrix} 1 & 0 \end{bmatrix}\xi(t), \tag{10.125}$$

for which a reduced-order Luenberger-type observer can be designed to estimate $\xi_2(t)$. Note that the structure of the designed observer contains multiple delayed output and input measurements (i.e., $y(t-\tau)$, $y(t-2\tau)$, $u(t-\tau)$). Then from (10.123), $\hat{x}_2(t)$ can be obtained as $\hat{x}_2(t) = \hat{\xi}_2(t) - y(t-\tau)$.

By this approach [9], we obtain an observer with no internal delay, but with multiple delayed output and input measurements. This observer has a simpler structure than (10.72) and (10.73) but a bit more complex than (10.1) and (10.2). On the other hand, in this example, condition (10.27) is not satisfied and thus an observer (10.1) and (10.2) does not exist. This highlights the merit and usefulness of the design method developed by the authors in [9]. On the other hand, we note that computing the left inverse of $Q_k(d)$ is not an easy task especially for high-order time-delay systems.

10.6 Concluding Remark

In this chapter, we have discussed several existing observer design methods for linear time-delay systems. Each method (as well as their observer structure) has their own advantages and disadvantages. Therefore, when faced with the task of designing an observer for a given linear time-delay system, one should go through a process of selection/elimination to choose a suitable observer design method that can result in a simple observer structure, hence a simple unknown input observer.

The reader can refer to the subsequent chapters for more results on the design of generalized observers for a wider class of time-delay systems than the one considered in this chapter. Firstly, in Chap. 11, we present a method for designing a functional observer to estimate m functions of the state vector of the time-delay system as described by (5.13) and (5.14). We show, via a simple numerical example, that our approach is less conservative than the Output Injection approach of [9].

References

1. Boyd, S., El Ghaoui, L., Feron, E., Balakrishnan, V.: Linear Matrix Inequalities in System and Control Theory. SIAM Studies in Applied Mathematics. Philadelphia, Pennsylvania (1994)
2. Darouach, M.: Linear functional observers for systems with delays in state variables. IEEE Trans. Autom. Control **46**(3), 491–496 (2001)
3. Darouach, M., Zasadzinski, M., Xu, S.J.: Full-order observers for linear systems with unknown inputs. IEEE Trans. Autom. Control **39**(3), 606–609 (1994)
4. Fernando, T.L., Trinh, H., Jennings, L.: Functional observability and the design of minimum order linear functional observers. IEEE Trans. Autom. Control **55**(5), 1268–1273 (2010)
5. Fiagbedzi, Y.A., Cherid, A.: Finite-dimensional observers for delay systems. IEEE Trans. Autom. Control **48**(11), 1986–1990 (2003)
6. Gu, G., Khargonekhar, P.P., Lee, E.B., Misra, P.: Finite-dimensional approximation of unstable infinite-dimensional systems. SIAM J. Control Optim. **30**(3), 704–716 (1992)
7. Hien, L.V., Trinh, H.: Refined jensen-based inequality approach to stability analysis of time-delay systems. IET Control Theory Appl. **9**(14), 2188–2194 (2015)
8. Hien, L.V., Trinh, H.: Exponential stability of time-delay systems via new weighted integral inequalities. Appl. Math. Comput. **275**, 335–344 (2016)
9. Hou, M., Zítek, P., Patton, R.J.: An observer design for linear time-delay systems. IEEE Trans. Autom. Control **47**(1), 121–125 (2002)
10. Leyva-Ramos, J., Pearson, A.E.: An asymptotic modal observer for linear autonomous time lag systems. IEEE Trans. Autom. Control **40**(7), 1291–1294 (1995)
11. Nam, P.T., Pathirana, P.N., Trinh, H.: ϵ-bounded state estimation for time-delay systems with bounded disturbances. Int. J. Control **87**(9), 1747–1756 (2014)
12. Trinh, H.: Linear functional state observers for time-delay systems. Int. J. Control **72**(18), 1642–1658 (1999)
13. Trinh, H., Aldeen, M.: Comments on "an asymptotic modal observer for linear autonomous time lag systems". IEEE Trans. Autom. Control **42**(5), 742–745 (1997)
14. Trinh, H., Fernando, T.: Functional Observers for Dynamical Systems. Springer, Berlin Heidelberg (2012)
15. Trinh, H., Fernando, T., Nahavandi, S.: Design of reduced-order functional observers for linear systems with unknown inputs. Asian J. Control **6**(4), 514–520 (2004)
16. Wu, Z., Michiels, W.: Reliably computing all characteristic roots of delay differential equations in a given right half plane using a spectral method. J. Comput. Appl. Math. **236**(9), 2499–2514 (2012)
17. Xu, S., Lam, J.: A survey of linear matrix inequality techniques in stability analysis of delay systems. Int. J. Syst. Sci. **39**(12), 1095–1113 (2008)

Chapter 11
Observers for Systems with a Time Delay in State and Output Vectors

11.1 Background

This chapter kick-starts the third part of the book, where we investigate the design of generalized observers to asymptotically estimate generalized functionals of a wide class of time-delay systems, including multiple time delays and/or non-commensurate time delays in both the state and output vectors. This chapter and the next several chapters will take the reader through different classes of time-delay systems, scenarios and illustrative examples to showcase the significance of the new generalized functional estimation problem.

Firstly, in this chapter, we design observers to estimate m functions of the state vector of time-delay system as described by (5.13) and (5.14), i.e.,

$$\dot{\bar{x}}_1(t) = \bar{A}_{11}\bar{x}_1(t) + \bar{A}_{d11}\bar{x}_1(t-\tau) + \bar{A}_{12}V_1 y(t) + \bar{A}_{d12}V_1 y(t-\tau),\ t \geq 0,$$
$$\tilde{y}(t) = \tilde{C}_1\bar{x}_1(t) + \tilde{C}_{d1}\bar{x}_1(t-\tau).$$

Note that the method in [1] can handle multiple time delays in both state and output vectors (see, (10.114) and (10.115)), and thus it can be used to design a reduced-order state observer to estimate $\bar{x}_1(t)$ of time-delay system (5.13) and (5.14). However, it is not clear to us how the method [1] can be extended to estimate a function of the state vector directly, i.e., $z(t) = F\bar{x}_1(t)$, without estimating the entire state vector $\bar{x}_1(t)$. In this chapter, we design functional observers with a range of complexities to estimate the function $z(t) = F\bar{x}_1(t)$.

11.2 Proposed Observer

To estimate the function $z(t) = F\bar{x}_1(t)$, $F \in \mathbb{R}^{m\times(n-1)}$, of the system (5.13) and (5.14) we consider the following functional observer

H. Trinh et al., *Unknown Inputs Estimation in Linear Time-Delay Systems Using Generalized Functional Observers*, Studies in Systems, Decision and Control 648,
https://doi.org/10.1007/978-3-032-11870-7_11

$$\hat{z}(t) = w(t) + M\tilde{y}(t) + M_d\tilde{y}(t-\tau), \tag{11.1}$$

$$\begin{aligned}\dot{w}(t) = {} & Nw(t) + N_d w(t-\tau) + G\tilde{y}(t) + G_d\tilde{y}(t-\tau) \\ & + G_{dd}\tilde{y}(t-2\tau) + Jy(t) + J_d y(t-\tau) \\ & + J_{dd}y(t-2\tau) + J_{ddd}y(t-3\tau),\end{aligned} \tag{11.2}$$

where $\tilde{y}(t) = V_2 y(t) - V_2 C_d D V_1 y(t-\tau)$, $w(\theta) = \rho(\theta)$, $\forall\theta \in [0, \tau]$, $w(t) \in \mathbb{R}^m$ and $\hat{z}(t) \in \mathbb{R}^m$ is the estimate of $z(t)$. Matrices $M, M_d, N, N_d, G, G_d, G_{dd}, J, J_d, J_{dd}$ and J_{ddd} are unknown and to be determined such that $\hat{z}(t)$ converges asymptotically to $z(t)$. Here, we consider the above observer with an internal delay and multiple delay output measurements, since it can cover a wider class of time-delay systems than an observer with no internal delay (see, Chap. 10). Also, the structure of the observer (11.1) and (11.2) is general enough to demonstrate the observer design procedure, where the unknown matrices can be derived in a systematic manner.

11.3 Observer Existence Conditions

To design observers (11.1) and (11.2) that asymptotically estimate $z(t) = F\bar{x}_1(t)$, we first need to derive some existence conditions by following similar analysis as in Section 10.3. For completeness, we derive their existence conditions and provide a systematic procedure for obtaining the observer matrices.

Firstly, with $e(t) = \hat{z}(t) - z(t)$ and from (11.1), we obtain the following equation

$$w(t) = e(t) + F\bar{x}_1(t) - M\tilde{y}(t) - M_d\tilde{y}(t-\tau). \tag{11.3}$$

The derivative of $e(t)$ is obtained as follows

$$\begin{aligned}\dot{e}(t) &= \dot{\hat{z}}(t) - \dot{z}(t) \\ &= \dot{w}(t) + M\dot{\tilde{y}}(t) + M_d\dot{\tilde{y}}(t-\tau) - F\dot{\bar{x}}_1(t),\end{aligned} \tag{11.4}$$

and after some substituting and rearranging, we obtain the following error dynamic equation

$$\begin{aligned}\dot{e}(t) = {} & Ne(t) + N_d e(t-\tau) + (M\tilde{C}_1\bar{A}_{12}V_1 - F\bar{A}_{12}V_1 + J)y(t) \\ & + (M\tilde{C}_1\bar{A}_{d12}V_1 + M_d\tilde{C}_1\bar{A}_{12}V_1 - F\bar{A}_{d12}V_1 + M\tilde{C}_{d1}\bar{A}_{12}V_1 + J_d)y(t-\tau) \\ & + (J_{dd} + M_d\tilde{C}_1\bar{A}_{d12}V_1 + M\tilde{C}_{d1}\bar{A}_{d12}V_1 + M_d\tilde{C}_{d1}\bar{A}_{12}V_1)y(t-2\tau) \\ & + (J_{ddd} + M_d\tilde{C}_{d1}\bar{A}_{d12}V_1)y(t-3\tau) + \big(NF + \tilde{T}\tilde{C}_1 + M\tilde{C}_1\bar{A}_{11} - F\bar{A}_{11}\big)\bar{x}_1(t) \\ & + \big(N_d F + \tilde{W}\tilde{C}_1 + \tilde{T}\tilde{C}_{d1} + M_d\tilde{C}_1\bar{A}_{11} + M(\tilde{C}_1\bar{A}_{d11} + \tilde{C}_{d1}\bar{A}_{11}) - F\bar{A}_{d11}\big)\bar{x}_1(t-\tau) \\ & + \big(\tilde{W}\tilde{C}_{d1} + \tilde{U}\tilde{C}_1 + M_d(\tilde{C}_1\bar{A}_{d11} + \tilde{C}_{d1}\bar{A}_{11}) + M\tilde{C}_{d1}\bar{A}_{d11}\big)\bar{x}_1(t-2\tau) \\ & + (\tilde{U}\tilde{C}_{d1} + M_d\tilde{C}_{d1}\bar{A}_{d11})\bar{x}_1(t-3\tau),\end{aligned} \tag{11.5}$$

where $\tilde{T}$, $\tilde{W}$ and $\tilde{U}$ are as defined in (10.84)–(10.86), respectively. Thus, from (11.5), observer (11.1) and (11.2) exists such that $\hat{z}(t)$ converges asymptotically to $z(t)$ for any initial condition $e(\theta)$, $\forall\theta \in [0, \tau]$, if the following conditions hold

$$\dot{e}(t) = Ne(t) + N_d e(t-\tau) \text{ is stable,} \tag{11.6}$$

$$J = F\bar{A}_{12}V_1 - M\tilde{C}_1\bar{A}_{12}V_1, \tag{11.7}$$

$$J_d = F\bar{A}_{d12}V_1 - M\tilde{C}_1\bar{A}_{d12}V_1 - M_d\tilde{C}_1\bar{A}_{12}V_1 - M\tilde{C}_{d1}\bar{A}_{12}V_1, \tag{11.8}$$

$$J_{dd} = -M_d\tilde{C}_1\bar{A}_{d12}V_1 - M\tilde{C}_{d1}\bar{A}_{d12}V_1 - M_d\tilde{C}_{d1}\bar{A}_{12}V_1, \tag{11.9}$$

$$J_{ddd} = -M_d\tilde{C}_{d1}\bar{A}_{d12}V_1, \tag{11.10}$$

$$NF + \tilde{T}\tilde{C}_1 + M\tilde{C}_1\bar{A}_{11} = F\bar{A}_{11}, \tag{11.11}$$

$$N_dF + \tilde{W}\tilde{C}_1 + \tilde{T}\tilde{C}_{d1} + M(\tilde{C}_1\bar{A}_{d11} + \tilde{C}_{d1}\bar{A}_{11}) + M_d\tilde{C}_1\bar{A}_{11} = F\bar{A}_{d11}, \tag{11.12}$$

$$\tilde{U}\tilde{C}_1 + \tilde{W}\tilde{C}_{d1} + M_d(\tilde{C}_1\bar{A}_{d11} + \tilde{C}_{d1}\bar{A}_{11}) + M\tilde{C}_{d1}\bar{A}_{d11} = 0, \tag{11.13}$$

$$\tilde{U}\tilde{C}_{d1} + M_d\tilde{C}_{d1}\bar{A}_{d11} = 0. \tag{11.14}$$

11.4 Design Procedure

With known matrices $F, \tilde{C}_1, \tilde{C}_{d1}, \bar{A}_{11}, \bar{A}_{d11}$, our objective is to solve the four coupled Eqs. (11.11)–(11.14) for the seven unknown matrices N, N_d, M, M_d, $\tilde{T}$, $\tilde{W}$ and $\tilde{U}$ such that the time-delay system (11.6) is asymptotically stable. Once $\tilde{T}$, $\tilde{W}$ and $\tilde{U}$ are found, then G, G_d and G_{dd} are obtained from (10.84)–(10.86), respectively. Finally, J, J_d, J_{dd} and J_{ddd} are obtained from (11.7)–(11.10), respectively. Thus, a straightforward observer design procedure.

Here, due to the presence of $\tilde{C}_{d1}$, the decoupling of Eqs. (11.11)–(11.14) into simpler equations cannot be conveniently carried out (see, (10.90)–(10.95) in Sect. 10.3). On the other hand, based on the solution approach reported in [2], we can express four Eqs. (11.11)–(11.14) into the following augmented form

$$\begin{bmatrix} N & N_d & \tilde{T} & \tilde{W} & \tilde{U} & M & M_d \end{bmatrix}\Theta = \Upsilon, \tag{11.15}$$

where $\Theta \in \mathbb{R}^{(2m+5(p-1))\times 4(n-1)}$ and $\Upsilon \in \mathbb{R}^{m\times 4(n-1)}$ are defined below

$$\Theta = \begin{bmatrix} F & 0 & 0 & 0 \\ 0 & F & 0 & 0 \\ \tilde{C}_1 & \tilde{C}_{d1} & 0 & 0 \\ 0 & \tilde{C}_1 & \tilde{C}_{d1} & 0 \\ 0 & 0 & \tilde{C}_1 & \tilde{C}_{d1} \\ \tilde{C}_1\bar{A}_{11} & \tilde{C}_{1a} & \tilde{C}_{d1}\bar{A}_{d11} & 0 \\ 0 & \tilde{C}_1\bar{A}_{11} & \tilde{C}_{1a} & \tilde{C}_{d1}\bar{A}_{d11} \end{bmatrix}, \tag{11.16}$$

$$\Upsilon = \begin{bmatrix} F\bar{A}_{11} & F\bar{A}_{d11} & 0 & 0 \end{bmatrix}, \tag{11.17}$$

and for brevity, we let $\tilde{C}_{1a} = \tilde{C}_1\bar{A}_{d11} + \tilde{C}_{d1}\bar{A}_{11}$. With regard to (11.15), it has a solution if and only if the following condition holds

$$\text{rank}\begin{bmatrix} \Theta \\ \Upsilon \end{bmatrix} = \text{rank}[\Theta]. \tag{11.18}$$

Note that Θ and Υ are known matrices and thus testing condition (11.18) is an easy task. Upon satisfying (11.18), the unknown matrices can be obtained as follows

$$\begin{bmatrix} N \ N_d \ \tilde{T} \ \tilde{W} \ \tilde{U} \ M \ M_d \end{bmatrix} = \Upsilon\Theta^+ + Z_a(I_{2m+5(p-1)} - \Theta\Theta^+), \tag{11.19}$$

where $\Theta^+ \in \mathbb{R}^{4(n-1)\times(2m+5(p-1))}$ is the Moore-Penrose inverse of Θ, and $Z_a \in \mathbb{R}^{m\times(2m+5(p-1))}$ is an arbitrary matrix.

Next, we look at the asymptotic stability requirement of the error system (11.6). Let us partition the identity matrix $I_{2m+5(p-1)} = \begin{bmatrix} \mathscr{I}^1 \ \mathscr{I}^2 \ \mathscr{I}^3 \ ... \ \mathscr{I}^7 \end{bmatrix}$, where $\mathscr{I}^i$ is a sub-matrix with an appropriate dimension. Thus, from (11.19), we can obtain N and N_d as follows

$$N = N_1 + Z_a N_2, \tag{11.20}$$

$$N_d = N_{d1} + Z_a N_{d2}, \tag{11.21}$$

where

$$N_1 = \Upsilon\Theta^+\mathscr{I}^1, \ N_2 = (I_{2m+5(p-1)} - \Theta\Theta^+)\mathscr{I}^1, \tag{11.22}$$

$$N_{d1} = \Upsilon\Theta^+\mathscr{I}^2, \ N_{d2} = (I_{2m+5(p-1)} - \Theta\Theta^+)\mathscr{I}^2, \tag{11.23}$$

$$\mathscr{I}^1 = \begin{bmatrix} I_m \\ 0_{(m+5(p-1))\times m} \end{bmatrix} \text{ and } \mathscr{I}^2 = \begin{bmatrix} 0_{m\times m} \\ I_m \\ 0_{5(p-1)\times m} \end{bmatrix}.$$

Thus, from (11.20) and (11.21), the error system (11.6) is expressed as

$$\dot{e}(t) = (N_1 + Z_a N_2)e(t) + (N_{d1} + Z_a N_{d2})e(t-\tau), \tag{11.24}$$

for which we can search for a Z_a to stabilize (11.24) (note that for the case where both N_2 and N_{d2} are zeros, it is necessary that the nominal system, $\dot{e}(t) = N_1 e(t) + N_{d1}e(t-\tau)$, is asymptotically stable). Once asymptotic stability of (11.24) is ensured, the rest of the observer matrices can be derived.

Remark 1 We consider an observer that has a simpler structure than (11.1) and (11.2), where

$$\hat{z}(t) = w(t) + M\tilde{y}(t), \tag{11.25}$$

$$\begin{aligned}\dot{w}(t) = {} & Nw(t) + N_d w(t-\tau) + G\tilde{y}(t) + G_d\tilde{y}(t-\tau) \\ & + Jy(t) + J_d y(t-\tau) + J_{dd}y(t-2\tau).\end{aligned} \tag{11.26}$$

Then, by following similar analysis as above, we can show that observer (11.25) and (11.26) exists such that $\hat{z}(t)$ converges asymptotically to $z(t)$ for any initial condition $e(\theta)$, $\forall\theta \in [0, \tau]$, if the following conditions hold

$$\dot{e}(t) = Ne(t) + N_d e(t-\tau) \text{ is stable,} \tag{11.27}$$
$$J = F\bar{A}_{12}V_1 - M\tilde{C}_1\bar{A}_{12}V_1, \tag{11.28}$$
$$J_d = F\bar{A}_{d12}V_1 - M\tilde{C}_1\bar{A}_{d12}V_1 - M\tilde{C}_{d1}\bar{A}_{12}V_1, \tag{11.29}$$
$$J_{dd} = -M\tilde{C}_{d1}\bar{A}_{d12}V_1, \tag{11.30}$$
$$NF + \tilde{T}\tilde{C}_1 + M\tilde{C}_1\bar{A}_{11} = F\bar{A}_{11}, \tag{11.31}$$
$$N_d F + \bar{W}\tilde{C}_1 + \tilde{T}\tilde{C}_{d1} + M\tilde{C}_{1a} = F\bar{A}_{d11}, \tag{11.32}$$
$$\bar{W}\tilde{C}_{d1} + M\tilde{C}_{d1}\bar{A}_{d11} = 0, \tag{11.33}$$

where

$$\bar{W} = G_d - N_d M. \tag{11.34}$$

The solvability of Eqs. (11.31)–(11.33) for the unknown matrix $\begin{bmatrix} N & N_d & \tilde{T} & \bar{W} & M \end{bmatrix}$ is subject to the following necessary and sufficient condition

$$\text{rank}\begin{bmatrix} \Theta_r \\ \Upsilon_r \end{bmatrix} = \text{rank}[\Theta_r], \tag{11.35}$$

where

$$\Theta_r = \begin{bmatrix} F & 0 & 0 \\ 0 & F & 0 \\ \tilde{C}_1 & \tilde{C}_{d1} & 0 \\ 0 & \tilde{C}_1 & \tilde{C}_{d1} \\ \tilde{C}_1\bar{A}_{11} & \tilde{C}_{1a} & \tilde{C}_{d1}\bar{A}_{d11} \end{bmatrix}, \tag{11.36}$$
$$\Upsilon_r = \begin{bmatrix} F\bar{A}_{11} & F\bar{A}_{d11} & 0 \end{bmatrix}. \tag{11.37}$$

Then the rest of the observer parameters can be obtained once asymptotic stability of (11.27) is confirmed.

Remark 2 We can also consider a functional observer with no internal delay, where

$$\hat{z}(t) = w(t) + M\tilde{y}(t), \tag{11.38}$$
$$\begin{aligned}\dot{w}(t) = {} & Nw(t) + G\tilde{y}(t) + G_d\tilde{y}(t-\tau) + Jy(t) \\ & + J_d y(t-\tau) + J_{dd}y(t-2\tau).\end{aligned} \tag{11.39}$$

We can show that observer (11.38) and (11.39) exists such that $\hat{z}(t)$ converges asymptotically to $z(t)$ for any initial condition $e(\theta)$, $\forall\theta \in [0, \tau]$, if the following conditions are met

$$J = F\bar{A}_{12}V_1 - M\tilde{C}_1\bar{A}_{12}V_1, \tag{11.40}$$

$$J_d = F\bar{A}_{d12}V_1 - M\tilde{C}_1\bar{A}_{d12}V_1 - M\tilde{C}_{d1}\bar{A}_{12}V_1, \tag{11.41}$$

$$J_{dd} = -M\tilde{C}_{d1}\bar{A}_{d12}V_1, \tag{11.42}$$

$$NF + \tilde{T}\tilde{C}_1 + M\tilde{C}_1\bar{A}_{11} = F\bar{A}_{11}, \ N \text{ is stable}, \tag{11.43}$$

$$G_d\tilde{C}_1 + \tilde{T}\tilde{C}_{d1} + M\tilde{C}_{1a} = F\bar{A}_{d11}, \tag{11.44}$$

$$G_d\tilde{C}_{d1} + M\tilde{C}_{d1}\bar{A}_{d11} = 0. \tag{11.45}$$

From (11.43)–(11.45), we can solve for the unknown matrix $\begin{bmatrix} N & \tilde{T} & G_d & M \end{bmatrix}$ if and only if the following condition holds

$$\operatorname{rank}\begin{bmatrix} \Theta_{rr} \\ \Upsilon_r \end{bmatrix} = \operatorname{rank}[\Theta_{rr}], \tag{11.46}$$

where

$$\Theta_{rr} = \begin{bmatrix} F & 0 & 0 \\ \tilde{C}_1 & \tilde{C}_{d1} & 0 \\ 0 & \tilde{C}_1 & \tilde{C}_{d1} \\ \tilde{C}_1\bar{A}_{11} & \tilde{C}_{1a} & \tilde{C}_{d1}\bar{A}_{d11} \end{bmatrix}. \tag{11.47}$$

11.4.1 A Numerical Example

To show the merit of the design method and observers (i.e., with a range of complexities) in this chapter, we consider time-delay system (5.13) and (5.14) with $\tau = 1s$, $\bar{A}_{11}$, $\bar{A}_{d11}$, $\tilde{C}_1$, $\tilde{C}_{d1}$ and F as given below
$\bar{A}_{11} = \begin{bmatrix} -2 & -3 \\ -1 & -3 \end{bmatrix}$, $\bar{A}_{d11} = \begin{bmatrix} 2 & -1 \\ 0 & 0 \end{bmatrix}$, $\tilde{C}_1 = \begin{bmatrix} 1 & 0 \end{bmatrix}$, $\tilde{C}_{d1} = \begin{bmatrix} 0 & 1 \end{bmatrix}$ and $F = \begin{bmatrix} 0 & 1 \end{bmatrix}$.

For this numerical example, if the method [1] is to be employed, we need to first check whether matrix $Q_2(s)$ as defined in (10.118) is unimodular or not. Based on the discussion in Section 10.5, we have $C(s) = [1 \ \ s]$, $A(s) = \begin{bmatrix} 2s-2 & -s-3 \\ -1 & -3 \end{bmatrix}$, and thus from (10.118), we obtain $Q_2(s) = \begin{bmatrix} 1 & s \\ s-2 & -4s-3 \end{bmatrix}$. It is clear that $Q_2(s)$ is not unimodular since its determinant is zero at $s = -1 \pm j\sqrt{2}$. Therefore, the design method [1] cannot be used to design a state observer.

We also find that a first-order observer without an internal delay in its structure does not exist. This is confirmed by checking condition (11.46) of observer (11.38) and (11.39), which does not hold. Therefore, we next consider an observer that has a more complex structure than observer (11.38) and (11.39). The obvious next choice is observer (11.25) and (11.26). Here, it is found that condition (11.35) is satisfied, and also matrices $N = -3$ and $N_d = 1$ which resulted in an asymptotic stable error time-delay system (11.27). Hence for this example, we can use observer (11.25) and (11.26) to estimate the function $z(t)$.

On the other hand, to estimate the entire state vector, we let $F = I_2$, i.e., $z(t) = F\bar{x}(t) = \bar{x}(t)$. Note that the method [1] cannot be used since $Q_2(s)$ is not unimodular. Based on our result, we can design a second-order observer (11.38) and (11.39). This is confirmed by checking condition (11.46). We obtain a stable second-order observer with $N = \begin{bmatrix} -5 & 3 \\ -1 & -3 \end{bmatrix}$, where $\lambda(N) = -4 \pm j\sqrt{2}$.

11.5 Concluding Remark

In this chapter, we have considered the design of various types of functional observers to estimate a function of the state vector for a class of time-delay systems with a state-delayed term in the output vector. This term can be easily dealt with with our approach which is an adapted solution approach in [2]. We have presented existence conditions for the proposed observers and outlined a procedure for determining their observer gain matrices. Via a numerical example, we have also shown that our approach is less conservative than the Output Injection approach in [1].

References

1. Hou, M., Zítek, P., Patton, R.J.: An observer design for linear time-delay systems. IEEE Trans. Automat. Control **47**(1), 121–125 (2002)
2. Nam, P.T., Pathirana, P.N., Trinh, H.: ϵ-bounded state estimation for time-delay systems with bounded disturbances. Int. J. Control **87**(9), 1747–1756 (2014)

Chapter 12
Observers for Systems with Multiple Delays in State and Output Vectors

12.1 Background

In this chapter, we take on an exploratory journey to solve the problem of designing various types of functional observers for time-delay system (5.28) and (5.29), i.e.,

$$\begin{aligned}\dot{\bar{x}}_1(t) &= \bar{A}_{11}\bar{x}_1(t) + \tilde{A}_{d11}\bar{x}_1(t-\tau) + \tilde{A}_{dd11}\bar{x}_1(t-2\tau) + \bar{A}_{12}V_1 y(t) + \bar{A}_{d12}V_1 y(t-\tau),\\ \tilde{y}(t) &= \tilde{C}_1\bar{x}_1(t) + \tilde{C}_{d1}\bar{x}_1(t-\tau) + \tilde{C}_{dd1}\bar{x}_1(t-2\tau).\end{aligned}$$

Due to the presence of multiple time delays in both the state and output vectors, we will employ more complex functional observers that contain multiple internal delays. Such observers have more complex structures than the ones considered in Chap. 11. On the other hand, by employing multiple internal delays, we can relax their existence conditions, thus a clear trade-off between simplicity and availability.

To start off, we first consider the problem of estimating the vector $z(t) = F\bar{x}_1(t)$ of system (5.28) and (5.29). We then consider a new functional estimation problem, where we directly estimate the vector $\bar{z}(t) = F\bar{x}_1(t) + F_d\bar{x}_1(t-\tau) + F_{dd}\bar{x}_1(t-2\tau)$ of system (5.28) and (5.29). This is a new functional estimation problem that has not yet been reported in the literature. Here, the vector $\bar{z}(t)$ contains multiple delayed terms, $F_d\bar{x}_1(t-\tau)$ and $F_{dd}\bar{x}_1(t-2\tau)$, instead of just the term $F\bar{x}_1(t)$, which is currently the case in the literature. The significance of estimating $\bar{z}(t)$ is that it opens a new horizon for the design of state observers of time-delay systems (5.28) and (5.29) (and more complex time-delay systems with mixed time delays in both the state and output vectors in which we will consider later in this book), where we can now expand the order of the designed observer beyond $(n-1)$ thanks to a more relaxed observer existence condition (to be presented in this chapter). Furthermore, the unknown input $d(t)$ can be directly estimated via this new type of generalized functional observers (see, Scenario 4 as discussed in Sect. 4.3.9, Chap. 4).

H. Trinh et al., *Unknown Inputs Estimation in Linear Time-Delay Systems Using Generalized Functional Observers*, Studies in Systems, Decision and Control 648,
https://doi.org/10.1007/978-3-032-11870-7_12

12.2 Functional Estimation Problem 1: $z(t) = F\bar{x}_1(t) \in \mathbb{R}^m$

To estimate the function $z(t) = F\bar{x}_1(t)$ of system (5.28) and (5.29), we consider the following observer with multiple internal delays

$$\begin{aligned} \hat{z}(t) &= w(t) + M\tilde{y}(t), && (12.1)\\ \dot{w}(t) &= Nw(t) + N_d w(t-\tau) + N_{dd} w(t-2\tau) + G\tilde{y}(t) \\ &\quad + G_d\tilde{y}(t-\tau) + G_{dd}\tilde{y}(t-2\tau) + Jy(t) + J_d y(t-\tau) \\ &\quad + J_{dd} y(t-2\tau) + J_{ddd} y(t-3\tau), && (12.2) \end{aligned}$$

where $\tilde{y}(t) = V_2 y(t) - V_2 C_d D V_1 y(t-\tau)$, $w(\theta) = \rho(\theta)$, $\forall \theta \in [0, 2\tau]$, $w(t) \in \mathbb{R}^m$ and $\hat{z}(t) \in \mathbb{R}^m$ is the estimate of $z(t)$. Matrices $M, N, N_d, N_{dd}, G, G_d, G_{dd}, J, J_d, J_{dd}$ and J_{ddd} are unknown and to be determined such that $\hat{z}(t)$ converges asymptotically to $z(t)$.

Compared to (11.1) and (11.2), observer (12.1) and (12.2) has an extra internal delay term, $N_{dd}w(t-2\tau)$. Its inclusion helps relax their observer existence conditions. Based on a similar methodology as in Chap. 11, we first derive their existence conditions and then a procedure for determining their parameters.

12.2.1 Observer Existence Conditions

From $e(t) = \hat{z}(t) - z(t)$ and (12.1), we obtain $w(t) = e(t) + F\bar{x}_1(t) - M\tilde{y}(t)$ and the following error dynamic equation

$$\begin{aligned} \dot{e}(t) &= Ne(t) + N_d e(t-\tau) + N_{dd} e(t-2\tau) + (J - F\bar{A}_{12}V_1 + M\tilde{C}_1\bar{A}_{12}V_1)y(t) \\ &+ (J_d - F\bar{A}_{d12}V_1 + M\tilde{C}_{d1}\bar{A}_{12}V_1 + M\tilde{C}_1\bar{A}_{d12}V_1)y(t-\tau) \\ &+ (J_{dd} + M\tilde{C}_{d1}\bar{A}_{d12}V_1 + M\tilde{C}_{dd1}\bar{A}_{12}V_1)y(t-2\tau) + (J_{ddd} + M\tilde{C}_{dd1}\bar{A}_{d12}V_1)y(t-3\tau) \\ &+ \big(NF + \tilde{T}\tilde{C}_1 + M\tilde{C}_1\bar{A}_{11} - F\bar{A}_{11}\big)\bar{x}_1(t) \\ &+ \big(N_d F + \bar{W}\tilde{C}_1 + \tilde{T}\tilde{C}_{d1} + M\bar{C}_1 - F\tilde{A}_{d11}\big)\bar{x}_1(t-\tau) \\ &+ \big(N_{dd} F + \bar{U}\tilde{C}_1 + \bar{W}\tilde{C}_{d1} + \tilde{T}\tilde{C}_{dd1} + M\bar{C}_{1a} - F\tilde{A}_{dd11}\big)\bar{x}_1(t-2\tau) \\ &+ (\bar{U}\tilde{C}_{d1} + \bar{W}\tilde{C}_{dd1} + M\hat{C}_1)\bar{x}_1(t-3\tau) + (\bar{U}\tilde{C}_{dd1} + M\tilde{C}_{dd1}\tilde{A}_{dd11})\bar{x}_1(t-4\tau), \end{aligned} \quad (12.3)$$

where

$$\begin{aligned} \bar{C}_1 &= \tilde{C}_1\tilde{A}_{d11} + \tilde{C}_{d1}\bar{A}_{11}, && (12.4)\\ \bar{U} &= G_{dd} - N_{dd}M, && (12.5)\\ \bar{C}_{1a} &= \tilde{C}_1\tilde{A}_{dd11} + \tilde{C}_{d1}\tilde{A}_{d11} + \tilde{C}_{dd1}\bar{A}_{11}, && (12.6)\\ \hat{C}_1 &= \tilde{C}_{d1}\tilde{A}_{dd11} + \tilde{C}_{dd1}\tilde{A}_{d11}. && (12.7) \end{aligned}$$

Here, $\tilde{T}$ and $\bar{W}$ are as defined in (10.84) and (11.34), respectively. Thus, from (12.3), observer (12.1) and (12.2) exists such that $\hat{z}(t)$ converges asymptotically to $z(t)$ for any initial condition $e(\theta),\ \forall\theta \in [0, 2\tau]$, if the following conditions hold

$$\dot{e}(t) = Ne(t) + N_d e(t-\tau) + N_{dd} e(t-2\tau) \text{ is asymptotically stable,} \tag{12.8}$$

$$J = F\bar{A}_{12}V_1 - M\tilde{C}_1\bar{A}_{12}V_1, \tag{12.9}$$

$$J_d = F\bar{A}_{d12}V_1 - M(\tilde{C}_{d1}\bar{A}_{12} + \tilde{C}_1\bar{A}_{d12})V_1, \tag{12.10}$$

$$J_{dd} = -M(\tilde{C}_{d1}\bar{A}_{d12} + \tilde{C}_{dd1}\bar{A}_{12})V_1, \tag{12.11}$$

$$J_{ddd} = -M\tilde{C}_{dd1}\bar{A}_{d12}V_1, \tag{12.12}$$

$$NF + \tilde{T}\tilde{C}_1 + M\tilde{C}_1\bar{A}_{11} = F\bar{A}_{11}, \tag{12.13}$$

$$N_d F + \bar{W}\tilde{C}_1 + \tilde{T}\tilde{C}_{d1} + M\bar{C}_1 = F\tilde{A}_{d11}, \tag{12.14}$$

$$N_{dd} F + \bar{U}\tilde{C}_1 + \bar{W}\tilde{C}_{d1} + \tilde{T}\tilde{C}_{dd1} + M\bar{C}_{1a} = F\tilde{A}_{dd11}, \tag{12.15}$$

$$\bar{U}\tilde{C}_{d1} + \bar{W}\tilde{C}_{dd1} + M\hat{C}_1 = 0, \tag{12.16}$$

$$\bar{U}\tilde{C}_{dd1} + M\tilde{C}_{dd1}\tilde{A}_{dd11} = 0. \tag{12.17}$$

Examining conditions (12.8)–(12.17), based on (12.9)–(12.12), we can obtain matrices J, J_d, J_{dd} and J_{ddd}, once M is found. Thus, by following a similar analytical approach in Chap. 11, our solution approach is to first solve the coupled matrix equations (12.13)–(12.17) for the seven unknown matrices, namely, N, N_d, N_{dd}, $\tilde{T}$, $\bar{W}$, $\bar{U}$ and M. Then, subject to a necessary and sufficient condition for the solvability of (12.13)–(12.17), we obtain a closed-form solution for matrices N, N_d and N_{dd}, and solve the stabilizability problem of the time-delay error system (12.8). Provided that the time-delay system (12.8) is asymptotically stable, from (10.84), (11.34) and (12.5), we can obtain matrices G, G_d, G_{dd} as follows

$$G = \tilde{T} + NM, \tag{12.18}$$

$$G_d = \bar{W} + N_d M, \tag{12.19}$$

$$G_{dd} = \bar{U} + N_{dd} M. \tag{12.20}$$

Now, a necessary and sufficient condition for the solvability of (12.13)–(12.17) can be readily obtained by expressing them in the following augmented form

$$\begin{bmatrix} N & N_d & N_{dd} & \tilde{T} & \bar{W} & \bar{U} & M \end{bmatrix}\Theta_a = \Upsilon_a, \tag{12.21}$$

where $\Theta_a \in \mathbb{R}^{(3m+4(p-1))\times 5(n-1)}$ and $\Upsilon_a \in \mathbb{R}^{m\times 5(n-1)}$ are defined below

$$\Theta_a = \begin{bmatrix} F & 0 & 0 & 0 & 0 \\ 0 & F & 0 & 0 & 0 \\ 0 & 0 & F & 0 & 0 \\ \tilde{C}_1 & \tilde{C}_{d1} & \tilde{C}_{dd1} & 0 & 0 \\ 0 & \tilde{C}_1 & \tilde{C}_{d1} & \tilde{C}_{dd1} & 0 \\ 0 & 0 & \tilde{C}_1 & \tilde{C}_{d1} & \tilde{C}_{dd1} \\ \tilde{C}_1\bar{A}_{11} & \bar{C}_1 & \bar{C}_{1a} & \hat{C}_1 & \tilde{C}_{dd1}\tilde{A}_{dd11} \end{bmatrix}, \tag{12.22}$$

$$\Upsilon_a = \begin{bmatrix} F\bar{A}_{11} & F\tilde{A}_{d11} & F\tilde{A}_{dd11} & 0 & 0 \end{bmatrix}. \tag{12.23}$$

From (12.21), we can solve for the seven unknown matrices if and only if the following condition holds

$$\text{rank}\begin{bmatrix} \Theta_a \\ \Upsilon_a \end{bmatrix} = \text{rank}\begin{bmatrix} \Theta_a \end{bmatrix}. \tag{12.24}$$

Since Θ_a and Υ_a are known matrices, testing condition (12.24) can be easily carried out. Upon the satisfaction of (12.24), the unknown matrices can be obtained as follows

$$\begin{bmatrix} N \; N_d \; N_{dd} \; \tilde{T} \; \bar{W} \; \bar{U} \; M \end{bmatrix} = \Upsilon_a\Theta_a^+ + \bar{Z}_a(I_{3m+4(p-1)} - \Theta_a\Theta_a^+), \tag{12.25}$$

where $\Theta_a^+ \in \mathbb{R}^{5(n-1)\times(3m+4(p-1))}$ is the Moore-Penrose inverse of Θ_a, and $\bar{Z}_a \in \mathbb{R}^{m\times(3m+4(p-1))}$ is an arbitrary matrix.

Next, we turn to the stabilizability problem of the time-delay error system (12.8). Denoting $\mathscr{I}^1$, $\mathscr{I}^2$ and $\mathscr{I}^3$ as the following selective matrices

$$\mathscr{I}^1 = \begin{bmatrix} I_m \\ 0_{(2m+4(p-1))\times m} \end{bmatrix}, \; \mathscr{I}^2 = \begin{bmatrix} 0_{m\times m} \\ I_m \\ 0_{(m+4(p-1))\times m} \end{bmatrix} \text{ and } \mathscr{I}^3 = \begin{bmatrix} 0_{2m\times m} \\ I_m \\ 0_{4(p-1)\times m} \end{bmatrix}.$$

Then, from (12.25), we obtain N, N_d and N_{dd} as follows

$$N = \bar{N}_1 + \bar{Z}_a\bar{N}_2, \tag{12.26}$$

$$N_d = \bar{N}_{d1} + \bar{Z}_a\bar{N}_{d2}, \tag{12.27}$$

$$N_{dd} = \bar{N}_{dd1} + \bar{Z}_a\bar{N}_{dd2}, \tag{12.28}$$

where

$$\bar{N}_1 = \Upsilon_a\Theta_a^+\mathscr{I}^1, \; \bar{N}_2 = (I_{3m+4(p-1)} - \Theta_a\Theta_a^+)\mathscr{I}^1, \tag{12.29}$$

$$\bar{N}_{d1} = \Upsilon_a\Theta_a^+\mathscr{I}^2, \; \bar{N}_{d2} = (I_{3m+4(p-1)} - \Theta_a\Theta_a^+)\mathscr{I}^2, \tag{12.30}$$

$$\bar{N}_{dd1} = \Upsilon_a\Theta_a^+\mathscr{I}^3, \; \bar{N}_{dd2} = (I_{3m+4(p-1)} - \Theta_a\Theta_a^+)\mathscr{I}^3. \tag{12.31}$$

Thus, from (12.26)–(12.28), the error system (12.8) is expressed as follows

$$\dot{e}(t) = (\bar{N}_1 + \bar{Z}_a\bar{N}_2)e(t) + (\bar{N}_{d1} + \bar{Z}_a\bar{N}_{d2})e(t-\tau) + (\bar{N}_{dd1} + \bar{Z}_a\bar{N}_{dd2})e(t-2\tau). \tag{12.32}$$

System (12.32) contains multiple time delays, and by employing the well-established Lyapunov-Krasovskii functional approach, we can establish stabilizability conditions of (12.32) in terms of linear matrix inequalities. Note that in some cases, we have $\Theta_a\Theta_a^+ = I_{3m+4(p-1)}$ and thus $\bar{N}_2$, $\bar{N}_{d2}$ and $\bar{N}_{dd2}$ are all zeros. In such case, $\bar{Z}_a$ (i.e., $\bar{Z}_a = 0$) plays no role in the stabilization of the time-delay (12.32). As such, we can only check whether the following time-delay system

$$\dot{e}(t) = \bar{N}_1 e(t) + \bar{N}_{d1} e(t-\tau) + \bar{N}_{dd1} e(t-2\tau), \tag{12.33}$$

is asymptotically stable or not. For completeness, stability conditions and stabilizability conditions for (12.33) and (12.32), respectively, will be derived in Chap. 20.

Finally, we consider some simpler observers for system (5.28) and (5.29) and present their existence conditions.

Remark 1 We consider the following observer which has one internal delay (i.e., $N_{dd} = 0$)

$$\hat{z}(t) = w(t) + M\tilde{y}(t), \tag{12.34}$$

$$\begin{aligned} \dot{w}(t) = {} & Nw(t) + N_d w(t-\tau) + G\tilde{y}(t) + G_d\tilde{y}(t-\tau) \\ & + G_{dd}\tilde{y}(t-2\tau) + Jy(t) + J_d y(t-\tau) + J_{dd} y(t-2\tau) \\ & + J_{ddd} y(t-3\tau). \end{aligned} \tag{12.35}$$

The above observer has only one internal delay, namely, $N_d w(t-\tau)$, and thus it has a simpler structure than (12.1) and (12.2). On the other hand, its existence conditions are expected to be more restrictive. By letting $N_{dd} = 0$ and hence $\bar{U} = G_{dd}$, the solvability of matrix equations (12.13)–(12.17) for the unknown matrix $\begin{bmatrix} N & N_d & \tilde{T} & \bar{W} & G_{dd} & M \end{bmatrix}$, where

$$\begin{bmatrix} N & N_d & \tilde{T} & \bar{W} & G_{dd} & M \end{bmatrix} \Theta_{ar} = \Upsilon_a, \tag{12.36}$$

is subject to the following necessary and sufficient condition

$$\operatorname{rank}\begin{bmatrix} \Theta_{ar} \\ \Upsilon_a \end{bmatrix} = \operatorname{rank}\begin{bmatrix} \Theta_{ar} \end{bmatrix}, \tag{12.37}$$

where

$$\Theta_{ar} = \begin{bmatrix} F & 0 & 0 & 0 & 0 \\ 0 & F & 0 & 0 & 0 \\ \tilde{C}_1 & \tilde{C}_{d1} & \tilde{C}_{dd1} & 0 & 0 \\ 0 & \tilde{C}_1 & \tilde{C}_{d1} & \tilde{C}_{dd1} & 0 \\ 0 & 0 & \tilde{C}_1 & \tilde{C}_{d1} & \tilde{C}_{dd1} \\ \tilde{C}_1\bar{A}_{11} & \bar{C}_1 & \bar{C}_{1a} & \hat{C}_1 & \tilde{C}_{dd1}\bar{A}_{dd11} \end{bmatrix}. \tag{12.38}$$

Then the rest of the observer parameters can be obtained, once asymptotic stability of (11.27) is confirmed.

Remark 2 We consider the following observer with no internal delay

$$\hat{z}(t) = w(t) + M\tilde{y}(t), \tag{12.39}$$

$$\begin{aligned}\dot{w}(t) = Nw(t) + G\tilde{y}(t) + G_d\tilde{y}(t-\tau) + G_{dd}\tilde{y}(t-2\tau) \\ + Jy(t) + J_d y(t-\tau) + J_{dd}y(t-2\tau) + J_{ddd}y(t-3\tau).\end{aligned} \tag{12.40}$$

By letting $N_d = 0$ and $N_{dd} = 0$, we have $\bar{U} = G_{dd}$ and $\bar{W} = G_d$. Accordingly, the solvability of matrix equations (12.13)–(12.17) for the unknown matrix $\begin{bmatrix} N & \tilde{T} & G_d & G_{dd} & M \end{bmatrix}$, where

$$\begin{bmatrix} N & \tilde{T} & G_d & G_{dd} & M \end{bmatrix}\Theta_{arr} = \Upsilon_a, \tag{12.41}$$

$$\Theta_{arr} = \begin{bmatrix} F & 0 & 0 & 0 & 0 \\ \tilde{C}_1 & \tilde{C}_{d1} & \tilde{C}_{dd1} & 0 & 0 \\ 0 & \tilde{C}_1 & \tilde{C}_{d1} & \tilde{C}_{dd1} & 0 \\ 0 & 0 & \tilde{C}_1 & \tilde{C}_{d1} & \tilde{C}_{dd1} \\ \tilde{C}_1\bar{A}_{11} & \bar{C}_1 & \bar{C}_{1a} & \hat{C}_1 & \tilde{C}_{dd1}\tilde{A}_{dd11} \end{bmatrix}, \tag{12.42}$$

is subject to the following necessary and sufficient condition

$$\operatorname{rank}\begin{bmatrix} \Theta_{arr} \\ \Upsilon_a \end{bmatrix} = \operatorname{rank}\begin{bmatrix} \Theta_{arr} \end{bmatrix}. \tag{12.43}$$

Provided that N is stable, the rest of the observer parameters can be obtained.

Remark 3 So far in our solution approach to matrix equations (12.13)–(12.17), we have exploited the free matrix $\bar{Z}_a$ and used it to solve the stabilizability problem of the error system (12.32). On the other hand, when the rate of asymptotic convergence of the observer error is specified a priori, for example, in the case of observer (12.39)–(12.40), matrix N and its the eigenvalues are given in advance. Then in such a case, the solvability of matrix equations (12.13)–(12.17) for the unknown matrix $\begin{bmatrix} \tilde{T} & G_d & G_{dd} & M \end{bmatrix}$ is subject to the following necessary and sufficient condition

$$\operatorname{rank}\begin{bmatrix} \bar{\Theta}_a \\ \bar{\Upsilon}_a \end{bmatrix} = \operatorname{rank}\begin{bmatrix} \bar{\Theta}_a \end{bmatrix}, \tag{12.44}$$

where

$$\bar{\Theta}_a = \begin{bmatrix} \tilde{C}_1 & \tilde{C}_{d1} & \tilde{C}_{dd1} & 0 & 0 \\ 0 & \tilde{C}_1 & \tilde{C}_{d1} & \tilde{C}_{dd1} & 0 \\ 0 & 0 & \tilde{C}_1 & \tilde{C}_{d1} & \tilde{C}_{dd1} \\ \tilde{C}_1\bar{A}_{11} & \bar{C}_1 & \bar{C}_{1a} & \hat{C}_1 & \tilde{C}_{dd1}\tilde{A}_{dd11} \end{bmatrix}, \tag{12.45}$$

$$\bar{\Upsilon}_a = \begin{bmatrix}(F\bar{A}_{11} - NF) & F\tilde{A}_{d11} & F\tilde{A}_{dd11} & 0 & 0\end{bmatrix}. \tag{12.46}$$

12.2.2 Numerical Examples

To implement the observers designed in this section, we consider the following numerical examples.

Numerical Example 1: We consider the following time-delay system with multiple delays and a known external input, $u(t)$,

$$\dot{\bar{x}}_1(t) = \bar{A}_{11}\bar{x}_1(t) + \tilde{A}_{d11}\bar{x}_1(t-\tau) + \tilde{A}_{dd11}\bar{x}_1(t-2\tau) + Bu(t), \tag{12.47}$$

$$\tilde{y}(t) = \tilde{C}_1\bar{x}_1(t) + \tilde{C}_{d1}\bar{x}_1(t-\tau) + \tilde{C}_{dd1}\bar{x}_1(t-2\tau), \tag{12.48}$$

$$z(t) = F\bar{x}_1(t), \tag{12.49}$$

where $\tau = 1s$, $\bar{A}_{11} = \begin{bmatrix} -3 & -3 \\ 1 & -5 \end{bmatrix}$, $\tilde{A}_{d11} = \begin{bmatrix} -4 & 0 \\ 0 & -2 \end{bmatrix}$, $\tilde{A}_{dd11} = \begin{bmatrix} -3 & 1 \\ 0 & -2 \end{bmatrix}$, $\tilde{C}_1 = I_2$, $\tilde{C}_{d1} = \begin{bmatrix} 0 & 1 \\ 0 & 1 \end{bmatrix}$, $\tilde{C}_{dd1} = \begin{bmatrix} 1 & 1 \\ 0 & 1 \end{bmatrix}$, $B = \begin{bmatrix} 1 \\ 2 \end{bmatrix}$ and $F = \begin{bmatrix} 0 & 1 \end{bmatrix}$.

Note that we include the external known input, $u(t)$, just for the purpose of showing via simulations the ability of the designed observers to asymptotically track the required function of the state vector, $z(t)$, to any arbitrary external inputs and initial conditions. Thus, with the term $Bu(t)$ in (12.47) instead of $\bar{A}_{12}V_1y(t)$ and $\bar{A}_{d12}V_1y(t-\tau)$ as in (5.28), we therefore replace $Jy(t)$, $J_dy(t-\tau)$, $J_{dd}y(t-2\tau)$ and $J_{ddd}y(t-3\tau)$ in (12.2) (also, (12.35) and (12.40)) with $Ju(t)$, $J_du(t-\tau)$, $J_{dd}u(t-2\tau)$ and $J_{ddd}u(t-3\tau)$, where

$$J = FB - M\tilde{C}_1B, \tag{12.50}$$

$$J_d = -M\tilde{C}_{d1}B, \tag{12.51}$$

$$J_{dd} = -M\tilde{C}_{dd1}B, \tag{12.52}$$

$$J_{ddd} = 0. \tag{12.53}$$

Now, as specified in (12.49), we are required to estimate the second state variable of the state vector $\bar{x}_1(t) = \begin{bmatrix} \bar{x}_{11}(t) \\ \bar{x}_{12}(t) \end{bmatrix}$, i.e., $\bar{x}_{12}(t)$. Note that there are two outputs in this system, namely, $\tilde{y}_1(t)$ and $\tilde{y}_2(t)$, i.e., from (12.48), we have

$$\tilde{y}_1(t) = \bar{x}_{11}(t) + \bar{x}_{11}(t-2\tau) + \bar{x}_{12}(t-\tau) + \bar{x}_{12}(t-2\tau), \tag{12.54}$$

$$\tilde{y}_2(t) = \bar{x}_{12}(t) + \bar{x}_{12}(t-\tau) + \bar{x}_{12}(t-2\tau). \tag{12.55}$$

However, from the above equations, it is not possible to explicitly derive $\bar{x}_{11}(t)$ and $\bar{x}_{12}(t)$ in terms of the outputs $\tilde{y}_1(t)$ and $\tilde{y}_2(t)$, due to the presence of multiple delayed terms in (12.54) and (12.55). This implies that even though the number of outputs is the same as the state variables, it is still necessary to design a dynamical observer to estimate the unknown state variables (in this case, $\bar{x}_{12}(t)$). Note that this situation does not happen for systems where their outputs do not contain state-delayed terms. This example serves to highlight an intriguing case as well as the complexity in designing observers for systems with multiple time delays in both state and output vectors.

Another way to show that it is not possible to derive $\bar{x}_{11}(t)$ and $\bar{x}_{12}(t)$ in terms of the output variables $\tilde{y}_1(t)$ and $\tilde{y}_2(t)$ is to check whether matrix $C(s)$, where $C(s) = \begin{bmatrix} s_2+1 & s_1+s_2 \\ 0 & s_1+s_2+1 \end{bmatrix}$, is unimodular or not (see, [1] and the discussion presented in Sect. 10.5 where $C(d) = \tilde{C}_1 + \tilde{C}_{d1}(d_1) + \tilde{C}_{dd1}(d_2)$). Here, $C(s)$ is not unimodular since its determinant is $(s_2+1)(s_1+s_2+1)$, which is zero at $s_2 = -1$ and/or $s_1 + s_2 = -1$.

Let us thus proceed with the design of a functional observer to estimate $z(t) = \bar{x}_{12}(t)$ of system (12.47) and (12.48). Obviously, we first seek the simplest functional observer as presented in Remark 1, i.e., observer (12.39) and (12.40). However, condition (12.43) is not satisfied, which implies that a first-order observer (12.39) and (12.40) does not exist in this example.

Our next option is to consider observer (12.34) and (12.35) with an internal delay as presented in Remark 1. It is found that condition (12.37) is satisfied. However, the solution to Eq. (12.36) results in $N = -1$ and $N_d = 2$, and hence the following time-delay error system is unstable (see, [2])

$$\dot{e}(t) = -e(t) + 2e(t-\tau).$$

Thus even though condition (12.37) is satisfied, we cannot ensure asymptotic stability of (11.27).

Thus, we next consider observer (12.1) and (12.2) with two internal delays. For this, condition (12.24) is satisfied (this is as expected since we know that condition (12.24) is generally less conservative than condition (12.37). Given that (12.37) is satisfied, (12.24) is satisfied). Accordingly, from (12.29)–(12.31), we obtain the following matrices

$$\bar{N}_1 = \Upsilon_a \Theta_a^+ \mathscr{I}^1 = -1.75,$$

$$\bar{N}_{d1} = \Upsilon_a \Theta_a^+ \mathscr{I}^2 = 1.25,$$

$$\bar{N}_{dd1} = \Upsilon_a \Theta_a^+ \mathscr{I}^3 = -0.75,$$

$$\bar{N}_2 = \bar{N}_{d2} = \bar{N}_{dd2} = \begin{bmatrix} \eta \\ 0_{6\times 1} \end{bmatrix},$$

where $\eta = [0.25\ \ 0.25\ \ 0.25\ \ 0\ \ -0.25]^T$. Since $\bar{N}_2 = \bar{N}_{d2} = \bar{N}_{dd2}$, we can simply let $\bar{Z}_a$ as $\bar{Z}_a = \begin{bmatrix} \bar{z}_a\ 0_{1\times 10} \end{bmatrix}$, where $\bar{z}_a$ is a scalar, and thus the error system (12.32) is obtained as follows

$$\begin{aligned} \dot{e}(t) &= (-1.75 + 0.25\bar{z}_a)e(t) + (1.25 + 0.25\bar{z}_a)e(t-\tau) \\ &\quad + (-0.75 + 0.25\bar{z}_a)e(t-2\tau). \end{aligned} \tag{12.56}$$

From (12.56), we can verify that its nominal system, i.e., when $\bar{z}_a = 0$, is asymptotically stable by simulations. On the other hand, we can search for a $\bar{z}_a$ such that the error, $e(t)$, converges to zero at a faster rate (this is a stabilizability problem as discussed in the section preceding Remark 1). Since this is a one-dimensional search problem, we can select a suitable $\bar{z}_a$. For $\tau = 1s$, Fig. 12.1 shows the trajectories of $e(t)$ for some values of $\bar{z}_a$. From the figure, we select $\bar{z}_a = 0.5$ as it gives a better response in terms of smaller overshoots and a fast convergence rate. Thus, with $\bar{z}_a$ selected as 0.5, we have $\bar{Z}_a = \begin{bmatrix} 0.5\ 0_{1\times 10} \end{bmatrix}$ and from (12.25), we solve for the unknown matrices and obtain $N = -1.625$, $N_d = 1.375$, $N_{dd} = -0.625$, $\tilde{T} = \begin{bmatrix} -3 & -1.375 \end{bmatrix}$, $\bar{W} = \begin{bmatrix} -4 & 3 \end{bmatrix}$, $\bar{U} = \begin{bmatrix} -3 & 4 \end{bmatrix}$ and $M = \begin{bmatrix} -1 & 1 \end{bmatrix}$. With the obtained matrices, we can check and confirm that conditions (12.13)–(12.17) are satisfied. Then G, G_d and G_{dd} are obtained from (12.18)–(12.20), respectively, where $G = \begin{bmatrix} -1.375 & -3 \end{bmatrix}$, $G_d = \begin{bmatrix} -5.375 & 4.375 \end{bmatrix}$ and $G_{dd} = \begin{bmatrix} -2.375 & 3.375 \end{bmatrix}$. Finally, from (12.50)–(12.52), we obtain $J = 1$, $J_d = 0$ and $J_{dd} = 1$. Accordingly, we have completed the design of a first-order observer to estimate the unknown state variable $\bar{x}_{12}(t)$, where

$$\hat{\bar{x}}_{12}(t) = w(t) + \begin{bmatrix} -1 & 1 \end{bmatrix}\tilde{y}(t), \tag{12.57}$$

$$\begin{aligned} \dot{w}(t) &= -1.625w(t) + 1.375w(t-1) - 0.625w(t-2) \\ &\quad + \begin{bmatrix} -1.375 & -3 \end{bmatrix}\tilde{y}(t) + \begin{bmatrix} -5.375 & 4.375 \end{bmatrix}\tilde{y}(t-\tau) \\ &\quad + \begin{bmatrix} -2.375 & 3.375 \end{bmatrix}\tilde{y}(t-2\tau) + u(t) + u(t-2). \end{aligned} \tag{12.58}$$

Figure 12.2 shows the trajectories of $\bar{x}_{12}(t)$ and its estimate, $\hat{\bar{x}}_{12}(t)$, for a step input, $u(t)$, with magnitudes 10, -10 and 10 occurring in the time intervals [0, 13], [13, 30] and [30, 40], respectively. The initial conditions are chosen as $\bar{x}_{11}(\theta) = 0.5$, $\bar{x}_{12}(\theta) = -0.5$ and $w(\theta) = 1.5$, $\forall\theta \in [-2, 0]$. We can see from Fig. 12.2 that $\hat{\bar{x}}_{12}(t)$ converges to $\bar{x}_{12}(t)$ after a finite time, and that $\hat{\bar{x}}_{12}(t)$ tracks the true trajectory of $\bar{x}_{12}(t)$.

Remark 4 In Numerical Example 1, we showed that to estimate the required state variable, $\bar{x}_{12}(t)$, a first-order observer of the form (12.39) and (12.40) does not exist due to the fact that condition (12.43) is not satisfied. Similar to the case discussed in Chap. 10, we can also have an option to increase the order of the observer (i.e., observer (12.39) and (12.40)) by increasing the number of functions to be estimated

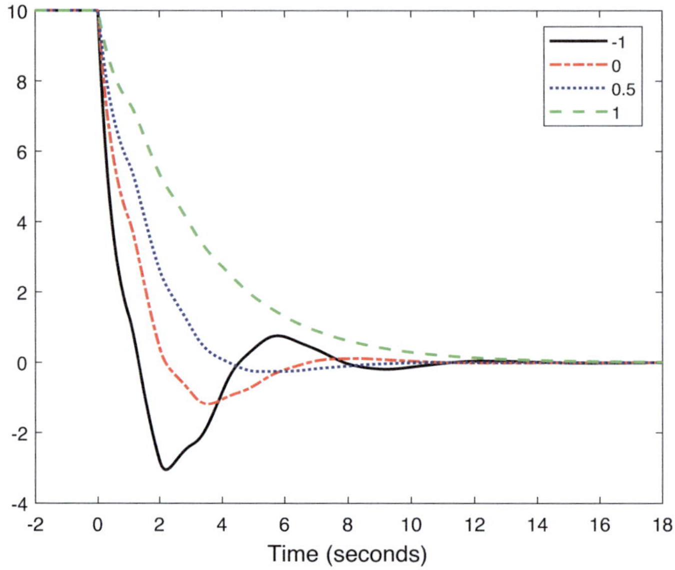

Fig. 12.1 Trajectories of $e(t)$ for $\bar{z}_a = -1, 0, 0.5, 1$ (Numerical Example 1, Chap. 12)

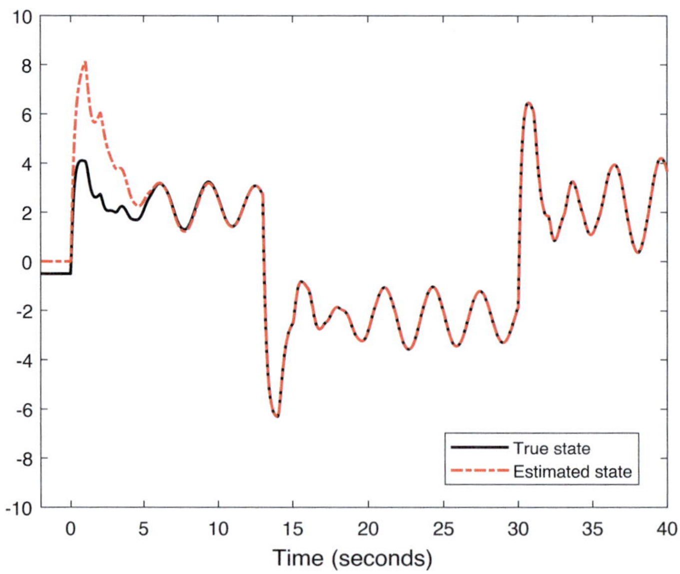

Fig. 12.2 Trajectories of $\bar{x}_{12}(t)$ and $\hat{\bar{x}}_{12}(t)$ (Numerical Example 1: First-order observer with internal delays, Chap. 12)

from one to two functions. For this, let us define $z_{ex}(t) = F_{ex}\bar{x}_1(t)$ as an additional function to be estimated, where $F_{ex} = \begin{bmatrix} 1 & 0 \end{bmatrix}$, and denote $z_a(t) \in \mathbb{R}^2$ as

$$z_a(t) = \begin{bmatrix} z(t) \\ z_{ex}(t) \end{bmatrix} = \begin{bmatrix} F \\ F_{ex} \end{bmatrix} \bar{x}_1(t) = F_a \bar{x}_1(t), \tag{12.59}$$

where $F_a = \begin{bmatrix} 0 & 1 \\ 1 & 0 \end{bmatrix}$. Thus, the two functions to be estimated are the two state variables, $\bar{x}_{11}(t)$ and $\bar{x}_{12}(t)$ of system (12.47) and (12.48). Now, by replacing F with F_a (where F_a is as given above) into condition (12.43), it is found that the condition is satisfied. Note that the solution to (12.41) results in a stable matrix N with fixed eigenvalues at $s = -1.75 \pm j0.6614$ (i.e., we have $\bar{N}_2 = 0_{2\times 2}$ and thus $\bar{Z}_a$ plays no role in the stabilization of N), and we obtain the following unknown matrix $\begin{bmatrix} N & \tilde{T} & G_d & G_{dd} & M \end{bmatrix}$, where $N = \begin{bmatrix} -3 & 1 \\ -2 & -0.5 \end{bmatrix}$, $\tilde{T} = \begin{bmatrix} 0 & -2 \\ -0.5 & -20 \end{bmatrix}$, $G_d = \begin{bmatrix} 0 & 0 \\ -2 & -6.5 \end{bmatrix}$, $G_{dd} = \begin{bmatrix} 0 & 0 \\ -1.5 & -6 \end{bmatrix}$ and $M = \begin{bmatrix} 0 & 0 \\ -0.5 & -3.5 \end{bmatrix}$. From (12.18) and (12.50)–(12.52), we obtain $G = \begin{bmatrix} -0.5 & -5.5 \\ -0.25 & -18.25 \end{bmatrix}$, $J = \begin{bmatrix} 2 \\ 8.5 \end{bmatrix}$, $J_d = \begin{bmatrix} 0 \\ 8 \end{bmatrix}$ and $J_{dd} = \begin{bmatrix} 0 \\ 8.5 \end{bmatrix}$. Thus far, we have completed the design of a second-order observer with no internal delay to estimate the two unknown state variables $\bar{x}_{12}(t)$ and $\bar{x}_{11}(t)$, where

$$\hat{z}_a(t) = \begin{bmatrix} \hat{\bar{x}}_{12}(t) \\ \hat{\bar{x}}_{11}(t) \end{bmatrix} = \begin{bmatrix} w_1(t) \\ w_2(t) \end{bmatrix} + \begin{bmatrix} 0 & 0 \\ -0.5 & -3.5 \end{bmatrix} \tilde{y}(t), \tag{12.60}$$

$$\begin{aligned} \dot{w}(t) &= Nw(t) + G\tilde{y}(t) + G_d\tilde{y}(t-\tau) + G_{dd}\tilde{y}(t-2\tau) \\ &\quad + Ju(t) + J_d u(t-\tau) + J_{dd}u(t-2\tau), \end{aligned} \tag{12.61}$$

where $w(t) = \begin{bmatrix} w_1(t) \\ w_2(t) \end{bmatrix}$, N, G, G_d, G_{dd}, J, J_d and J_{dd} are given above.

Comparing observer (12.60) and (12.61) with observer (12.57) and (12.58), we have the following observations. Observer (12.60) and (12.61) can estimate both unknown state variables $\bar{x}_{12}(t)$ and $\bar{x}_{11}(t)$. It is a second-order observer with no internal delay, and we expect that the error converges to zero at a faster convergence rate than the error system (12.56). On the other hand, (12.57) and (12.58) is a first-order observer but with the two added internal delays. Thus, in this section, we have provided some available options for the design of state/functional observers with various structures. Overall, the state/functional observer design problem for systems with multiple delays in the state and output vectors is an intriguing problem.

Figure 12.3 shows the trajectories of $\bar{x}_{12}(t)$ and its estimate, $\hat{\bar{x}}_{12}(t)$, for a step input, $u(t)$, with magnitudes 10, –10 and 10 in the time intervals [0, 13], [13, 30] and [30, 40], respectively. The initial conditions are chosen as $\bar{x}_{11}(\theta) = 0.5$, $\bar{x}_{12}(\theta) =$

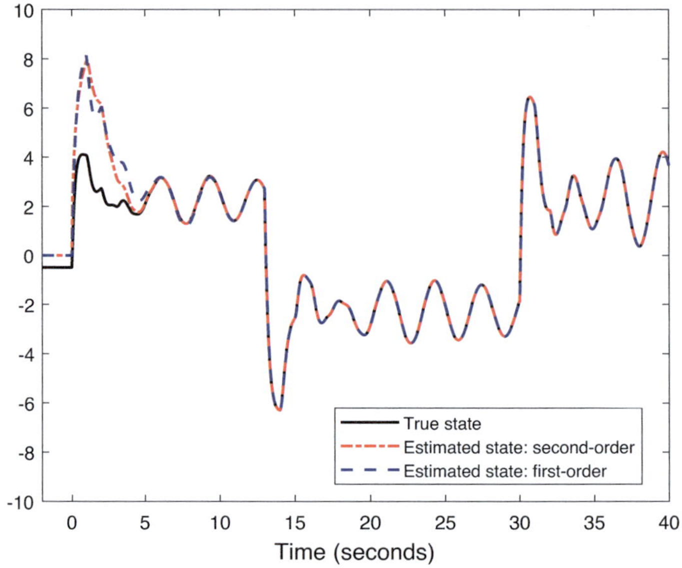

Fig. 12.3 Trajectories of $\bar{x}_{12}(t)$ and $\hat{\bar{x}}_{12}(t)$ (Numerical Example 1: Second-order observer with no internal delay, Chap. 12)

-0.5 and $w(\theta) = \begin{bmatrix} 1.5 \\ 0 \end{bmatrix}$, $\forall\theta \in [-2, 0]$. Superimposed on Fig. 12.3 is the trajectory of $\hat{\bar{x}}_{12}(t)$, which is obtained from observer (12.57) and (12.58). We can see from Fig. 12.3 that $\hat{\bar{x}}_{12}(t)$ converges asymptotically to $\bar{x}_{12}(t)$.

Finally, Fig. 12.4 shows the trajectories of $\bar{x}_{11}(t)$ and its estimate, $\hat{\bar{x}}_{11}(t)$ with the same input and intial conditions as stated above. Again, we can see from Fig. 12.4 that $\hat{\bar{x}}_{11}(t)$ converges asymptotically to $\bar{x}_{11}(t)$.

12.3 Functional Estimation Problem 2: $\bar{z}(t) = F\bar{x}_1(t) + F_d\bar{x}_1(t-\tau) + F_{dd}\bar{x}_1(t-2\tau)$

In Remark 4, we showed an interesting case where we increased the order of the observer (12.39) and (12.40) by adding an additional row, $F_{ex} = \begin{bmatrix} 1 & 0 \end{bmatrix}$, into F_a. In doing so, matrix Θ_{arr} becomes full rank and condition (12.43) is satisfied. As a result, we can estimate both $\bar{x}_{11}(t)$ and $\bar{x}_{12}(t)$ by a second-order observer without any internal delay. On the other hand, inspecting Θ_{arr} reveals that there is a limited number of additional rows that can be added to F, since any additional row must be linearly independent to F and the maximum number of additional rows is therefore $(n-m-1)$, i.e., rank $\begin{bmatrix} F \\ F_{ex} \end{bmatrix} = (n-1)$. This case has motivated us to look further

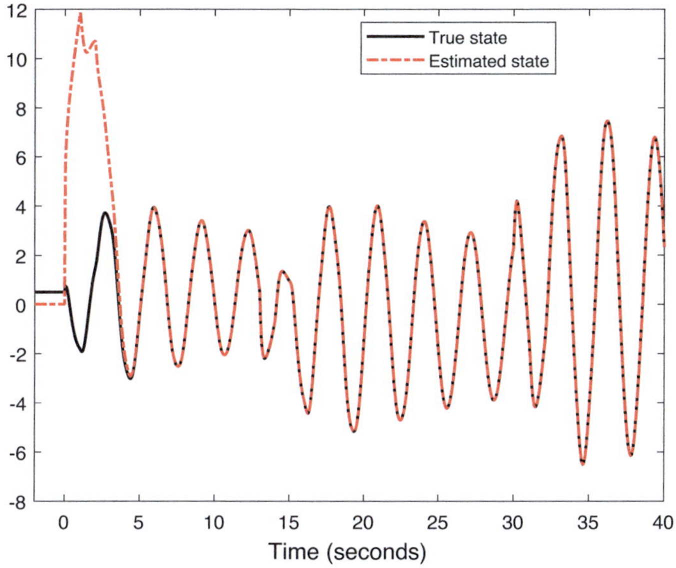

Fig. 12.4 Trajectories of $\bar{x}_{11}(t)$ and $\hat{\bar{x}}_{11}(t)$ (Numerical Example 1: Second-order observer with no internal delay, Chap. 12)

into the problem of estimating the following generalized functional of time-delay system (5.28) and (5.29)

$$\bar{z}(t) = F\bar{x}_1(t) + F_d\bar{x}_1(t-\tau) + F_{dd}\bar{x}_1(t-2\tau), \tag{12.62}$$

where F_d, $F_{dd} \in \mathbb{R}^{m\times(n-1)}$ are known matrices. The problem of estimating $\bar{z}(t)$ opens a very interesting situation where it is now possible to add more rows via F_d and F_{dd} to relax condition (12.43) (also, (12.24) and (12.37)). Here, we note that this estimation problem has not yet been reported in the literature. The rest of this chapter will discuss this new estimation problem and show its attractive features and generality compared to the problem of estimating $z(t) = F\bar{x}_1(t)$. Furthermore, the reader can refer to Chaps. 13–19, where we look at the advantages and applications of this new estimation problem to a wide range of time-delay systems.

12.3.1 Observer Existence Conditions

To estimate $\bar{z}(t)$ defined by (12.62), let us employ observer (12.1) and (12.2), i.e.,

$$\hat{\bar{z}}(t) = w(t) + M\tilde{y}(t),$$
$$\dot{w}(t) = Nw(t) + N_d w(t-\tau) + N_{dd} w(t-2\tau) + G\tilde{y}(t)$$

$$+ G_d\tilde{y}(t-\tau) + G_{dd}\tilde{y}(t-2\tau) + Jy(t) + J_d y(t-\tau) \\ + J_{dd}y(t-2\tau) + J_{ddd}y(t-3\tau).$$

In the following, we present their existence conditions by following a similar logic as in Sect. 12.2.1. Accordingly, from $e(t) = \hat{\bar{z}}(t) - \bar{z}(t)$, we have

$$e(t) = w(t) - M\tilde{y}(t) - F\bar{x}_1(t) - F_d\bar{x}_1(t-\tau) - F_{dd}\bar{x}_1(t-2\tau). \qquad (12.63)$$

Then, we obtain the following error dynamic equation

$$\begin{aligned}
\dot{e}(t) &= Ne(t) + N_d e(t-\tau) + N_{dd}e(t-2\tau) + \big(J - F\bar{A}_{12}V_1 + M\tilde{C}_1\bar{A}_{12}V_1\big)y(t) \\
&+ \big(J_d - F\bar{A}_{d12}V_1 + M\tilde{C}_{d1}\bar{A}_{12}V_1 + M\tilde{C}_1\bar{A}_{d12}V_1 - F_d\bar{A}_{12}V_1\big)y(t-\tau) \\
&+ \big(J_{dd} + M\tilde{C}_{d1}\bar{A}_{d12}V_1 + M\tilde{C}_{dd1}\bar{A}_{12}V_1 - F_d\bar{A}_{d12}V_1 - F_{dd}\bar{A}_{12}V_1\big)y(t-2\tau) \\
&+ \big(J_{ddd} + M\tilde{C}_{dd1}\bar{A}_{d12}V_1 - F_{dd}\bar{A}_{d12}V_1\big)y(t-3\tau) \\
&+ \big(NF - F\bar{A}_{11} + M\tilde{C}_1\bar{A}_{11} + \tilde{T}\tilde{C}_1\big)\bar{x}_1(t) \\
&+ \big(NF_d + N_d F + \bar{W}\tilde{C}_1 + \tilde{T}\tilde{C}_{d1} + M\bar{C}_1 - F\tilde{A}_{d11} - F_d\bar{A}_{11}\big)\bar{x}_1(t-\tau) \\
&+ \big(NF_{dd} + N_d F_d + N_{dd}F + \bar{U}\tilde{C}_1 + \bar{W}\tilde{C}_{d1} + \tilde{T}\tilde{C}_{dd1} + M\bar{C}_{1a} - F\tilde{A}_{dd11} \\
&- F_d\tilde{A}_{d11} - F_{dd}\bar{A}_{11}\big)\bar{x}_1(t-2\tau) \\
&+ \big(N_d F_{dd} + N_{dd}F_d + \bar{U}\tilde{C}_{d1} + \bar{W}\tilde{C}_{dd1} + M\hat{C}_1 - F_d\tilde{A}_{dd11} - F_{dd}\tilde{A}_{d11}\big)\bar{x}_1(t-3\tau) \\
&+ \big(N_{dd}F_{dd} + \bar{U}\tilde{C}_{dd1} + M\tilde{C}_{dd1}\tilde{A}_{dd11} - F_{dd}\tilde{A}_{dd11}\big)\bar{x}_1(t-4\tau). \qquad (12.64)
\end{aligned}$$

Thus, from (12.64), observer (12.1) and (12.2) exists such that $\hat{\bar{z}}(t)$ converges asymptotically to $\bar{z}(t)$ for any initial condition $e(\theta)$, $\forall\theta \in [0, 2\tau]$, if conditions (12.8), (12.9), (12.13) and the following conditions hold

$$J_d = \big(F\bar{A}_{d12} + F_d\bar{A}_{12} - M(\tilde{C}_1\bar{A}_{d12} + \tilde{C}_{d1}\bar{A}_{12})\big)V_1, \qquad (12.65)$$

$$J_{dd} = \big(F_d\bar{A}_{d12} + F_{dd}\bar{A}_{12} - M(\tilde{C}_{d1}\bar{A}_{d12} + \tilde{C}_{dd1}\bar{A}_{12})\big)V_1, \qquad (12.66)$$

$$J_{ddd} = F_{dd}\bar{A}_{d12}V_1 - M\tilde{C}_{dd1}\bar{A}_{d12}V_1, \qquad (12.67)$$

$$NF_d + N_d F + \tilde{T}\tilde{C}_{d1} + \bar{W}\tilde{C}_1 + M\bar{C}_1 = F\tilde{A}_{d11} + F_d\bar{A}_{11}, \qquad (12.68)$$

$$NF_{dd} + N_d F_d + N_{dd}F + \bar{U}\tilde{C}_1 + \bar{W}\tilde{C}_{d1} + \tilde{T}\tilde{C}_{dd1} + M\bar{C}_{1a} \\ = F\tilde{A}_{dd11} + F_d\tilde{A}_{d11} + F_{dd}\bar{A}_{11}, \qquad (12.69)$$

$$N_d F_{dd} + N_{dd}F_d + \bar{U}\tilde{C}_{d1} + \bar{W}\tilde{C}_{dd1} + M\hat{C}_1 = F_d\tilde{A}_{dd11} + F_{dd}\tilde{A}_{d11}, \qquad (12.70)$$

$$N_{dd}F_{dd} + \bar{U}\tilde{C}_{dd1} + M\tilde{C}_{dd1}\tilde{A}_{dd11} = F_{dd}\tilde{A}_{dd11}. \qquad (12.71)$$

The solution to Eqs. (12.13) and (12.68)–(12.71) for the seven unknown matrices, $[N \;\; N_d \;\; N_{dd} \;\; \tilde{T} \;\; \bar{W} \;\; \bar{U} \;\; M]$, is subject to the following necessary and sufficient condition

$$\text{rank}\begin{bmatrix} \tilde{\Theta}_a \\ \tilde{\Upsilon}_a \end{bmatrix} = \text{rank}\big[\tilde{\Theta}_a\big], \tag{12.72}$$

where

$$\tilde{\Theta}_a = \begin{bmatrix} F & F_d & F_{dd} & 0 & 0 \\ 0 & F & F_d & F_{dd} & 0 \\ 0 & 0 & F & F_d & F_{dd} \\ \tilde{C}_1 & \tilde{C}_{d1} & \tilde{C}_{dd1} & 0 & 0 \\ 0 & \tilde{C}_1 & \tilde{C}_{d1} & \tilde{C}_{dd1} & 0 \\ 0 & 0 & \tilde{C}_1 & \tilde{C}_{d1} & \tilde{C}_{dd1} \\ \tilde{C}_1\bar{A}_{11} & \bar{C}_1 & \bar{C}_{1a} & \hat{C}_1 & \tilde{C}_{dd1}\tilde{A}_{dd11} \end{bmatrix}, \tag{12.73}$$

$$\tilde{\Upsilon}_a = \big[\tilde{\Upsilon}_{a1}\ \tilde{\Upsilon}_{a2}\ \tilde{\Upsilon}_{a3}\ \tilde{\Upsilon}_{a4}\ \tilde{\Upsilon}_{a5}\big], \tag{12.74}$$

and

$$\begin{cases} \tilde{\Upsilon}_{a1} = F\bar{A}_{11}, \\ \tilde{\Upsilon}_{a2} = F\tilde{A}_{d11} + F_d\bar{A}_{11}, \\ \tilde{\Upsilon}_{a3} = F\tilde{A}_{dd11} + F_d\tilde{A}_{d11} + F_{dd}\bar{A}_{11}, \\ \tilde{\Upsilon}_{a4} = F_d\tilde{A}_{dd11} + F_{dd}\tilde{A}_{d11}, \\ \tilde{\Upsilon}_{a5} = F_{dd}\tilde{A}_{dd11}. \end{cases} \tag{12.75}$$

Thus, in the same way, we can show that to design observers (12.34) and (12.35), (12.39) and (12.40) which estimate the function (12.62), we need to solve, respectively, the following matrix equations

$$\big[N\ N_d\ \tilde{T}\ \bar{W}\ G_{dd}\ M\big]\tilde{\Theta}_{ar} = \tilde{\Upsilon}_a, \tag{12.76}$$

$$\big[N\ \tilde{T}\ G_d\ G_{dd}\ M\big]\tilde{\Theta}_{arr} = \tilde{\Upsilon}_a, \tag{12.77}$$

where

$$\tilde{\Theta}_{ar} = \begin{bmatrix} F & F_d & F_{dd} & 0 & 0 \\ 0 & F & F_d & F_{dd} & 0 \\ \tilde{C}_1 & \tilde{C}_{d1} & \tilde{C}_{dd1} & 0 & 0 \\ 0 & \tilde{C}_1 & \tilde{C}_{d1} & \tilde{C}_{dd1} & 0 \\ 0 & 0 & \tilde{C}_1 & \tilde{C}_{d1} & \tilde{C}_{dd1} \\ \tilde{C}_1\bar{A}_{11} & \bar{C}_1 & \bar{C}_{1a} & \hat{C}_1 & \tilde{C}_{dd1}\tilde{A}_{dd11} \end{bmatrix}, \tag{12.78}$$

and

$$\tilde{\Theta}_{arr} = \begin{bmatrix} F & F_d & F_{dd} & 0 & 0 \\ \tilde{C}_1 & \tilde{C}_{d1} & \tilde{C}_{dd1} & 0 & 0 \\ 0 & \tilde{C}_1 & \tilde{C}_{d1} & \tilde{C}_{dd1} & 0 \\ 0 & 0 & \tilde{C}_1 & \tilde{C}_{d1} & \tilde{C}_{dd1} \\ \tilde{C}_1\bar{A}_{11} & \tilde{C}_1 & \bar{C}_{1a} & \hat{C}_1 & \tilde{C}_{dd1}\tilde{A}_{dd11} \end{bmatrix}. \tag{12.79}$$

The solution to Eqs. (12.76) and (12.77) is subject to the following necessary and sufficent conditions, respectively,

$$\text{rank}\begin{bmatrix} \tilde{\Theta}_{ar} \\ \tilde{\Upsilon}_a \end{bmatrix} = \text{rank}\big[\tilde{\Theta}_{ar}\big], \tag{12.80}$$

$$\text{rank}\begin{bmatrix} \tilde{\Theta}_{arr} \\ \tilde{\Upsilon}_a \end{bmatrix} = \text{rank}\big[\tilde{\Theta}_{arr}\big]. \tag{12.81}$$

Remark 5 The significance of estimating $\bar{z}(t)$ (see, (12.62)) is that we can now expand the order of the designed observer beyond $(n-1)$ to meet condition (12.72) (or (12.80), (12.81)) as well as the required convergence rate of the error vector $e(t)$.

12.3.2 A Numerical Example

Numerical Example 2: To illustrate the significance of this new finding as remarked in Remark 5, let us reconsider Numerical Example 1. As discussed in Remark 4, we increased the order of the observer to $n = 2$ and estimated both state variables $\bar{x}_{11}(t)$ and $\bar{x}_{12}(t)$ by a second-order observer with no internal delays. Note that the designed second-order observer has two stable fixed eigenvalues at $s = -1.75 \pm j0.6614$ (i.e., $\bar{Z}_a$ plays no role in the stabilization of N). Thus the convergence rate of the error vector, $e(t)$, cannot be arbitrarily assigned.

Let us now change $\bar{A}_{11}$ in Numerical Example 1 to $\bar{A}_{11} = \begin{bmatrix} -3 & -3 \\ 1 & -1.4 \end{bmatrix}$. However, it is found that by increasing the order of the observer (12.39) and (12.40) to $n = 2$ (as discussed in Remark 4) does not work. This is due to the fact that the solution to (12.41) results in an unstable matrix N with fixed eigenvalues at $s = \mathbf{0.05} \pm j1.8702$ (i.e., here we have $\bar{N}_2 = 0_{2\times 2}$ and thus $\bar{Z}_a$ plays no role in the stabilization of N). Thus, a stable second-order observer (12.39) and (12.40) does not exist.

As discussed in Sect. 12.3, we now utilize F_d and F_{dd} by selecting one additional function, namely, $z_e(t)$, to be estimated, where

$$z_e(t) = \bar{x}_{11}(t-\tau) + \bar{x}_{12}(t-2\tau). \tag{12.82}$$

Now, the vector $\bar{z}(t)$ to be estimated is

$$\bar{z}(t) = \begin{bmatrix} z_a(t) \\ z_e(t) \end{bmatrix} = \begin{bmatrix} \bar{x}_{12}(t) \\ \bar{x}_{11}(t) \\ \bar{x}_{11}(t-\tau) + \bar{x}_{12}(t-2\tau) \end{bmatrix}, \tag{12.83}$$

and according to (12.62), we have $F = \begin{bmatrix} 0 & 1 \\ 1 & 0 \\ 0 & 0 \end{bmatrix}$, $F_d = \begin{bmatrix} 0 & 0 \\ 0 & 0 \\ 1 & 0 \end{bmatrix}$ and $F_{dd} = \begin{bmatrix} 0 & 0 \\ 0 & 0 \\ 0 & 1 \end{bmatrix}$.

With F, F_d and F_{dd} given above, condition (12.81) is satisfied. Most significantly, the solution to Eq. (12.77) results in matrix $\bar{N}_2 \neq 0$ and the matrix pair $(\bar{N}_1, \bar{N}_2)$ is observable. This means that we can arbitrarily assign the eigenvalues of matrix N to be at anywhere in the left-half of the complex s-plane. By selecting the eigenvalues of $N = (\bar{N}_1 + \bar{Z}_a\bar{N}_2)$ at, say, $\{-3, -4, -5\}$ and using the pole-placement method, we can easily obtain $\bar{Z}_a = [\bar{z}_a \;\; 0_{3\times 10}]$, where $\bar{z}_a = \begin{bmatrix} 103.7702 \\ -72.7457 \\ -587.2017 \end{bmatrix}$. With $\bar{Z}_a$ given above, the solution to Eq. (12.77) yields the following matrices

$$N = \begin{bmatrix} 12.9383 & 3.2033 & 4.4065 \\ -15.0214 & -2.5038 & -4.0076 \\ -70.6964 & -11.5172 & -22.4344 \end{bmatrix},\; G = \begin{bmatrix} -32.4682 & 68.3719 \\ 20.8337 & -47.0630 \\ 129.7740 & -311.7497 \end{bmatrix},$$

$$G_d = \begin{bmatrix} 8.8130 & -15.4228 \\ -10.0153 & 7.5267 \\ -61.0689 & 92.4205 \end{bmatrix},\; G_{dd} = \begin{bmatrix} 6.6098 & -17.6261 \\ -7.5115 & 10.0305 \\ -43.5517 & 100.9378 \end{bmatrix} \text{ and}$$

$$M = \begin{bmatrix} 2.2033 & -6.6098 \\ -2.5038 & 2.5115 \\ -14.5172 & 36.9517 \end{bmatrix}.$$

With $N_d = 0$, $N_{dd} = 0$ and the above obtained matrices, we can verify that conditions (12.13) and (12.68)–(12.71) are satisfied. Here, matrix N is stable with the eigenvalues at $\{-3, -4, -5\}$. Finally, from (12.50)–(12.52), we obtain $J = \begin{bmatrix} 13.0163 \\ -1.5191 \\ -59.3861 \end{bmatrix}$, $J_d = \begin{bmatrix} 8.8130 \\ -0.0153 \\ -43.8689 \end{bmatrix}$ and $J_{dd} = \begin{bmatrix} 6.6098 \\ 2.4885 \\ -28.3517 \end{bmatrix}$.

The significance of the result presented in this section is thus very clearly demonstrated in this Example. By a novel idea of estimating a generalized functional $\bar{z}(t)$ as defined in (12.62), we can now expand the order of the observer (12.39) and (12.40) thanks to the addition of F_d and F_{dd} in (12.62) to satisfy condition (12.81) and asymptotic stability of the error vector $e(t)$. Finally, Figs. 12.5 and 12.6 show the trajectories of $\bar{x}_{12}(t)$, $\hat{\bar{x}}_{12}(t)$, and $\bar{x}_{11}(t)$, $\hat{\bar{x}}_{11}(t)$, respectively.

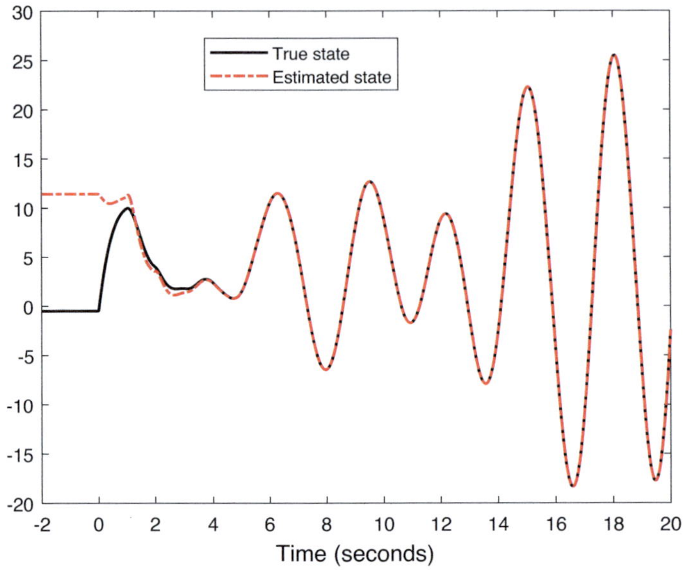

Fig. 12.5 Trajectories of $\bar{x}_{12}(t)$ and $\hat{\bar{x}}_{12}(t)$ (Numerical Example 2: Third-order observer with no internal delay, Chap. 12)

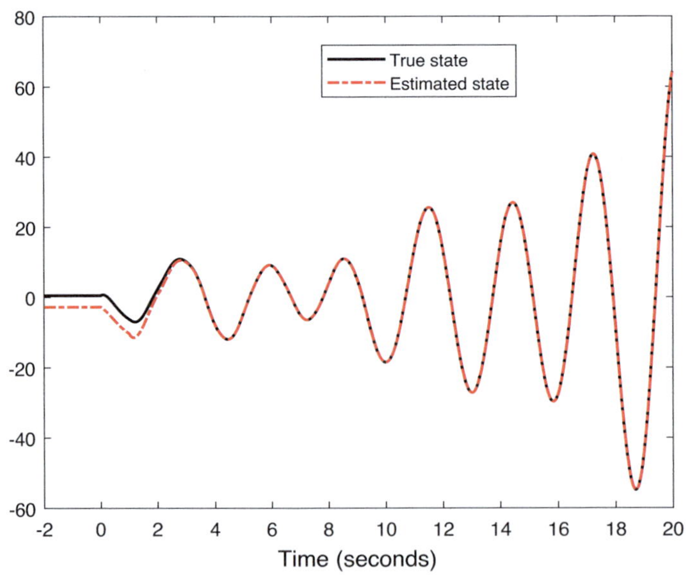

Fig. 12.6 Trajectories of $\bar{x}_{11}(t)$ and $\hat{\bar{x}}_{11}(t)$ (Numerical Example 2: Third-order observer with no internal delay, Chap. 12)

12.4 Concluding Remark

In this chapter, we have, for the first time, reported the problem of estimating a generalized functional observer for a class of time-delay systems with multiple delays in the state and output vectors. By a novel idea of estimating a generalized functional that has multiple state-delayed terms, we have relaxed their observer existence conditions. Via a numerical example, we have shown the significance of the new estimation problem, where we can expand the dimension of the designed observers beyond $(n-1)$. This finding has enabled us to design state observers for a wide class of time-delay systems. This will be further investigated in the subsequent chapters.

References

1. Hou, M., Zítek, P., Patton, R.J.: An observer design for linear time-delay systems. IEEE Trans. Automat. Control **47**(1), 121–125 (2002)
2. Mori, T.: Criteria for asymptotic stability of linear time-delay systems. IEEE Trans. Automat. Control **30**(2), 158–161 (1985)

Chapter 13
Instantaneous and Delayed Functional Estimation of Time-Delay Systems

13.1 Background and Motivations

In this chapter, inspired by the new finding presented in Sect. 12.3, Remark 5 and the numerical example therein, we relook at time-delay system (4.3) and (4.4), i.e.,

$$\begin{aligned}\dot{\bar{x}}_1(t) &= \bar{A}_{11}\bar{x}_1(t) + \bar{A}_{d11}\bar{x}_1(t-\tau) + \bar{A}_{12}V_1 y(t) + \bar{A}_{d12}V_1 y(t-\tau),\ t \geq 0,\\ \bar{y}(t) &= \tilde{C}_1\bar{x}_1(t),\end{aligned}$$

and we consider a new problem of estimating the following generalized functional

$$\bar{z}(t) = F\bar{x}_1(t) + F_d\bar{x}_1(t-\tau), \tag{13.1}$$

where $F, F_d \in \mathbb{R}^{m\times(n-1)}$ are known and constant matrices. The motivation for considering the above problem stemmed from Eq. (4.5). Here, if the following functional

$$\bar{z}(t) = -\bar{A}_{21}\bar{x}_1(t) - \bar{A}_{d21}\bar{x}_1(t-\tau) \tag{13.2}$$

can be estimated (instead of $\bar{x}_1(t)$ as in Design Option 1 of Chap. 4), then from (4.5) the unknown input $d(t)$ can be estimated, according to the following equation

$$\hat{d}(t) = V_1\dot{y}(t) - \bar{A}_{22}V_1 y(t) - \bar{A}_{d22}V_1 y(t-\tau) + \hat{\bar{z}}(t). \tag{13.3}$$

If such a functional observer can be designed (i.e., $\hat{\bar{z}}(t)$ converges asymptotically to $\bar{z}(t)$), then it can lead to a simpler unknown input observer. This is the reason for the inclusion of Scenario 4 in Chap. 4.

Furthermore, the functional estimation problem (13.1) enables the design of *delayed* functional/state observers in the absence of instantaneous functional/state observers. This situation arises, for example, when the non-minimum phase condi-

H. Trinh et al., *Unknown Inputs Estimation in Linear Time-Delay Systems Using Generalized Functional Observers*, Studies in Systems, Decision and Control 648,
https://doi.org/10.1007/978-3-032-11870-7_13

tion (10.47) is not satisfied, or the stability of the designed observers (10.48) and (10.49), (10.72) and (10.73) cannot be met for a range of time delay, τ, etc.

In the literature, the problem of designing delayed state observers for time-delay systems has been well reported (see, [2–4, 6]). In these studies, via some coordinate state transformations but with tedious algebraic manipulations, time-delay systems are transformed into new coordinates, where all the time-delay terms in the system are associated with the output and input vectors only.

13.2 Results: The Case Where $\tilde{C}_{d1} = \mathbf{0}_{(p-1)\times(n-1)}$

To estimate $\bar{z}(t)$ as defined in (13.1) for system (4.3) and (4.4), let us consider observer (10.72) and (10.73), i.e.,

$$\begin{aligned}\hat{\bar{z}}(t) &= w(t) + M\bar{y}(t) + M_d\bar{y}(t-\tau),\\ \dot{w}(t) &= Nw(t) + N_dw(t-\tau) + G\bar{y}(t) + G_d\bar{y}(t-\tau) + G_{dd}\bar{y}(t-2\tau)\\ &\quad + Jy(t) + J_dy(t-\tau) + J_{dd}y(t-2\tau).\end{aligned}$$

By following similar lines as in Sect. 10.3, we derive their existence conditions and show that they can be applied to a wider class of time-delay systems.

Consider the error vector $e(t) = \hat{\bar{z}}(t) - \bar{z}(t)$ and from (10.72) where $\hat{z}(t)$ is replaced with $\hat{\bar{z}}(t)$, we have

$$w(t) = e(t) + F\bar{x}_1(t) + F_d\bar{x}_1(t-\tau) - M\bar{y}(t) - M_d\bar{y}(t-\tau). \tag{13.4}$$

The derivative of $e(t)$, i.e., $\dot{e}(t) = \dot{w}(t) + M\dot{\bar{y}}(t) + M_d\dot{\bar{y}}(t-\tau) - F\dot{\bar{x}}_1(t) - F_d\dot{\bar{x}}_1(t-\tau)$ is obtained as follows

$$\begin{aligned}\dot{e}(t) &= Ne(t) + N_de(t-\tau) + \big(J - F\bar{A}_{12}V_1 + M\tilde{C}_1\bar{A}_{12}V_1\big)y(t)\\ &\quad + \big(J_d - F\bar{A}_{d12}V_1 - F_d\bar{A}_{12}V_1 + M\tilde{C}_1\bar{A}_{d12}V_1 + M_d\tilde{C}_1\bar{A}_{12}V_1\big)y(t-\tau)\\ &\quad + \big(J_{dd} + M_d\tilde{C}_1\bar{A}_{d12}V_1 - F_d\bar{A}_{d12}V_1\big)y(t-2\tau)\\ &\quad + \big(NF + \tilde{T}\tilde{C}_1 + M\tilde{C}_1\bar{A}_{11} - F\bar{A}_{11}\big)\bar{x}_1(t)\\ &\quad + \big(NF_d + N_dF + \tilde{W}\tilde{C}_1 + M\tilde{C}_1\bar{A}_{d11} + M_d\tilde{C}_1\bar{A}_{11} - F\bar{A}_{d11} - F_d\bar{A}_{11}\big)\bar{x}_1(t-\tau)\\ &\quad + \big(N_dF_d + \tilde{U}\tilde{C}_1 + M_d\tilde{C}_1\bar{A}_{d11} - F_d\bar{A}_{d11}\big)\bar{x}_1(t-2\tau),\end{aligned} \tag{13.5}$$

where $\tilde{T}$, $\tilde{W}$ and $\tilde{U}$ are defined in (10.84)–(10.86), respectively.

Without repeating, we skip a few steps, where conditions for ensuring $e(t)$ converges asymptotically to zeros are obtained (they can be readily obtained from (13.5) and thus are ommitted here). Subsequently, the seven unknown matrices $[N \;\; N_d \;\; \tilde{T} \;\; \tilde{W} \;\; \tilde{U} \;\; M \;\; M_d]$ from the following equation

$$\begin{bmatrix} N & N_d & \tilde{T} & \tilde{W} & \tilde{U} & M & M_d \end{bmatrix} \tilde{\Theta}_d = \tilde{\Upsilon}_d, \tag{13.6}$$

can be derived if and only if the following necessary and sufficient condition holds

$$\operatorname{rank}\begin{bmatrix} \tilde{\Theta}_d \\ \tilde{\Upsilon}_d \end{bmatrix} = \operatorname{rank}\begin{bmatrix} \tilde{\Theta}_d \end{bmatrix}, \tag{13.7}$$

where

$$\tilde{\Theta}_d = \begin{bmatrix} F & F_d & 0 \\ 0 & F & F_d \\ \tilde{C}_1 & 0 & 0 \\ 0 & \tilde{C}_1 & 0 \\ 0 & 0 & \tilde{C}_1 \\ \tilde{C}_1\bar{A}_{11} & \tilde{C}_1\bar{A}_{d11} & 0 \\ 0 & \tilde{C}_1\bar{A}_{11} & \tilde{C}_1\bar{A}_{d11} \end{bmatrix}, \tag{13.8}$$

and

$$\tilde{\Upsilon}_d = \begin{bmatrix} F\bar{A}_{11} & F\bar{A}_{d11} + F_d\bar{A}_{11} & F_d\bar{A}_{d11} \end{bmatrix}. \tag{13.9}$$

Once condition (13.7) is satisfied and its solution resulted in a stable error system (10.77), the rest of the observer parameters can be obtained.

Finally, for observer (10.72) and (10.73) with $N_d = 0$, equation (13.6) is reduced to

$$\begin{bmatrix} N & \tilde{T} & \tilde{W} & \tilde{U} & M & M_d \end{bmatrix} \tilde{\Theta}_{dr} = \tilde{\Upsilon}_d, \tag{13.10}$$

and condition (13.7) is reduced to the following condition

$$\operatorname{rank}\begin{bmatrix} \tilde{\Theta}_{dr} \\ \tilde{\Upsilon}_d \end{bmatrix} = \operatorname{rank}\begin{bmatrix} \tilde{\Theta}_{dr} \end{bmatrix}, \tag{13.11}$$

where

$$\tilde{\Theta}_{dr} = \begin{bmatrix} F & F_d & 0 \\ \tilde{C}_1 & 0 & 0 \\ 0 & \tilde{C}_1 & 0 \\ 0 & 0 & \tilde{C}_1 \\ \tilde{C}_1\bar{A}_{11} & \tilde{C}_1\bar{A}_{d11} & 0 \\ 0 & \tilde{C}_1\bar{A}_{11} & \tilde{C}_1\bar{A}_{d11} \end{bmatrix}. \tag{13.12}$$

13.2.1 Numerical Examples

To demonstrate the significance of estimating the generalized functional $\bar{z}(t)$ as defined by (13.1) for system (4.3) and (4.4), we consider the following examples.

Numerical Example 1: We consider time-delay system (4.3) and (4.4) but with an external input, $u(t)$, replacing $y(t)$ and $y(t-\tau)$, i.e.,

$$\dot{\bar{x}}_1(t) = \bar{A}_{11}\bar{x}_1(t) + \bar{A}_{d11}\bar{x}_1(t-\tau) + Bu(t), \tag{13.13}$$

$$\bar{y}(t) = \tilde{C}_1\bar{x}_1(t), \tag{13.14}$$

$$\bar{z}(t) = F\bar{x}_1(t) + F_d\bar{x}_1(t-\tau), \tag{13.15}$$

where $\bar{A}_{11} = \begin{bmatrix} 0 & 0 \\ 0 & 1 \end{bmatrix}$, $\bar{A}_{d11} = \begin{bmatrix} -1 & 2 \\ 0 & 0 \end{bmatrix}$, $\tilde{C}_1 = \begin{bmatrix} 1 & 0 \end{bmatrix}$, $B = \begin{bmatrix} 0 \\ 1 \end{bmatrix}$, F and F_d are to be given later in the subsequent discussion.

As is the case in Numerical Example 1 (Chap. 12), for simplicity and simulation purpose, we include the external input, $u(t)$, instead of both terms $\bar{A}_{12}V_1y(t)$ and $\bar{A}_{d12}V_1y(t-\tau)$. Thus, to ensure $e(t)$ converges to zeros for any arbitrary input, $u(t)$, we replace $Jy(t)$, $J_dy(t-\tau)$ and $J_{dd}y(t-2\tau)$ in (10.73) with $Ju(t)$, $J_du(t-\tau)$ and $J_{dd}u(t-2\tau)$, where

$$J = FB - M\tilde{C}_1B, \tag{13.16}$$

$$J_d = F_dB - M_d\tilde{C}_1B, \tag{13.17}$$

$$J_{dd} = 0. \tag{13.18}$$

Now, let say that we want to estimate the unmeasurable state variable $\bar{x}_{12}(t)$, i.e., $\bar{z}(t) = \begin{bmatrix} 0 & 1 \end{bmatrix}\bar{x}_1(t)$. For this, we would first check if observer (10.1) and (10.2) can be designed or not since it is the simplest observer. Here, condition (10.22) is satisfied. However, from (10.25) and (10.26), we have $N_1 = 1$ and $N_2 = 0$, and thus $N = 1$. Clearly, the matrix pair (N_1, N_2) is unobservable (it is also undetectable), as there is an unstable fixed pole at $s = 1 > 0$. This fixed pole is also called the unstable invariant zero of the system. This implies that observer (10.1) and (10.2) does not exist in this example.

Next, as for observer (10.48) and (10.49) (or (10.72) and (10.73)), we need to solve the stabilization problem of system (10.68) (or (10.112)), i.e., finding $\bar{Z}$ such that the following time-delay system is asymptotically stable for any given value of τ and any initial condition $e(\theta)$, $\forall\theta \in [-\tau, 0]$,

$$\dot{e}(t) = e(t) + 2\bar{Z}e(t-\tau).$$

For the stability of the above one-dimensional time-delay system, an exact stability condition [5] requires that $1 + 2\bar{Z} < 0$, with $2\bar{Z} \geq -\frac{1}{\tau}$. Its stability can only be satisfied for $0 \leq \tau \leq -\frac{1}{2\bar{Z}}$. Thus, for an arbitrary large τ, its stability condition is not met.

For the method based on the Output Injection approach [1] as discussed in Sect. 10.5, one has $A(d) = \begin{bmatrix} -d & 2d \\ 0 & 1 \end{bmatrix}$ and hence $Q_2(d) = \begin{bmatrix} C \\ CA(d) \end{bmatrix} = \begin{bmatrix} 1 & 0 \\ -d & 2d \end{bmatrix}$. For arbitrary $s \in \mathbb{C}$ and consider the matrix $Q_2(s) = \begin{bmatrix} 1 & 0 \\ -s & 2s \end{bmatrix}$ and thus $|Q_2(s)| = 2s$. Clearly, if $s = 0$ then $\text{rank}(Q_2(s)) = 1 \neq 2$. Hence, $Q_2(s)$ is not unimodular. Thus, the design method by [1] cannot be applied to design an observer for this example.

On the other hand, with the state transformation approach developed by [3, 4], where we introduce the following coordinate changes

$$\tilde{x}_{11}(t) = \bar{x}_{11}(t), \tag{13.19}$$
$$\tilde{x}_{12}(t) = -\bar{x}_{11}(t) + 2\bar{x}_{12}(t-\tau). \tag{13.20}$$

By taking the derivatives of (13.19) and (13.20), we obtain

$$\begin{aligned} \dot{\tilde{x}}_{11}(t) &= \dot{\bar{x}}_{11}(t) = -\bar{x}_{11}(t-\tau) + 2\bar{x}_{12}(t-\tau) \\ &= \tilde{x}_{12}(t) + \bar{y}(t) - \bar{y}(t-\tau), \end{aligned} \tag{13.21}$$
$$\begin{aligned} \dot{\tilde{x}}_{12}(t) &= -\dot{\bar{x}}_{11}(t) + 2\dot{\bar{x}}_{12}(t-\tau) \\ &= \bar{x}_{11}(t-\tau) - 2\bar{x}_{12}(t-\tau)) + 2(\bar{x}_{12}(t-\tau) + u(t-\tau)) \\ &= \bar{y}(t-\tau) + 2u(t-\tau), \end{aligned} \tag{13.22}$$

which can be expressed into the following $\tilde{x}_1$-coordinate form

$$\begin{aligned} \begin{bmatrix} \dot{\tilde{x}}_{11}(t) \\ \dot{\tilde{x}}_{12}(t) \end{bmatrix} &= \begin{bmatrix} 0 & 1 \\ 0 & 0 \end{bmatrix} \begin{bmatrix} \tilde{x}_{11}(t) \\ \tilde{x}_{12}(t) \end{bmatrix} + \begin{bmatrix} 0 \\ 2 \end{bmatrix} u(t-\tau) \\ &\quad + \begin{bmatrix} 1 \\ 0 \end{bmatrix} \bar{y}(t) + \begin{bmatrix} -1 \\ 1 \end{bmatrix} y(t-\tau), \end{aligned} \tag{13.23}$$
$$\bar{y}(t) = \tilde{x}_{11}(t). \tag{13.24}$$

With the coordinate changes as defined by (13.19) and (13.20), all the delay terms are now associated with the output and input of the transformed system (13.23) and (13.24). As a result, we can design a reduced-order state observer to estimate $\tilde{x}_{12}(t)$ in the $\tilde{x}_1$-coordinate system. Note that the structure of the designed observer also contains delayed output and input measurements, i.e., $\bar{y}(t-\tau)$ and $u(t-\tau)$. Then from (13.20), the following *delayed* observer for $\bar{x}_{12}(t-\tau)$ is obtained

$$\hat{\bar{x}}_{12}(t-\tau) = 0.5\hat{\tilde{x}}_{12}(t) + 0.5\bar{y}(t), \tag{13.25}$$

where $\hat{\bar{x}}_{12}(t-\tau)$ is an estimate of $\bar{x}_{12}(t-\tau)$. In this example, instantaneous estimation of the state variable $\bar{x}_{12}(t)$ cannot be achieved. In the context of estimating the unknown input $d(t)$ (see, (13.3)), we can see that the ability to estimate a delayed state vector or a delayed function of the state vector plays a crucial role.

It should be noted here that the state transformation approach ([2–4, 6]) involves very tedious algebraic manipulations, particularly for high-order systems with multiple outputs and delays. Via this example, we will show that the design of non-delayed and delayed functional/state observers for time-delay systems can be realized in a unified framework.

From (13.1), we let $F = \begin{bmatrix} 0 & 0 \end{bmatrix}$ and $F_d = \begin{bmatrix} 0 & 1 \end{bmatrix}$. Then, we have $\bar{z}(t) = \bar{x}_{12}(t-\tau)$, and therefore our objective here is to estimate $\bar{x}_{12}(t-\tau)$.

To proceed with observer design, we first check condition (13.11) as we aim for a simple observer with $N_d = 0$ (i.e., no internal delay). It is found that condition (13.11) is satisfied. Furthermore, the solution to Eq. (13.10) results in the matrix pair (N_1, N_2) being observable. By letting the eigenvalue of N to be at, for example, -5, the solution to (13.10) yields the following matrices: $N = -5$, $\tilde{T} = 0$, $\tilde{W} = 3$, $\tilde{U} = 0$, $M = 3$ and $M_d = 0$. From (10.84)–(10.86), we obtain $G = -15$, $G_d = 3$ and $G_{dd} = 0$. Finally, from (13.16) and (13.17), we obtain $J = 0$ and $J_d = 1$. Accordingly, we have systematically designed a first-order functional observer to estimate the function $\bar{z}(t) = \bar{x}_{12}(t-\tau)$, where

$$\hat{\bar{z}}(t) = w(t) + 3\bar{y}(t),$$
$$\dot{w}(t) = -5w(t) - 15\bar{y}(t) + 3\bar{y}(t-\tau) + u(t-\tau).$$

The above discussion clearly demonstrates the simplicity and effectiveness of our design approach based on the estimation of the function (13.1).

For $\tau = 1s$, Fig. 13.1 shows the trajectories of $\bar{x}_{12}(t)$, $\bar{x}_{12}(t-\tau)$ and $\hat{\bar{x}}_{12}(t-\tau)$.

Finally, Fig. 13.2 shows the trajectory of $e(t)$, with $e(\theta) = -120$, $\forall \theta \in [-1, 0]$. It is clear from both figures that the error converges asymptotically to zero.

Numerical Example 2: In this example, we consider a numerical example reported in [3], where matrices $\bar{A}_{11}$, $\bar{A}_{d11}$, $\tilde{C}_1$ and B are as given below

$$\bar{A}_{11} = \begin{bmatrix} -1 & 3 & 0 & 0 & 0 \\ 2 & -5 & 0 & 0 & 0 \\ -4 & 1 & 0 & 0 & -1 \\ -3 & 7 & 1 & 0 & 0 \\ -6 & 2 & 0 & -1 & 0 \end{bmatrix}, \bar{A}_{d11} = \begin{bmatrix} -7 & 0 & 0 & 0 & 0 \\ 2 & -3 & 0 & -3 & 0 \\ 3 & -2 & 0 & 4 & 0 \\ 4 & -4 & 0 & -9 & 0 \\ 5 & 1 & 0 & 3 & 0 \end{bmatrix}, B = \begin{bmatrix} 1 \\ 2 \\ 3 \\ 4 \\ 5 \end{bmatrix} \text{ and } \tilde{C}_1 = \begin{bmatrix} I_2 & 0_{2\times 3} \end{bmatrix}.$$

It was shown in [3] that it is not possible to design an observer to estimate the three unmeasurable state variables (namely, $\bar{x}_{13}(t)$, $\bar{x}_{14}(t)$ and $\bar{x}_{15}(t)$) by the approach in [1] or other observer design methods that employ an observer without internal delays. Instead, based on their proposed state transformation approach [3], it is still possible to design an observer to estimate the delayed state variables $\bar{x}_{13}(t-\tau)$, $\bar{x}_{14}(t-\tau)$ and $\bar{x}_{15}(t-\tau)$.

Now, thanks to our unified framework via the estimation of the new generalized functional $\bar{z}(t)$ as defined in (13.1), we can now readily design a delayed state observer for this example without the need to compute any complicated state transformations.

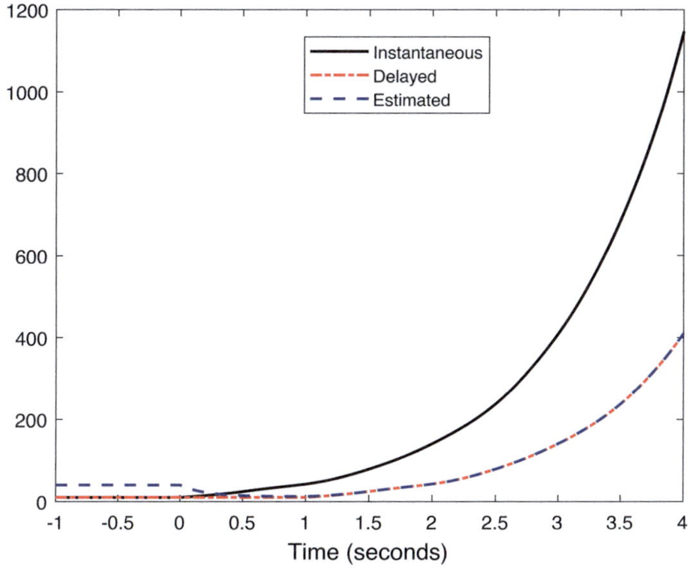

Fig. 13.1 Trajectories of $\bar{x}_{12}(t)$, $\bar{x}_{12}(t-\tau)$ and $\hat{\bar{x}}_{12}(t-\tau)$ (Numerical Example 1: First-order delayed observer, Chap. 13)

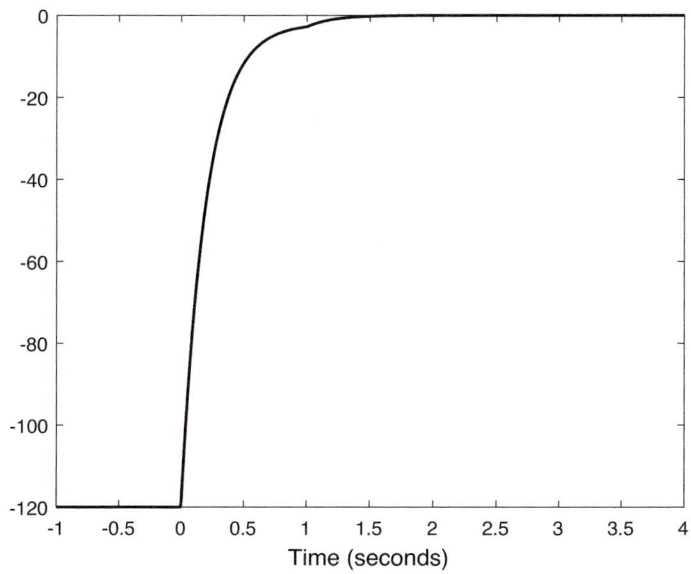

Fig. 13.2 Trajectory of $e(t) = \bar{x}_{12}(t-\tau) - \hat{\bar{x}}_{12}(t-\tau)$. (Numerical Example 1: First-order delayed observer, Chap. 13)

Firstly, let us provide a further analysis. We start with the case where we select F and F_d in (13.1) as $F = \begin{bmatrix} 0_{3\times2} & I_3 \end{bmatrix}$ and $F_d = 0_{3\times5}$. Thus, this is the case where we wish to estimate all three unmeasurable state variables $\bar{x}_{13}(t)$, $\bar{x}_{14}(t)$ and $\bar{x}_{15}(t)$ of the state vector $\bar{x}_1(t)$. By checking condition (13.11), it is found that it is satisfied, but the solution to Eq. (13.10) does not give a stable matrix N as there is a fixed eigenvalue at $s = 1$, which is an unstable pole. Accordingly, a third-order observer of the form (10.72) and (10.73) with $N_d = 0$ does not exist. Thus, this analysis reaffirms the finding reported in [3] that it is not possible to estimate $\bar{x}_{13}(t)$, $\bar{x}_{14}(t)$ and $\bar{x}_{15}(t)$ using observer (10.72) and (10.73) without internal delays.

Secondly, by letting F and F_d in (13.1) as $F = 0_{3\times5}$ and $F_d = \begin{bmatrix} 0_{3\times2} & I_3 \end{bmatrix}$, i.e., in this case, we wish to estimate the three delayed state variables $\bar{x}_{13}(t-\tau)$, $\bar{x}_{14}(t-\tau)$ and $\bar{x}_{15}(t-\tau)$. Note that this is exactly the problem considered in [3] via the use of a tedious state transformation approach. Instead, we can now readily design a delayed state observer for this example. For this, we simply check condition (13.11), and it is found that the condition is satisfied. Then the solution to Eq. (13.10) indeed gives an observable matrix pair (N_1, N_2). Therefore, a stable matrix N can always be found such that its eigenvalues can be placed at anywhere in the left-half complex s-plane. The rest of the observer parameters can then be easily obtained and thus we omit them here.

Numerical Example 3: This example is given to show the significance of the new state estimation problem of the function (13.1). We consider time-delay system (13.13) and (13.14), where matrices $\bar{A}_{11}$, $\bar{A}_{d11}$, $\tilde{C}_1$ and B are as given below

$$\bar{A}_{11} = \begin{bmatrix} -2 & 3 & 0 & 0 \\ 2 & 0 & 0 & -2 \\ -4 & 1 & 0 & -1 \\ -3 & 7 & 1 & 0 \end{bmatrix}, \bar{A}_{d11} = \begin{bmatrix} -7 & 0 & -10 & 0 \\ 2 & -3 & 0 & -3 \\ 3 & -2 & 0 & 4 \\ 4 & -4 & 0 & -9 \end{bmatrix}, B = \begin{bmatrix} 1 \\ 2 \\ 3 \\ 4 \end{bmatrix} \text{ and } \tilde{C}_1 = \begin{bmatrix} I_2 & 0_{2\times2} \end{bmatrix}.$$

For this example, the two outputs of the system are $\bar{x}_{11}(t)$ and $\bar{x}_{12}(t)$. Thus, there are two unmeasurable state variables, namely, $\bar{x}_{13}(t)$ and $\bar{x}_{14}(t)$. Let us now consider the problem of designing an observer, for instance, to estimate both the unmeasurable state variables, $\bar{x}_{13}(t)$ and $\bar{x}_{14}(t)$. According to (13.1), this is equivalent to the estimation of the vector $\bar{z}(t)$, where $F = \begin{bmatrix} 0_2 & I_2 \end{bmatrix}$ and $F_d = 0_{2\times4}$.

From a practical perspective, it is desirable to design a low-order observer with a simple structure. Thus, to estimate $\bar{z}(t)$, we first pick a second-order observer (10.72) and (10.73) with $N_d = 0_{2\times2}$. This is a practical choice since the order of the observer is the minimum, i.e., it is the same as the two state variables to be estimated. Moreover, this observer does not involve an internal delay, i.e., $N_d = 0_{2\times2}$, and thus its stability is independent of the size of the time delay. To proceed with the design of this observer, we first check condition (13.11). It is found that the condition is satisfied; however, the solution to equation (13.10) results in an unstable matrix N (here, $N_1 = \begin{bmatrix} 0 & -3.6667 \\ 1 & 6 \end{bmatrix}$ and $N_2 = 0$, i.e., the matrix pair (N_1, N_2) is not detectable nor observable). Hence our preferred observer does not exist.

Normally, we would look for a second-order observer (10.72) and (10.73) with an internal delay, $N_d w(t-\tau)$, and then solve the stabilization problem (see, (10.112)) to ensure asymptotic stability of the error time-delay system for a range of time delays.

Thanks to our unified observer design framework, we now have another viable option to design a practical observer for time-delay systems. In this option, we still consider an observer (10.72) and (10.73) without N_d, but we expand its dimension from two to three. This is possible by utilizing F_d in (13.1). Therefore, we select an additional function, $z_{ex}(t) = F_{ex}\bar{x}_1(t-\tau)$, to be estimated.

In this example, we pick the additional function to be estimated as

$$z_{ex}(t) = \bar{x}_{13}(t-\tau). \tag{13.26}$$

Thus, from (13.1), the vector $\bar{z}(t)$ to be estimated is now

$$\bar{z}(t) = \begin{bmatrix} z(t) \\ z_{ex}(t) \end{bmatrix} = \begin{bmatrix} \bar{x}_{13}(t) \\ \bar{x}_{14}(t) \\ \bar{x}_{13}(t-\tau) \end{bmatrix}, \tag{13.27}$$

and according to (13.1), we have $F = \begin{bmatrix} 0\,0\,1\,0 \\ 0\,0\,0\,1 \\ 0\,0\,0\,0 \end{bmatrix}$ and $F_d = \begin{bmatrix} 0\,0\,0\,0 \\ 0\,0\,0\,0 \\ 0\,0\,1\,0 \end{bmatrix}$. Note that, here we have rank $\begin{bmatrix} \tilde{C}_1 & 0 \\ F & F_d \end{bmatrix} = 5$, which implies that the vector $\bar{z}(t)$ is linearly independent from $\bar{y}(t)$.

With F and F_d given above, we proceed to design a third-order observer of the form (10.72) and (10.73) with $N_d = 0_{3\times 3}$. Condition (13.11) is satisfied and the solution to Eq. (13.10) also results in an observable matrix pair (N_1, N_2), and thus the eigenvalues of N can be assigned at anywhere in the left-half of the complex s-plane. By selecting the eigenvalues of N at, e.g., $\{-3, -4, -5\}$, the rest of the parameters of the observer can be obtained as follows

$$N = \begin{bmatrix} 0 & -3.6667 & 2.4545 \\ 1 & 6 & -61.9091 \\ 0 & 2.4444 & -18 \end{bmatrix}, N_d = 0_{3\times 3}, G = \begin{bmatrix} 17.4394 & -7.7364 \\ 53.1545 & -33.4273 \\ 11.2222 & -9.2667 \end{bmatrix}, G_d = \begin{bmatrix} 7.3848 & -9.2727 \\ -45.3364 & 87.5455 \\ -16.3778 & 28.6667 \end{bmatrix}, \quad G_{dd} = \begin{bmatrix} 0 & 0 \\ 0 & 0 \\ 5.6667 & -6 \end{bmatrix}, \quad M = \begin{bmatrix} 0.2455 & -1.3333 \\ -6.1909 & 3 \\ -1.8000 & 1.2222 \end{bmatrix},$$

$$M_d = \begin{bmatrix} 0 & 0 \\ 0 & 0 \\ 0 & -1.3333 \end{bmatrix}, J = \begin{bmatrix} 5.4212 \\ 4.1909 \\ -0.6444 \end{bmatrix}, J_d = \begin{bmatrix} 0 \\ 0 \\ 5.6667 \end{bmatrix} \text{ and } J_{dd} = 0_{3\times 1}.$$

Remark 1 We also found that condition (13.11) is satisfied and N is stable for other cases of F_{ex}, for example, $F_{ex} = [0\ 0\ 0\ 1]$. Through this numerical example, we can see the significance of the new functional estimation problem (13.1) for time-delay system (4.3) and (4.4). For the first time, by a novel idea of estimating an additional delayed function of the state vector, $z_{ex}(t) = F_{ex}\bar{x}_1(t-\tau)$, we can expand

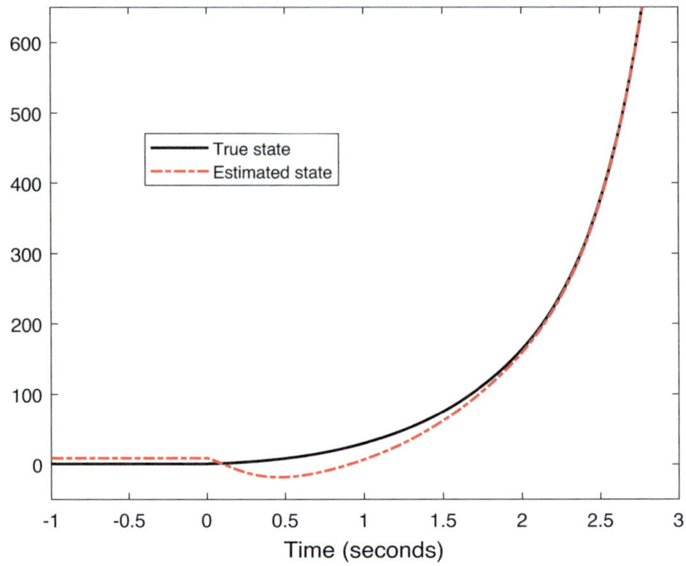

Fig. 13.3 Trajectories of $\bar{x}_{13}(t)$ and $\hat{\bar{x}}_{13}(t)$ (Numerical Example 3: Third-order observer without internal delay, Chap. 13)

the dimension of the observer (10.72) and (10.73) to achieve asymptotic estimation of the original required function, $z(t)$.

Here, we note that a key open research problem that needs to be addressed in the near future is how to systematically find a matrix $F_{ex} \in \mathbb{R}^{\nu\times(n-1)}$ with the minimum dimension, i.e., the smallest integer ν, to ensure the satisfaction of condition (13.11) (or (13.7)) and stability of N (or asymptotic stability of the error time-delay system $\dot{e}(t) = Ne(t) + N_d e(t-\tau)$).

Figures 13.3 and 13.4 show the trajectories of $\bar{x}_{13}(t)$ and $\bar{x}_{14}(t)$ and their estimates, $\hat{\bar{x}}_{13}(t)$ and $\hat{\bar{x}}_{14}(t)$.

Figure 13.5 depicts the trajectories of the errors $e_1(t) = (\bar{x}_{13}(t) - \hat{\bar{x}}_{13}(t))$ and $e_2(t) = (\bar{x}_{14}(t) - \hat{\bar{x}}_{14}(t))$. It is clear that the errors converge asymptotically to zero.

13.3 Results: The Case Where $\tilde{C}_{d1} \neq 0_{(p-1)\times(n-1)}$

For the remainder of this chapter, we consider time-delay system (5.13) and (5.14) with $\tilde{C}_{d1} \neq 0_{(p-1)\times(n-1)}$, i.e.,

$$\dot{\bar{x}}_1(t) = \bar{A}_{11}\bar{x}_1(t) + \bar{A}_{d11}\bar{x}_1(t-\tau) + \bar{A}_{12}V_1 y(t) + \bar{A}_{d12}V_1 y(t-\tau),\ t \geq 0,$$
$$\tilde{y}(t) = \tilde{C}_1\bar{x}_1(t) + \tilde{C}_{d1}\bar{x}_1(t-\tau),$$

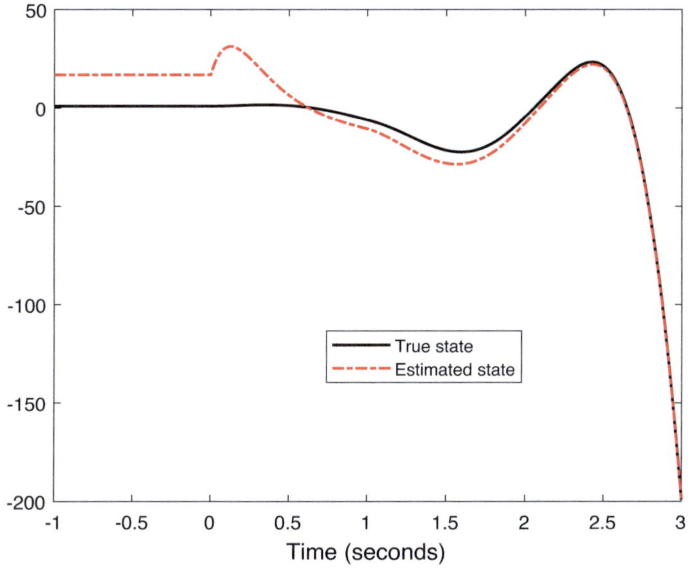

Fig. 13.4 Trajectories of $\bar{x}_{14}(t)$ and $\hat{\bar{x}}_{14}(t)$ (Numerical Example 3: Third-order observer without internal delay, Chap. 13)

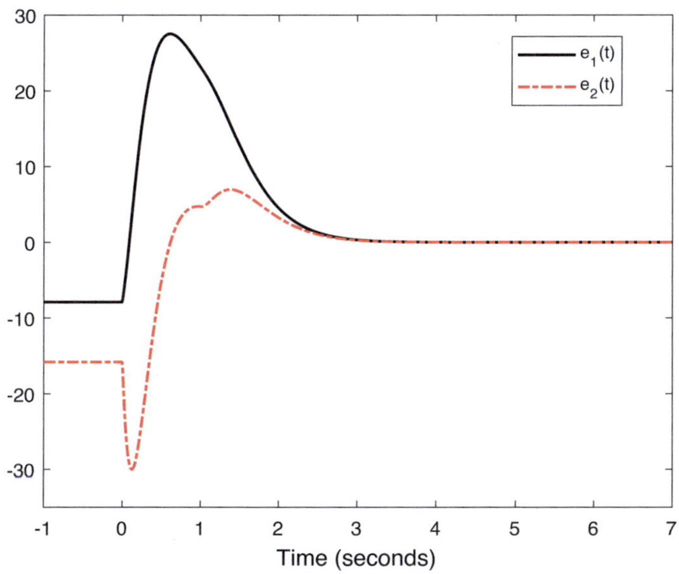

Fig. 13.5 Trajectories of $e_1(t)$ and $e_2(t)$ (Numerical Example 3: Third-order observer without internal delay, Chap. 13)

and we look at the problem of estimating the vector (13.1). As discussed in Sect. 5.2.1, decomposition (5.11) resulted in the disturbance-free time-delay system (5.13) and (5.14), and the unknown input is given by (4.5). Thus, if a functional observer can be designed to estimate $\bar{z}(t)$ directly, then the unknown input $d(t)$ can be estimated according to (13.3).

Now, to estimate $\bar{z}(t)$ defined in (13.1) for system (5.13) and (5.14), we consider a general observer of the form (11.1) and (11.2), where $\hat{z}(t)$ is replaced with $\hat{\bar{z}}(t)$ in (11.1), i.e.,

$$\begin{aligned}\hat{\bar{z}}(t) &= w(t) + M\tilde{y}(t) + M_d\tilde{y}(t-\tau),\\ \dot{w}(t) &= Nw(t) + N_dw(t-\tau) + G\tilde{y}(t) + G_d\tilde{y}(t-\tau) + G_{dd}\tilde{y}(t-2\tau)\\ &\quad + Jy(t) + J_dy(t-\tau) + J_{dd}y(t-2\tau) + J_{ddd}y(t-3\tau).\end{aligned}$$

In the following, we omit some detailed derivations and only present their existence conditions. Accordingly, a functional observer (11.1) and (11.2) for system (5.13) and (5.14) exists such that the error vector $e(t) = \hat{\bar{z}}(t) - \bar{z}(t)$ converges asymptotically to zeros if conditions (11.6), (11.7), (11.10), (11.11), (11.14) and the following conditions hold

$$J_d = F\bar{A}_{d12}V_1 + F_d\bar{A}_{12}V_1 - M\tilde{C}_1\bar{A}_{d12}V_1 - M\tilde{C}_{d1}\bar{A}_{12}V_1 - M_d\tilde{C}_1\bar{A}_{12}V_1, \tag{13.28}$$

$$J_{dd} = F_d\bar{A}_{d12}V_1 - M\tilde{C}_{d1}\bar{A}_{d12}V_1 - M_d\tilde{C}_1\bar{A}_{d12}V_1 - M_d\tilde{C}_{d1}\bar{A}_{12}V_1, \tag{13.29}$$

$$\begin{aligned}&NF_d + N_dF + \tilde{T}\tilde{C}_{d1} + \tilde{W}\tilde{C}_1 + M\big(\tilde{C}_1\bar{A}_{d11} + \tilde{C}_{d1}\bar{A}_{11}\big)\\ &\quad + M_d\tilde{C}_1\bar{A}_{11} = F\bar{A}_{d11} + F_d\bar{A}_{11},\end{aligned} \tag{13.30}$$

$$N_dF_d + \tilde{U}\tilde{C}_1 + \tilde{W}\tilde{C}_{d1} + M_d\big(\tilde{C}_1\bar{A}_{d11} + \tilde{C}_{d1}\bar{A}_{11}\big) + M\tilde{C}_{d1}\bar{A}_{d11} = F_d\bar{A}_{d11}, \tag{13.31}$$

where $\tilde{T}$, $\tilde{W}$ and $\tilde{U}$ are as defined in (10.84)–(10.86), respectively.

As for the four coupled matrix equations (11.11), (11.14), (13.30) and (13.31), there are seven unknown matrices, $[N \ N_d \ \tilde{T} \ \tilde{W} \ \tilde{U} \ M \ M_d]$, for which they can be expressed by the following equation

$$\big[N \ N_d \ \tilde{T} \ \tilde{W} \ \tilde{U} \ M \ M_d\big]\tilde{\Theta}_{dd} = \tilde{\Upsilon}_{dd}, \tag{13.32}$$

where

$$\tilde{\Theta}_{dd} = \begin{bmatrix} F & F_d & 0 & 0 \\ 0 & F & F_d & 0 \\ \tilde{C}_1 & \tilde{C}_{d1} & 0 & 0 \\ 0 & \tilde{C}_1 & \tilde{C}_{d1} & 0 \\ 0 & 0 & \tilde{C}_1 & \tilde{C}_{d1} \\ \tilde{C}_1\bar{A}_{11} & \tilde{C}_{1a} & \tilde{C}_{d1}\bar{A}_{d11} & 0 \\ 0 & \tilde{C}_1\bar{A}_{11} & \tilde{C}_{1a} & \tilde{C}_{d1}\bar{A}_{d11} \end{bmatrix}, \tag{13.33}$$

$$\tilde{\Upsilon}_{dd} = \begin{bmatrix} F\bar{A}_{11} & F\bar{A}_{d11} + F_d\bar{A}_{11} & F_d\bar{A}_{d11} & 0 \end{bmatrix}, \tag{13.34}$$

and $\tilde{C}_{1a} = \tilde{C}_1\bar{A}_{d11} + \tilde{C}_{d1}\bar{A}_{11}$.

A solution to (13.32) exists if and only if the following necessary and sufficient condition holds

$$\text{rank}\begin{bmatrix} \tilde{\Theta}_{dd} \\ \tilde{\Upsilon}_{dd} \end{bmatrix} = \text{rank}\big[\tilde{\Theta}_{dd}\big]. \tag{13.35}$$

Once condition (13.35) is satisfied and its solution resulted in a stable error system (11.6), the rest of the observer parameters can be obtained.

Finally, in the case where $N_d = 0$, i.e., observer (11.1) and (11.2) has no internal delay, and then there are only six unknown matrices in (13.32), and matrix $\tilde{\Theta}_{dd}$ is replaced with $\tilde{\Theta}_{ddr}$, where

$$\tilde{\Theta}_{ddr} = \begin{bmatrix} F & F_d & 0 & 0 \\ \tilde{C}_1 & \tilde{C}_{d1} & 0 & 0 \\ 0 & \tilde{C}_1 & \tilde{C}_{d1} & 0 \\ 0 & 0 & \tilde{C}_1 & \tilde{C}_{d1} \\ \tilde{C}_1\bar{A}_{11} & \tilde{C}_{1a} & \tilde{C}_{d1}\bar{A}_{d11} & 0 \\ 0 & \tilde{C}_1\bar{A}_{11} & \tilde{C}_{1a} & \tilde{C}_{d1}\bar{A}_{d11} \end{bmatrix}. \tag{13.36}$$

13.3.1 Numerical Examples

In the following, we provide some numerical examples to show the design of functional observers for time-delay systems (5.13) and (5.14), which will be used to design unknown input observers in Sect. 5.3.

Numerical Example 4: In this example, we consider time-delay system (5.13) and (5.14) with $\tau = 0.5s$, and matrices $\bar{A}_{11}$, $\bar{A}_{d11}$, $\bar{A}_{12}$, $\bar{A}_{d12}$, V_1, $\tilde{C}_1$ and $\tilde{C}_{d1}$ are as given in Numerical Example 6 in Section 4.3.10 and Numerical Example 1 in Sect. 5.3.1. Our objective is to design a functional observer to estimate the following required function of their state vector

$$\bar{z}(t) = F\bar{x}_1(t) + F_d\bar{x}_1(t-\tau), \tag{13.37}$$

where $F = \begin{bmatrix}0 & 0\end{bmatrix}$ and $F_d = \begin{bmatrix}0 & 1\end{bmatrix}$. Such a functional observer is required so that the unknown input, $d(t)$, of the system can be estimated based on (5.17).

To proceed, we first consider a first-order observer (11.1) and (11.2) with no internal delay, i.e., $N_d = 0$, as it is the simplest observer. For this, we first check condition (13.35) where $\tilde{\Theta}_{dd}$ is replaced with $\tilde{\Theta}_{ddr}$. It is found that condition (13.35) holds; however, the solution to Eq. (13.32) results in an unstable matrix with $N = 1$. In this case, $N_1 = 1$ and $N_2 = 0$, i.e., the matrix pair (N_1, N_2) is unobservable and also undetectable as there is an unstable fixed pole at $s = 1 > 0$. This implies that a first-order observer (11.1) and (11.2) with $N_d = 0$ does not exist for this example. In contrast, as reported in Numerical Example 1, for the case where $\tilde{C}_{d1} = 0_{1\times 2}$, it was possible to design a stable first-order functional observer (the eigenvalue of the observer can be arbitrarily assigned) with $N_d = 0$. In general, it is expected that the presence of $C_{d1} \neq 0_{1\times 2}$ in (5.14) causes further complexities into the observer design, which leads to more complex structured observers.

Next, we look at a first-order observer (11.1) and (11.2) with an internal delay, i.e., $N_d \neq 0$. In this case, condition (13.35) holds. Furthermore, the solution to (13.32) results in the following time-delay error system

$$\dot{e}(t) = (N_1 + ZN_2)e(t) + (N_{d1} + ZN_{d2})e(t-\tau), \tag{13.38}$$

where $N_1 = 0.55$, $N_{d1} = -0.45$ and $N_2 = N_{d2} = \begin{bmatrix}0.45 & 0.45 & 0 & -0.15 & 0 & -0.15 & 0\end{bmatrix}^T$. We can simply let $Z = [a \;\; 0_{1\times 6}]$ and find a scalar a such that the below error time-delay system is asymptotically stable for a given value of τ and any initial condition $e(\theta)$, $\forall\theta \in [-\tau, 0]$,

$$\dot{e}(t) = (0.55 + 0.45a)e(t) + (-0.45 + 0.45a)e(t-\tau). \tag{13.39}$$

Based on the exact stability condition as given by Mori [5], for $\tau = 0.5s$ and by letting $a = -3$, we have the following stable time-delay system, $\dot{e}(t) = -0.8e(t) - 1.8e(t-0.5)$. We can see that by employing an internal delay term in the observer, we can achieve asymptotic stability for $e(t)$. It should be noted here that the asymptotic stability of (13.39) is dependent on the size of the time-delay, τ.

Since we have found a scalar a to ensure asymptotic stability of $e(t)$, all parameters of observer (11.1) and (11.2) are now obtained as follows

$N = -0.8$, $N_d = -1.8$, $M = 0.6$, $M_d = 0$, $G = -0.48$, $G_d = -0.48$, $G_{dd} = 0$, $J = [0.6 \;\; 0]$, $J_d = [5 \;\; 0]$, $J_{dd} = [-0.4 \;\; 0]$ and $J_{ddd} = [0 \;\; 0]$.

The above designed observer to estimate the delayed state variable $\bar{x}_{12}(t-0.5)$ is then used to estimate the unknown input (5.17) for Numerical Example 1 in Chap. 5, where $\hat{d}(t)$ asymptotically tracks $d(t)$.

Numerical Example 5: In this example, we consider time-delay system (5.13) and (5.14) with $\tau = 1s$, matrices $\bar{A}_{11}$, $\bar{A}_{d11}$, $\bar{A}_{12}$, $\bar{A}_{d12}$, V_1, $\tilde{C}_1$ and $\tilde{C}_{d1}$ given as follows

$\bar{A}_{11} = \begin{bmatrix} -10 & -2 \\ 0 & -7 \end{bmatrix}$, $\bar{A}_{d11} = \begin{bmatrix} -3 & 2 \\ -3 & -6 \end{bmatrix}$, $\bar{A}_{12} = \begin{bmatrix} -27 \\ -22 \end{bmatrix}$, $\bar{A}_{d12} = \begin{bmatrix} -10 \\ -34 \end{bmatrix}$, $V_1 = \begin{bmatrix} 1 & 0 \end{bmatrix}$, $\tilde{C}_1 = \begin{bmatrix} 1 & 0 \end{bmatrix}$ and $\tilde{C}_{d1} = \begin{bmatrix} 1 & 1 \end{bmatrix}$.

Our objective is to design a functional observer to estimate the following function of their state vector

$$\bar{z}(t) = F\bar{x}_1(t) + F_d\bar{x}_1(t-\tau), \tag{13.40}$$

where $F = \begin{bmatrix} 0 & 1 \end{bmatrix}$ and $F_d = \begin{bmatrix} 1 & 1 \end{bmatrix}$. Note the function, $\bar{z}(t)$, that we aim to estimate is a combination of the instantaneous and delayed state variables $\bar{x}_{12}(t)$, $\bar{x}_{11}(t-\tau)$ and $\bar{x}_{12}(t-\tau)$, i.e., $\bar{z}(t) = \bar{x}_{11}(t-\tau) + \bar{x}_{12}(t) + \bar{x}_{12}(t-\tau)$.

In the following, we will explain and show the design of functional observers (11.1) and (11.2) for this example, by exploring different available options via successively expanding the order of the designed observer.

Firstly, we consider a first-order observer (11.1) and (11.2) with no internal delay, i.e., $N_d = 0$, as it is the simplest observer. For this, we check condition (13.35) where $\tilde{\Theta}_{dd}$ is replaced with $\tilde{\Theta}_{ddr}$. However, it is found that condition (13.35) does not hold since $\text{rank}[\tilde{\Theta}_{ddr}] = 6$, but $\text{rank}\begin{bmatrix} \tilde{\Theta}_{ddr} \\ \tilde{\Upsilon}_{dd} \end{bmatrix} = 7$. Thus, a first-order observer (11.1) and (11.2) with $N_d = 0$ does not exist for this example.

Secondly, we consider a first-order observer (11.1) and (11.2) with $N_d \neq 0$. Nevertheless, condition (13.35) still does not hold since $\text{rank}[\tilde{\Theta}_{dd}] = 7$, but $\text{rank}\begin{bmatrix} \tilde{\Theta}_{dd} \\ \tilde{\Upsilon}_{dd} \end{bmatrix} = 8$, and thus a first-order observer (11.1) and (11.2) with $N_d \neq 0$ does not exist for this example. As a remark, the left-hand side of (13.35) has reached its maximum value, i.e., 8, and thus any subsequent increase in the order of the observer will only increase, if any, the value of the right-hand side of (13.35). This opens a possibility of satisfying condition (13.35) by expanding the order of the observer, which will be discussed in the following. So far, checking the existence of a first-order functional observer is a simple task, as it involves little computational efforts. Both sides of (13.35) involve simple computations of the rank of known matrices.

Next, since a first-order functional observer with an internal delay does not exist, our next available option is to increase the order of the functional observer. Thus, we increase the order of the observer to $m = 2$ by increasing the number of functions to be estimated from one to two functions. For this, we denote $\bar{z}^1(t)$ as an additional function to be estimated, where

$$\bar{z}^1(t) = F^1\bar{x}_1(t) + F_d^1\bar{x}_1(t-\tau), \tag{13.41}$$

$F^1 = [f_1^1 \;\; f_2^1]$ and $F_d^1 = [f_{d1}^1 \;\; f_{d2}^1]$ are row vectors to be found, such that condition (13.35) is satisfied. Augmenting (13.40) with (13.41), we obtain the following

$$\begin{aligned} \bar{z}_a(t) = \begin{bmatrix} \bar{z}(t) \\ \bar{z}^1(t) \end{bmatrix} &= \begin{bmatrix} F \\ F^1 \end{bmatrix} \bar{x}_1(t) + \begin{bmatrix} F_d \\ F_d^1 \end{bmatrix} \bar{x}_1(t-\tau) \\ &= F_a\bar{x}_1(t) + F_{da}\bar{x}_1(t-\tau), \end{aligned} \tag{13.42}$$

where $F_a = \begin{bmatrix} 0 & 1 \\ f_1^1 & f_2^1 \end{bmatrix}$ and $F_{da} = \begin{bmatrix} 1 & 1 \\ f_{d1}^1 & f_{d2}^1 \end{bmatrix}$. Note that $[F_a \;\; F_{da}]$ must be a full-row rank matrix in order for $\bar{z}^1(t)$ to be linearly independent of $\bar{z}(t)$. In other words, for additional functions to be estimated, they must be linearly independent among themselves and of $\bar{z}(t)$ in order to provide additional degrees of freedom to enable condition (13.35) to be satisfied.

Let us now design a second-order functional observer (11.1) and (11.2) with $N_d = 0_{2\times 2}$ to estimate the function $\bar{z}_a(t)$ defined by (13.42). To achieve this objective, we need to find additional rows F^1 and F_d^1 such that condition (13.35) holds, where for $\tilde{\Theta}_{ddr}$ and $\tilde{\Upsilon}_{dd}$, F and F_d are now replaced with F_a and F_{da}, respectively. The aim is to find F^1 and F_d^1 so that the value of the right-hand side of (13.35) equals 7, i.e., $\text{rank}[\tilde{\Theta}_{ddr}] = 7$, while also ensuring that the value of the left-hand side of (13.35) equals 7 as well, i.e., $\text{rank}\begin{bmatrix} \tilde{\Theta}_{ddr} \\ \tilde{\Upsilon}_{dd} \end{bmatrix} = 7$.

In general, finding F^1 and F_d^1 to satisfy condition (13.35) is a difficult and computationally extensive problem, which is currently still an open research problem. In this example, however, it is found that F^1 and F_d^1 do not exist.

There are still two options available. Option (i): We consider a second-order functional observer (11.1) and (11.2) with $N_d \neq 0_{2\times 2}$; or Option (ii): We consider a third-order functional observer (11.1) and (11.2) with $N_d = 0_{3\times 3}$. Both options are viable as far as condition (13.35) is concerned for this example. In particular, in Option (i), $\tilde{\Theta}_{dd}$ is now a column matrix and it is possible to find F^1 and F_d^1 to satisfy condition (13.35), i.e., $\text{rank}[\tilde{\Theta}_{dd}] = 8$. The same logic applies to Option (ii). In the following, for completeness, we look at the design of two observers based on Options (i) and (ii).

Option (i): Second-order functional observer (11.1) and (11.2) with $N_d \neq 0_{2\times 2}$. By letting, say, $F^1 = [1 \;\; 0]$ and $F_d^1 = \begin{bmatrix} 1 & -1 \end{bmatrix}$, we have $F_a = \begin{bmatrix} 0 & 1 \\ 1 & 0 \end{bmatrix}$ and $F_{da} = \begin{bmatrix} 1 & 1 \\ 1 & -1 \end{bmatrix}$, which ensures matrix $[F_a \;\; F_{da}]$ has full-row rank.

Next, we check condition (13.35), where F and F_d are replaced with F_a and F_{da}, respectively. It is found that matrix $\tilde{\Theta}_{dd}$ is now a full-column rank matrix and condition (13.35) is satisfied. Solving (13.32) for N and N_d, and by letting $Z = 0$ (i.e., without solving the stabilization problem), we obtain the following error time-delay system

$$\dot{e}(t) = N_1 e(t) + N_{d1} e(t-\tau), \tag{13.43}$$

where $N_1 = \begin{bmatrix} -6.6296 & 4.9259 \\ -0.6914 & -5.7284 \end{bmatrix}$ and $N_{d1} = \begin{bmatrix} 0.9259 & -0.8148 \\ 4.2716 & -3.3457 \end{bmatrix}$.

With $\tau = 1s$ and by using the MATLAB package developed by Wu and Michiels [7], it is found that the error time-delay system (13.43) is asymptotically stable. Furthermore, there is a pair of dominant poles at $-1.1453 \pm j4.0742$, which ensures that the error vector $e(t)$ coverges to zeros within about four seconds. Figure 13.6 shows the trajectories of $e_1(t)$ and $e_2(t)$ with the initial condition

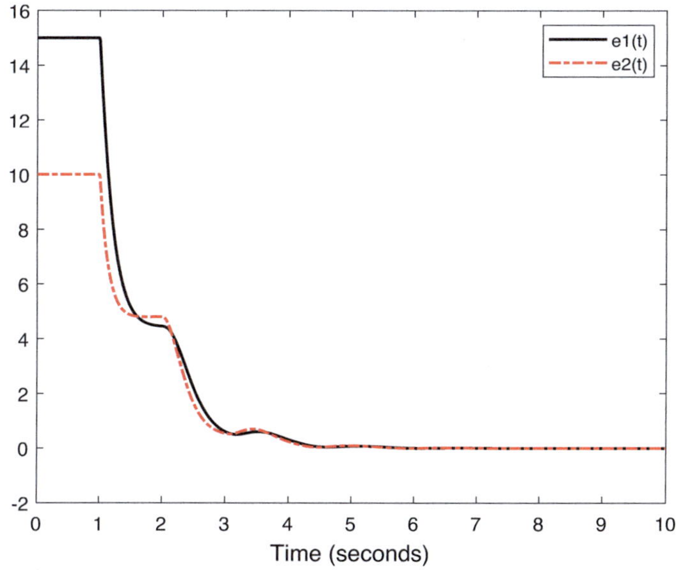

Fig. 13.6 Trajectories of $e_1(t)$ and $e_2(t)$ (Numerical Example 5, Chap. 13)

$e(\theta) = \begin{bmatrix} 15 \\ 10 \end{bmatrix}$, $\forall\theta \in [-1, 0]$. From the figure, $e(t)$ converges asymptotically to zero within about four seconds as expected.

Note that in this example, matrices $N_2 \neq 0$ and $N_{d2} \neq 0$, and thus we can solve the stabilization problem (see, Chap. 20) of finding a Z to make the error vector of the following time-delay system

$$\dot{e}(t) = (N_1 + ZN_2)e(t) + (N_{d1} + ZN_{d2})e(t - \tau), \tag{13.44}$$

converge to zero within a shorter time than for the case where $Z = 0$. Let us for now be content with the case $Z = 0$ and proceed to determine the rest of the observer parameters. With $Z = 0$, the solution to (13.32) gives the following matrices

$$N = N_1, N_d = N_{d1}, \tilde{T} = \begin{bmatrix} -3.0741 \\ 2.2716 \end{bmatrix}, \tilde{W} = \begin{bmatrix} -5 \\ 3 \end{bmatrix}, \tilde{U} = \begin{bmatrix} 0 \\ 0 \end{bmatrix}, M = \begin{bmatrix} 0.1852 \\ 0.6543 \end{bmatrix}$$

and $M_d = \begin{bmatrix} 0 \\ 0 \end{bmatrix}$. From (10.84)–(10.86), (11.7), (13.28), (13.29) and (11.10), the rest of the observer matrices, G, G_d, G_{dd}, J, J_d, J_{dd} and J_{ddd} are obtained. Accordingly, we obtain the following second-order observer for $\bar{z}_a(t)$

$$\hat{\bar{z}}_a(t) = \begin{bmatrix} \hat{\bar{z}}(t) \\ \hat{\bar{z}}^1(t) \end{bmatrix} = \begin{bmatrix} w_1(t) \\ w_2(t) \end{bmatrix} + \begin{bmatrix} 0.1852 \\ 0.6543 \end{bmatrix} \tilde{y}(t), \tag{13.45}$$

$$\begin{aligned} \begin{bmatrix} \dot{w}_1(t) \\ \dot{w}_2(t) \end{bmatrix} &= \begin{bmatrix} -6.6296 & 4.9259 \\ -0.6914 & -5.7284 \end{bmatrix} \begin{bmatrix} w_1(t) \\ w_2(t) \end{bmatrix} + \\ &\begin{bmatrix} 0.9259 & -0.8148 \\ 4.2716 & -3.3457 \end{bmatrix} \begin{bmatrix} w_1(t-\tau) \\ w_2(t-\tau) \end{bmatrix} + \begin{bmatrix} -1.0786 \\ -1.6046 \end{bmatrix} \tilde{y}(t) \\ &+ \begin{bmatrix} -5.3617 \\ 1.6019 \end{bmatrix} \tilde{y}(t-\tau) + \begin{bmatrix} -17 \\ -9.3333 \end{bmatrix} y_1(t) \\ &+ \begin{bmatrix} -72.0741 \\ 23.6049 \end{bmatrix} y_1(t-\tau) + \begin{bmatrix} -35.8519 \\ 52.7901 \end{bmatrix} y_1(t-2\tau). \end{aligned} \tag{13.46}$$

From (13.45) and (13.46), $\hat{\bar{z}}(t)$ is obtained, and

$$\hat{\bar{z}}(t) = w_1(t) + 0.1852\tilde{y}(t). \tag{13.47}$$

This observer is used in Numerical Example 2 (Chap. 5) to estimate the unknown input, $d(t)$, of a time-delay system described by (5.1) and (5.2) (please see Numerical Example 2 in Chap. 5).

Option (ii): Third-order functional observer (11.1) and (11.2) with $N_d = 0_{3\times 3}$. Thanks to our unified approach to solve the new functional estimation problem (13.1), it is now possible to design a higher-order functional observer to estimate the required function $\bar{z}(t)$ with no internal delay. To design a third-order observer for this example, we let $\bar{z}_{ex}(t)$ be an additional function to be estimated, where

$$\bar{z}_{ex}(t) = F_{ex}\bar{x}_1(t) + F_{dex}\bar{x}_1(t-\tau), \tag{13.48}$$

Augmenting (13.48) with (13.42), we obtain the following vector

$$\begin{bmatrix} \bar{z}_a(t) \\ \bar{z}_{ex}(t) \end{bmatrix} = \begin{bmatrix} F_a \\ F_{ex} \end{bmatrix} \bar{x}_1(t) + \begin{bmatrix} F_{da} \\ F_{dex} \end{bmatrix} \bar{x}_1(t-\tau), \tag{13.49}$$

where $\begin{bmatrix} F_a & F_{da} \\ F_{ex} & F_{dex} \end{bmatrix}$ is a full-row rank matrix.

In this example, we let $F^1 = [1 \;\; 0]$ and $F_d^1 = \begin{bmatrix} 1 & -1 \end{bmatrix}$ as in Option (i). Also, we choose $F_{ex} = [1 \;\; 1]$ and $F_{dex} = [0 \;\; 0]$. With these selections, the following matrix has full-row rank

$$\begin{bmatrix} F_a & F_{da} \\ F_{ex} & F_{dex} \end{bmatrix} = \begin{bmatrix} 0 & 1 & 1 & 1 \\ 1 & 0 & 1 & -1 \\ 1 & 1 & 0 & 0 \end{bmatrix}.$$

To proceed with the design of a third-order with $N_d = 0_{3\times 3}$, we check condition (13.35) where for $\tilde{\Theta}_{ddr}$ and $\tilde{\Upsilon}_{dd}$, F and F_d are now replaced with $\begin{bmatrix} F_a \\ F_{ex} \end{bmatrix}$ and $\begin{bmatrix} F_{da} \\ F_{dex} \end{bmatrix}$, respectively. It is worth mentioning that with the new replacements, $\tilde{\Theta}_{ddr}$ is now a square matrix and has full rank. Thus, condition (13.35) is satisfied. This justifies the reason for increasing the order of the designed observer and the requirement that $\begin{bmatrix} F_a & F_{da} \\ F_{ex} & F_{dex} \end{bmatrix}$ has full-row rank.

We then solve (13.32), where $N_d = 0_{3\times 3}$ and $\tilde{\Theta}_{dd}$ is replaced with $\tilde{\Theta}_{ddr}$. The solution to (13.32) gives a stable matrix N ($N_2 = 0$ and thus $N = N_1$), where $N = \begin{bmatrix} -7.5 & 4 & 2.5 \\ -4.5 & -10 & 10.5 \\ -2.5 & -1 & -6.5 \end{bmatrix}$, $\lambda(N) = \{-6.949 \pm j5.9003, -10.1020\}$. Note that the negative real parts indicate stable eigenvalues of N, and thus we expect that the error $e(t)$ converges to zeros (about one second) in a much shorter time than in the case of Option (i). The rest of the observer matrices are obtained as follows

$$M = \begin{bmatrix} 1 \\ 4 \\ 0 \end{bmatrix}, \quad M_d = \begin{bmatrix} 0 \\ 0 \\ 0 \end{bmatrix}, \quad G = \begin{bmatrix} 12 \\ -15 \\ -9 \end{bmatrix}, \quad G_d = \begin{bmatrix} 0 \\ 24 \\ 0 \end{bmatrix}, \quad G_{dd} = \begin{bmatrix} 0 \\ 0 \\ 0 \end{bmatrix},$$

$J = \begin{bmatrix} 5 & 0 \\ 81 & 0 \\ -49 & 0 \end{bmatrix}$, $J_d = \begin{bmatrix} -24 & 0 \\ 221 & 0 \\ -44 & 0 \end{bmatrix}$, $J_{dd} = \begin{bmatrix} 0 & 0 \\ 200 & 0 \\ 0 & 0 \end{bmatrix}$ and $J_{ddd} = 0_{3\times 2}$.

From the above design process, we can see the significance of our unified observer design approach for time-delay systems by solving the novel functional estimation problem (13.1). This observer is used in Numerical Example 2 (Chap. 5) to estimate the unknown input, $d(t)$, of a time-delay system described by (5.1) and (5.2).

13.4 Concluding Remark

In this chapter we have presented a comprehensive analysis on the problem of estimating the vector $\bar{z}(t) = F\bar{x}_1(t) + F_d\bar{x}_1(t - \tau)$ for both time-delay systems (4.3) and (4.4), (5.13) and (5.14). Thanks to this unified framework, it is now possible to design a wider range of observers than currently available in the literature. The results derived in this chapter are used for the estimation of some unknown inputs as discussed in Chaps. 4 and 5. Finally, we have shown that we can design delayed observers without using any complicated state transformations such as the method in [3].

In the next chapter, Chap. 14, as a byproduct of the results presented so far, we look at the problem of designing dynamic output feedback controllers for time-delay systems via the novel application of using $\hat{z}(t)$ to realize a general control law, $u(t) = Fx(t) + F_dx(t - \tau)$, for a time-delay system. The objective of the controller is to ensure that one of the selected outputs of the system asymptotically tracks any step (or ramp) reference command input, $r(t)$.

References

1. Hou, M., Zítek, P., Patton, R.J.: An observer design for linear time-delay systems. IEEE Trans. Automat. Control **47**(1), 121–125 (2002)
2. Huong, D.C., Huynh, V.T., Trinh, H.: Integral outputs-based robust state observers design for time-delay systems. SIAM J. Control Optim. **57**(3), 2214–2239 (2019)
3. Huong, D.C., Trinh, H.: Method for computing state transformations of time-delay systems. IET Control Theory Appl. **9**(16), 2405–2413 (2015)
4. Huong, D.C., Trinh, H.: New state transformations of time-delay systems with multiple delays and their applications to state observer design. J. Franklin Inst. **353**(14), 3487–3523 (2016)
5. Mori, T.: Criteria for asymptotic stability of linear time-delay systems. IEEE Trans. Automat. Control **30**(2), 158–161 (1985)
6. Trinh, H., Huong, D.C.: A new method for designing distributed reduced-order functional observers of interconnected time-delay systems. J. Franklin Inst. **355**(3), 1411–1451 (2018)
7. Wu, Z., Michiels, W.: Reliably computing all characteristic roots of delay differential equations in a given right half plane using a spectral method. J. Comput. Appl. Math. **236**(9), 2499–2514 (2012)

Chapter 14
Dynamic Output Feedback Controllers for Time-Delay Systems

14.1 Background, Motivations and Problem Statement

In this chapter, we apply the design of instantaneous and delayed functional observers (see, Chap. 13) to solve the problem of designing dynamic output feedback controllers for time-delay systems. In particular, we consider the problem of designing dynamic output feedback controllers such that one of the selected outputs of the system asymptotically tracks any step (or ramp) reference command input, $r(t)$.

Let us first consider the following time-delay system with a single control input $u(t)$

$$\dot{x}(t) = Ax(t) + A_d x(t-\tau) + Bu(t), \tag{14.1}$$

$$y(t) = Cx(t), \tag{14.2}$$

where $x(t) \in \mathbb{R}^n$, $y(t) \in \mathbb{R}^p$, $p \geq 1$ and $u(t) \in \mathbb{R}$ are, respectively, the state vector, output vector and control input. Matrices A, $A_d \in \mathbb{R}^{n\times n}$, $C \in \mathbb{R}^{p\times n}$, and $B \in \mathbb{R}^{n\times 1}$ is a column matrix. The positive constant τ is a known time delay and $x(\theta) = \phi(\theta)$, $\forall\theta \in [-\tau, 0]$, is a continuous initial function.

For the above time-delay system, let $y_i(t)$ be the output required to track asymptotically any constant step reference input, $r(t)$. To solve this problem, we propose a novel control scheme that contains an integral controller integrated with an instantaneous/delayed functional observer that we have developed earlier in Chap. 13. Let us first define

$$e(t) = r(t) - y_i(t) \tag{14.3}$$

as the error between the step reference input $r(t)$ and the output $y_i(t)$. We then introduce the following measurable state variable, $x_I(t)$, where

$$\dot{x}_I(t) = e(t) = r(t) - y_i(t) = r(t) - C_i x(t), \tag{14.4}$$

H. Trinh et al., *Unknown Inputs Estimation in Linear Time-Delay Systems Using Generalized Functional Observers*, Studies in Systems, Decision and Control 648,
https://doi.org/10.1007/978-3-032-11870-7_14

and $C_i \in \mathbb{R}^{1\times n}$ is the i-th row of C. Augmenting (14.4) with (14.1) and (14.2), we obtain the following augmented system

$$\dot{x}_a(t) = A_a x_a(t) + A_{ad} x_a(t-\tau) + B_a u(t) + Rr(t), \tag{14.5}$$

$$y(t) = C_a x_a(t), \tag{14.6}$$

where $x_a(t) = \begin{bmatrix} x(t) \\ x_I(t) \end{bmatrix}$, $A_a = \begin{bmatrix} A & 0_{n\times 1} \\ -C_i & 0 \end{bmatrix}$, $A_{ad} = \begin{bmatrix} A_d & 0_{n\times 1} \\ 0_{1\times n} & 0 \end{bmatrix}$, $B_a = \begin{bmatrix} B \\ 0 \end{bmatrix}$, $R = \begin{bmatrix} 0_{n\times 1} \\ 1 \end{bmatrix}$ and $C_a = \begin{bmatrix} C & 0 \end{bmatrix}$.

For time-delay system (14.5) and (14.6), we first design the following state feedback controller

$$u(t) = F_a x_a(t) + F_d x(t-\tau), \tag{14.7}$$

where $F_a \in \mathbb{R}^{1\times(n+1)}$ and $F_d \in \mathbb{R}^{1\times n}$ are feedback matrices, to be found such that the below closed-loop system is asymptotically stable

$$\dot{x}_{ac}(t) = (A_a + B_a F_a) x_{ac}(t) + (A_{ad} + B_a F_{ad}) x_{ac}(t-\tau), \tag{14.8}$$

where $F_{ad} = [F_d \;\; 0]$. Note that controller (14.7) contains a delayed state feedback term, i.e., $F_d x(t-\tau)$. In some cases, we can find F_d to make the closed-loop system (14.8) delay-free, i.e., $A_{ad} + B_a F_{ad} = 0$. This can be achieved if and only if the following condition is satisfied

$$\text{rank}\begin{bmatrix} A_d & B \end{bmatrix} = \text{rank}[B]. \tag{14.9}$$

When the above condition is not met, we find F_a and F_d by solving a stabilization problem involving some LMIs (see, Chap. 20). In general, $F_d x(t-\tau)$ adds some extra degrees of freedom and thus it offers a more relaxed controller design problem than the case where $F_d = 0$ (see, Numerical Example 1 in this chapter).

14.2 Main Results: $C_d = 0$, Step/Ramp Tracking Using $\hat{z}(t)$ Where $z(t) = Fx(t) + F_d x(t-\tau)$

Once F_a and F_d are found, we further partition F_a as $F_a = [F \;\; F_I]$. Then the control law $u(t)$ is expressed as follows

$$u(t) = u_p(t) + u_I(t), \tag{14.10}$$

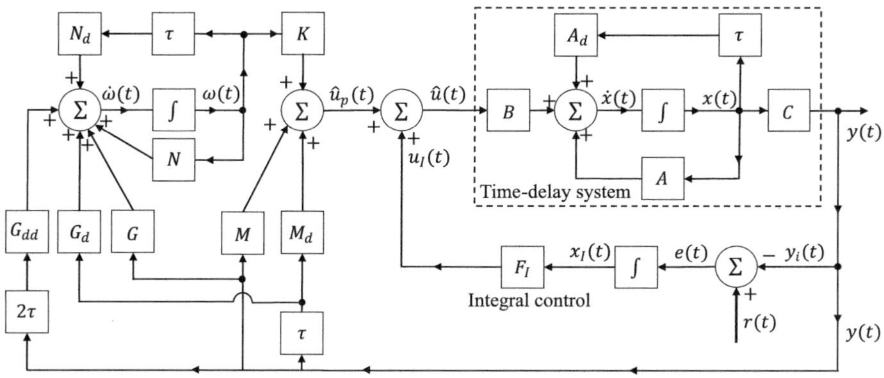

Fig. 14.1 Schematic diagram of a dynamic output feedback controller (14.15), Chap. 14

where

$$u_p(t) = Fx(t) + F_d x(t-\tau), \tag{14.11}$$
$$u_I(t) = F_I x_I(t). \tag{14.12}$$

From (14.12), it is clear that $u_I(t)$ can be realized, as it involves the integration of the error $e(t)$, which is constructed from the known reference step input $r(t)$ and the output $y_i(t)$ (see, Fig. 14.1). However, $u_p(t)$ cannot be directly realized, given that it is a state feedback controller that contains unmeasurable state variables and unmeasurable delayed state variables. Thanks to the new finding reported in Chap. 13, we now propose the following novel generalized observer to estimate $u_p(t)$

$$\hat{u}_p(t) = Kw(t) + My(t) + M_d y(t-\tau), \tag{14.13}$$
$$\dot{w}(t) = Nw(t) + N_d w(t-\tau) + Gy(t) + G_d y(t-\tau) + G_{dd} y(t-2\tau), \tag{14.14}$$

where $w(t) \in \mathbb{R}^m$, and $K, M, M_d, N, N_d, G, G_d$ and G_{dd} are matrices of appropriate dimensions to be found such that $\hat{u}_p(t)$ converges asymptotically to $u_p(t)$. The design of observer (14.13) and (14.14) will be discussed later in this chapter. Therefore, if such an observer (14.13) and (14.14) exists, the output tracking problem can be solved by using the following dynamic output feedback controller

$$\hat{u}(t) = \hat{u}_p(t) + u_I(t), \tag{14.15}$$

where $\hat{u}_p(t)$ is defined in (14.13) and (14.14). Figure 14.1 shows a schematic diagram of the implementation of the dynamic output feedback controller (14.15).

The scheme is novel as, for the first time, a generalized observer is designed to directly realize a state feedback controller that contains both delayed and non-delay unmeasurable state variables. Since observer (14.13) and (14.14) directly estimates

$u_p(t)$ (i.e., instead of estimating the unmeasurable state variables as is currently the case), the resulted controller scheme is simpler with fewer computation blocks (see, Numerical Example 1, Remark 1).

14.2.1 Existence Conditions

Next, we derive existence conditions for observer (14.13) and (14.14) and their parameters. Here, we note that (14.14) does not contain $u(t)$ as it is now used to realize the control signal $u_p(t)$ (see, Fig. 14.1). As such, there will be additional constraints imposed on their existence conditions. The following development is for the case where $m = 1$, i.e., first-order functional observers to estimate $u_p(t)$. In this case, we let $K = 1$. By following similar lines of reasoning as in Chap. 13, we can readily derive existence conditions for observer (14.13) and (14.14) with $m = 1$ and $K = 1$. Accordingly, let $e(t) = \hat{u}_p(t) - u_p(t)$ and from (14.13) we have

$$w(t) = e(t) + Fx(t) + F_d x(t-\tau) - My(t) - M_d y(t-\tau). \tag{14.16}$$

The derivative of $e(t)$, i.e., $\dot{e}(t) = \dot{w}(t) + M\dot{y}(t) + M_d\dot{y}(t-\tau) - F\dot{x}(t) - F_d\dot{x}(t-\tau)$ is obtained as follows

$$\begin{aligned}\dot{e}(t) &= Ne(t) + N_d e(t-\tau) + \big(MCB - FB\big)u(t) + \big(M_dCB - F_dB\big)u(t-\tau)\\ &\quad + \big(NF + \tilde{T}C + MCA - FA\big)x(t) + \big(NF_d + N_dF + \tilde{W}C + MCA_d + M_dCA\\ &\quad - FA_d - F_dA\big)x(t-\tau) + \big(N_dF_d + \tilde{U}C + M_dCA_d - F_dA_d\big)x(t-2\tau),\end{aligned} \tag{14.17}$$

where $\tilde{T}$, $\tilde{W}$ and $\tilde{U}$ are defined by (10.84)–(10.86), respectively.

From (14.17), conditions for ensuring that $e(t)$ converges asymptotically to zeros are as given below

$$\dot{e}(t) = Ne(t) + N_d e(t-\tau) \text{ is stable,} \tag{14.18}$$

$$MCB = FB, \tag{14.19}$$

$$M_dCB = F_dB, \tag{14.20}$$

$$NF + \tilde{T}C + MCA = FA, \tag{14.21}$$

$$NF_d + N_dF + \tilde{W}C + MCA_d + M_dCA = FA_d + F_dA, \tag{14.22}$$

$$N_dF_d + \tilde{U}C + M_dCA_d = F_dA_d. \tag{14.23}$$

Note that due to the structural constraint of (14.14), i.e., without $u(t)$, their existence conditions include two additional conditions (14.19) and (14.20). Now, there are seven unknown matrices to be solved for in (14.19)–(14.23). They can be expressed in a compact form as shown below

$$\begin{bmatrix} N & N_d & \tilde{T} & \tilde{W} & \tilde{U} & M & M_d \end{bmatrix} \bar{\Theta} = \bar{\Upsilon}, \tag{14.24}$$

where

$$\bar{\Theta} = \begin{bmatrix} F & F_d & 0 & 0 & 0 \\ 0 & F & F_d & 0 & 0 \\ C & 0 & 0 & 0 & 0 \\ 0 & C & 0 & 0 & 0 \\ 0 & 0 & C & 0 & 0 \\ CA & CA_d & 0 & CB & 0 \\ 0 & CA & CA_d & 0 & CB \end{bmatrix}, \tag{14.25}$$

and

$$\bar{\Upsilon} = \begin{bmatrix} FA & FA_d + F_dA & F_dA_d & FB & F_dB \end{bmatrix}. \tag{14.26}$$

The unknown matrices can be obtained if and only if the following necessary and sufficient condition holds

$$\text{rank} \begin{bmatrix} \bar{\Theta} \\ \bar{\Upsilon} \end{bmatrix} = \text{rank} \begin{bmatrix} \bar{\Theta} \end{bmatrix}. \tag{14.27}$$

Then, the rest of the observer parameters can be obtained once condition (14.27) is satisfied and its solution leads to a stable error system (14.18).

14.2.2 Numerical Examples

Numerical Example 1: In this example, we design a dynamic output feedback controller to achieve set-point tracking for the following single input, single output (SISO) time-delay system

$$\dot{x}(t) = \begin{bmatrix} 1 & 1 \\ 1 & -1 \end{bmatrix} x(t) + \begin{bmatrix} 30 & 20 \\ 30 & 20 \end{bmatrix} x(t-\tau) + \begin{bmatrix} 1 \\ 1 \end{bmatrix} u(t),$$
$$y(t) = [1 \;\; 0] x(t).$$

For this example, our objective is to design a controller to make the output $y(t)$ asymptotically track any step change in the reference input $r(t)$. We note here that the system is an unstable time-delay system. Also, without F_d, it is not possible to find F_a to achieve asymptotic stability of (14.8) for any given large time delay, τ. On the other hand, with $F_d \neq 0$, we can now readily design a stabilizing state feedback controller (14.7) for the augmented system (14.5). The controller gains F_a and F_d are selected as below

$$F_a = [-9.5 \;\; -2.5 \;\; 30], \;\; F_d = [-30 \;\; -20].$$

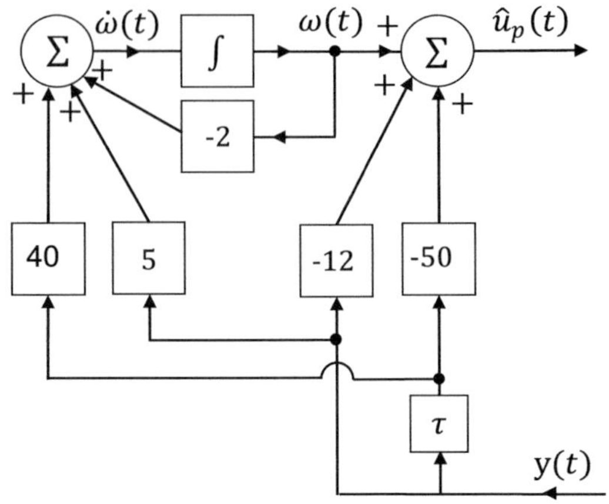

Fig. 14.2 Schematic diagram of the implementation of $\hat{u}_p(t)$ as described by (14.30) and (14.31) (Numerical Example 1, Chap. 14)

With the above controller gains, we have $(A_{ad} + B_a F_{ad}) = 0_{3\times 3}$ and $(A_a + B_a F_a) = \begin{bmatrix} -8.5 & -1.5 & 30 \\ -8.5 & -3.5 & 30 \\ -1 & 0 & 0 \end{bmatrix}$. Hence, the closed-loop system (14.8) is stable (independent of the time delay τ) with eigenvalues at $\{-3, -4, -5\}$. From (14.11) and (14.12), we have

$$u_p(t) = [-9.5 \quad -2.5]x(t) + [-30 \quad -20]x(t-\tau), \tag{14.28}$$

$$u_I(t) = 30x_I(t). \tag{14.29}$$

From (14.28), $u_p(t)$ contains two terms, namely, $-2.5x_2(t)$ and $-20x_2(t-\tau)$, which are associated with the unmeasurable state variables $x_2(t)$ and $x_2(t-\tau)$. We therefore design an observer (14.13) and (14.14) to estimate $u_p(t)$. It is found that a first-order observer exists and is given as below

$$\hat{u}_p(t) = w(t) - 12y(t) - 50y(t-\tau), \tag{14.30}$$

$$\dot{w}(t) = -2w(t) + 5y(t) + 40y(t-\tau). \tag{14.31}$$

Figure 14.2 shows a schematic diagram of the implementation of $\hat{u}_p(t)$. From the Figure, to realize $\hat{u}_p(t)$, two summers, one integrator, one delay unit and five proportional gains are required.

Figure 14.3 shows the response of the output $y(t)$ to step changes in $r(t)$. The time-delay is selected as $\tau = 5s$. The output asymptotically tracks the step changes in $r(t)$ as required.

To show how well controller scheme (14.15) performs in comparison with controller (14.7), Fig. 14.4 shows the trajectories of $y(t)$ from using both controllers $\hat{u}(t)$ and $u(t)$ to step changes in $r(t)$. Apart from some differences due to initial errors in the estimation, both trajectories are then the same, and thus highlighting the effectiveness of controller scheme (14.15).

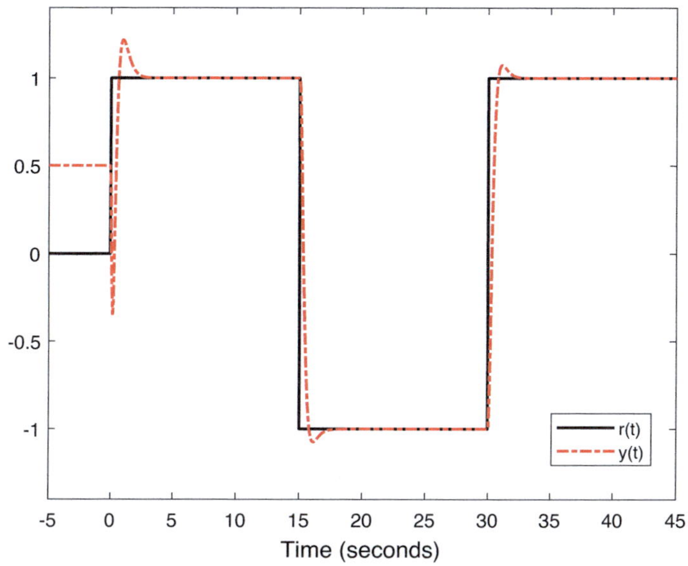

Fig. 14.3 Trajectories of $y(t)$ and $r(t)$ (Numerical Example 1, Chap. 14)

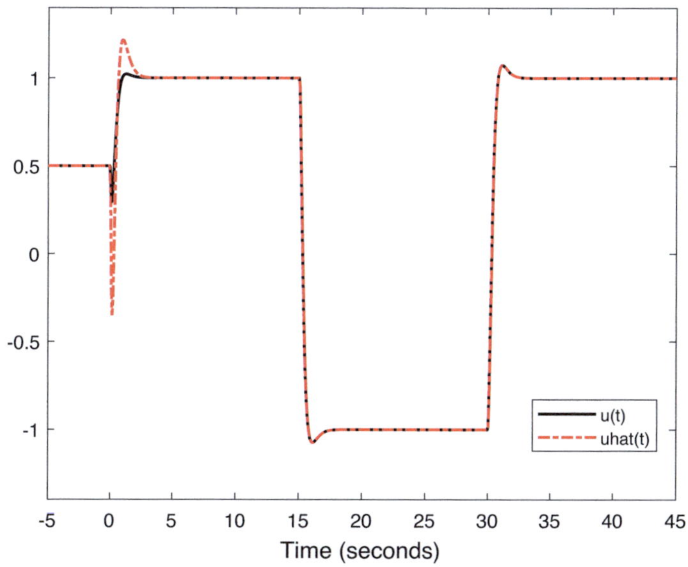

Fig. 14.4 Trajectories of $y(t)$ using $\hat{u}(t)$ and $u(t)$ (Numerical Example 1, Chap. 14).

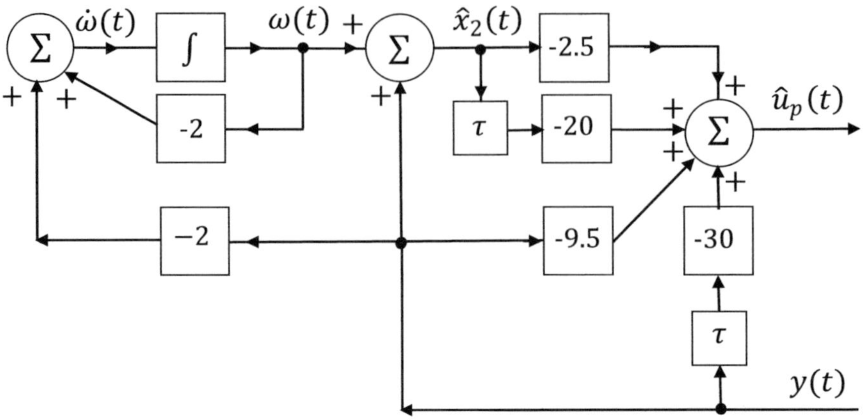

Fig. 14.5 Schematic diagram of the implementation of $\hat{u}_p(t)$ described by (14.32) (Numerical Example 1, Chap. 14)

Remark 1 In general, controller scheme (14.15) where $u_p(t)$ is directly estimated by the functional observer (14.13) and (14.14) resulted in fewer hardware components and hence a simpler controller scheme than by estimating the unmeasurable state variables, as is currently the case.

For this example, we have $y(t) = x_1(t)$ and thus from (14.28), $u_p(t)$ is the same as

$$u_p(t) = -9.5y(t) - 30y(t-\tau) - 2.5x_2(t) - 20x_2(t-\tau).$$

As is currently the case, to be able to implement the above control law, we design an observer to estimate $x_2(t)$, then $u_p(t)$ is realized as follows

$$\hat{u}_p(t) = -9.5y(t) - 30y(t-\tau) - 2.5\hat{x}_2(t) - 20\hat{x}_2(t-\tau). \tag{14.32}$$

The above observer design problem to estimate $x_2(t)$ is equivalent to the design of an observer (14.13) and (14.14) for the case where $F_a = [0 \;\; 1]$ and $F_d = [0 \;\; 0]$. For this, it is found that we can also design a first-order observer to estimate $x_2(t)$, where

$$\hat{x}_2(t) = w(t) + y(t), \tag{14.33}$$

$$\dot{w}(t) = -2w(t) - 2y(t). \tag{14.34}$$

With the above observer, i.e., $\hat{x}_2(t)$, the implementation of control law (14.32) is shown in Fig. 14.5. From the Figure, the hardware components for realizing $\hat{u}_p(t)$ now require three summers, one integrator, two delay units and six proportional gains. Clearly, this scheme resulted in more hardware components than the scheme based on observer (14.30) and (14.31).

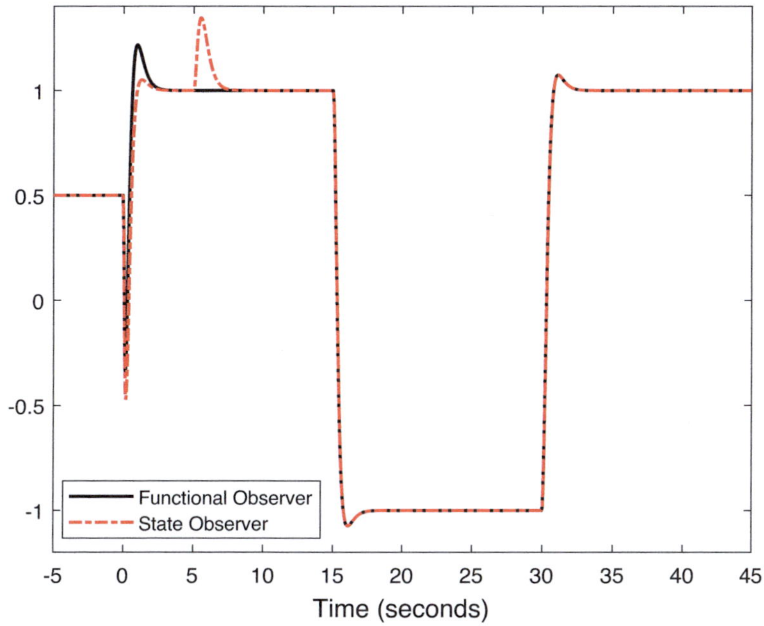

Fig. 14.6 Trajectories of $y(t)$ using control schemes (14.30), (14.31) and (14.32) (Numerical Example 1, Chap. 14)

Furthermore, Fig. 14.6 shows the trajectories of $y(t)$ from using two control schemes (14.30), (14.31) and (14.32). It is clear that control schemes based on generalized functional observer result in a better transient response. The spike occurs at $t = 5s$ associated with the scheme (14.32) is due to the initial estimated error of the delayed state variable $\hat{x}_2(t - \tau)$. Overall, via this numerical example, we can see the advantages of employing a generalized functional observer to directly realize a state feedback controller containing both delayed and instantaneous unmeasurable state variables.

Remark 2 We can design a controller to make a selected output of the controlled system track the step input as well as the ramp input with zero steady-state errors. For this, we consider a scheme with double integral control. With an abuse of notation, let $x_I^2(t)$ be defined as follows

$$\dot{x}_I^2(t) = x_I(t). \tag{14.35}$$

Then the following augmented system with double integral control is obtained

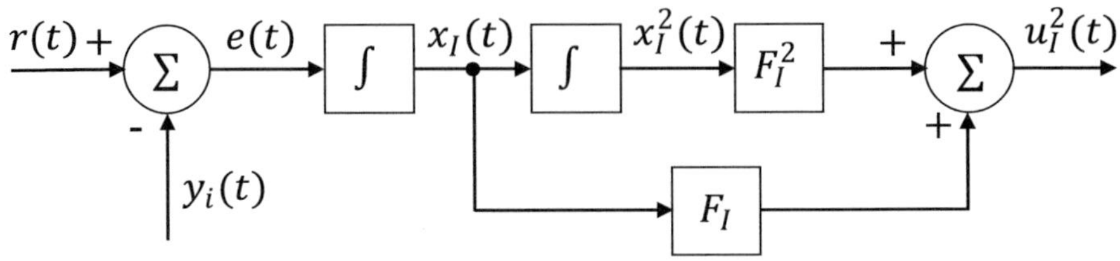

Fig. 14.7 Schematic diagram of the implementation of $u_I^2(t)$

$$\begin{bmatrix} \dot{x}(t) \\ \dot{x}_I(t) \\ \dot{x}_I^2(t) \end{bmatrix} = \begin{bmatrix} A & 0 & 0 \\ -C_i & 0 & 0 \\ 0 & 1 & 0 \end{bmatrix} \begin{bmatrix} x(t) \\ x_I(t) \\ x_I^2(t) \end{bmatrix} + \begin{bmatrix} B \\ 0 \\ 0 \end{bmatrix} u(t)$$
$$+ \begin{bmatrix} A_d & 0 & 0 \\ 0 & 0 & 0 \\ 0 & 0 & 0 \end{bmatrix} \begin{bmatrix} x(t-\tau) \\ x_I(t-\tau) \\ x_I^2(t-\tau) \end{bmatrix} + \begin{bmatrix} 0 \\ 1 \\ 0 \end{bmatrix} r(t). \tag{14.36}$$

For the above system, we design a state feedback controller of the form

$$u(t) = [F \quad F_I \quad F_I^2] \begin{bmatrix} x(t) \\ x_I(t) \\ x_I^2(t) \end{bmatrix} + F_d x(t-\tau), \tag{14.37}$$

to stabilize system (14.36). Then, the implementation of the control law is realized with double integral control, $u_I^2(t) = F_I x_I(t) + F_I^2 x_I^2(t)$, and an instantaneous/delayed functional observer to estimate $u_p(t)$, where $u_p(t) = Fx(t) + F_d x(t-\tau)$. Figure 14.7 shows a schematic diagram of the implementation of $u_I^2(t) = F_I x_I(t) + F_I^2 x_I^2(t)$ with double integral control, and two gains F_I and F_I^2.

Now, in Numerical Example 1, we obtain $[F \quad F_I \quad F_I^2] = [-20 \quad 2 \quad 81 \quad 180]$ and $F_d = -[30 \quad 20]$ such that the closed-loop system is stable (independent of the time delay τ) with the eigenvalues at $\{-3, -4, -5, -6\}$. To estimate the control law $u_p(t) = [-20 \quad 2]x(t) - [30 \quad 20]x(t-\tau)$, a first-order observer exists and is given below

$$\hat{u}_p(t) = w(t) - 18y(t) - 50y(t-\tau), \tag{14.38}$$
$$\dot{w}(t) = -2w(t) - 4y(t) + 40y(t-\tau). \tag{14.39}$$

Figure 14.8 shows the response of the output $y(t)$ subject to ramp-inputs in $r(t)$. The time-delay is selected as $\tau = 5s$. The output asymptotically tracks ramp-inputs in $r(t)$ as required. On the other hand, Fig. 14.9 shows the response of $y(t)$ without double integral control, and as a result, zero steady-state error cannot be achieved.

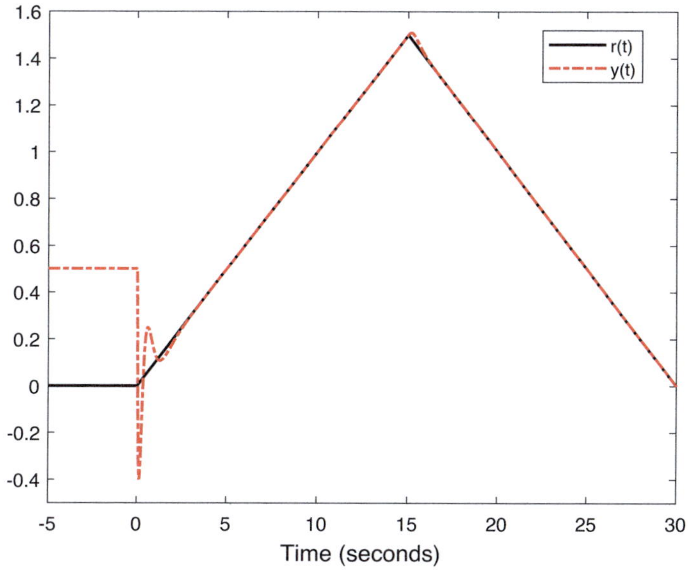

Fig. 14.8 Trajectories of $y(t)$ and $r(t)$ (Numerical Example 1 with double integral control, Remark 2, Chap. 14)

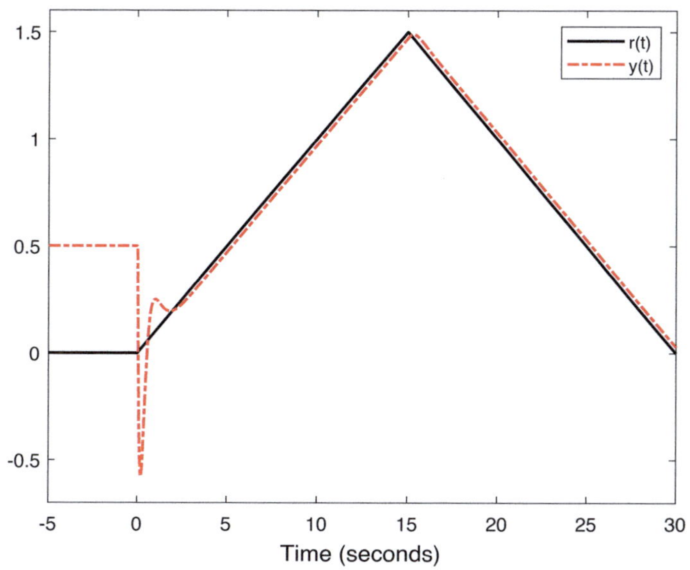

Fig. 14.9 Trajectories of $y(t)$ and $r(t)$ (Numerical Example 1 without double integral control, Remark 2, Chap. 14)

14.3 Main Results: $C_d \neq 0$, Step Change Tracking Using $\hat{z}(t)$ Where $z(t) = Fx(t) + F_d x(t - \tau)$

In this section, for completeness, we consider time-delay system (14.1) and (14.2), where the output vector $y(t)$ is now of the form

$$y(t) = Cx(t) + C_d x(t - \tau).$$

Our objective is to design an integral controller in combination with an instantaneous/delayed functional observer to ensure that a selected output, $y_i(t) = C_i x(t) + C_{di} x(t - \tau)$, of the system asymptotically tracks any constant step reference input, $r(t)$. The results are omitted here, since they can be derived by following similar lines of reasoning presented earlier in this chapter and Chap. 13. Instead, we present the following example to illustrate this case.

14.3.1 A Numerical Example

Numerical Example 2: In this example, matrices A, A_d, B, C and C_d are given as follows

$$A = \begin{bmatrix} 1 & 2 \\ 1 & -1 \end{bmatrix}, A_d = \begin{bmatrix} -30 & 20 \\ -30 & 20 \end{bmatrix}, B = \begin{bmatrix} 1 \\ 1 \end{bmatrix}, C = \begin{bmatrix} 1 & 0 \\ 0 & 1 \end{bmatrix} \text{ and } C_d = \begin{bmatrix} 0 & 1 \\ 1 & 1 \end{bmatrix}.$$

With $\tau = 1s$, the open-loop time-delay system is unstable since the solution to the equation $\det(\lambda I - A - A_d e^{-\tau\lambda}) = 0$ results in four unstable eigenvalues, where $\lambda = \{2.5580 \pm j1.7685, 0.8024 \pm j7.1237\}$ (see, [1]). Our objective is to design a controller to ensure that the first output of the system, i.e., $y_1(t) = x_1(t) + x_2(t - \tau)$, asymptotically tracks any constant step reference input, $r(t)$. Based on a similar analysis presented in this chapter (see, (14.4)–(14.10)), we design the following proportional-integral controller

$$u(t) = u_p(t) + u_I(t), \tag{14.40}$$

where

$$u_p(t) = Fx(t) + F_d x(t - \tau), \tag{14.41}$$

$$u_I(t) = F_I x_I(t), \tag{14.42}$$

$F = \begin{bmatrix} -2.8333 & -1.6667 \end{bmatrix}$, $F_d = \begin{bmatrix} 30 & -20 \end{bmatrix}$, $F_I = 1$ and $\dot{x}_I(t) = r(t) - y_1(t)$. Note that feedback matrix $F_a = \begin{bmatrix} F & F_I \end{bmatrix}$ is obtained by placing the eigenvalues of matrix $(A_a + B_a F_a)$ at $\{-1, -1.5, -2\}$. With the above controller, the closed-loop system (14.8) is asymptotically stable with three stable eigenvalues at $\{-0.4858, -0.9481 \pm j1.5339\}$. Figure 14.10 shows the responses of the output $y_1(t)$ to step changes in

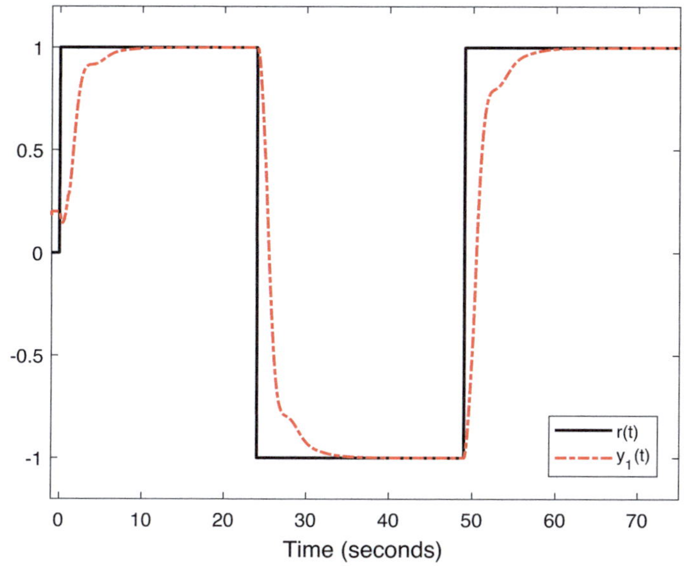

Fig. 14.10 Trajectories of $y_1(t)$ and $r(t)$ (Example 2, without an observer, Chap. 14)

$r(t)$. The time-delay is $\tau = 1s$. It is clear that $y_1(t)$ asymptotically tracks step changes in $r(t)$ as required.

However, control law (14.41) requires $Fx(t)$ and $F_d x(t-\tau)$ to be available for the feedback control. Even though there are two outputs for this system, namely, $y_1(t)$ and $y_2(t)$, i.e.,

$$y_1(t) = x_1(t) + x_2(t-\tau), \tag{14.43}$$

$$y_2(t) = x_2(t) + x_1(t-\tau) + x_2(t-\tau), \tag{14.44}$$

it is still necessary to design an observer to realize control law (14.41). In this regard, we propose a functional observer, $\hat{u}_p(t)$, to directly estimate $u_p(t)$ such that $\hat{u}_p(t)$ asymptotically converges to $u_p(t)$. This is done based on the conditions derived in this chapter and the design method presented Chap. 13. Accordingly, we can design two functional observers to directly estimate $u_p(t)$. One is a first-order functional observer with an internal delay, and the other is a second-order functional observer without an internal delay. For illustrative purpose, we design a second-order functional observer without an internal delay to estimate the following function,

$$u_a(t) = \begin{bmatrix} u_p(t) \\ z_a(t) \end{bmatrix}, \tag{14.45}$$

where the additional function to be estimated, $z_a(t)$, is given as folows

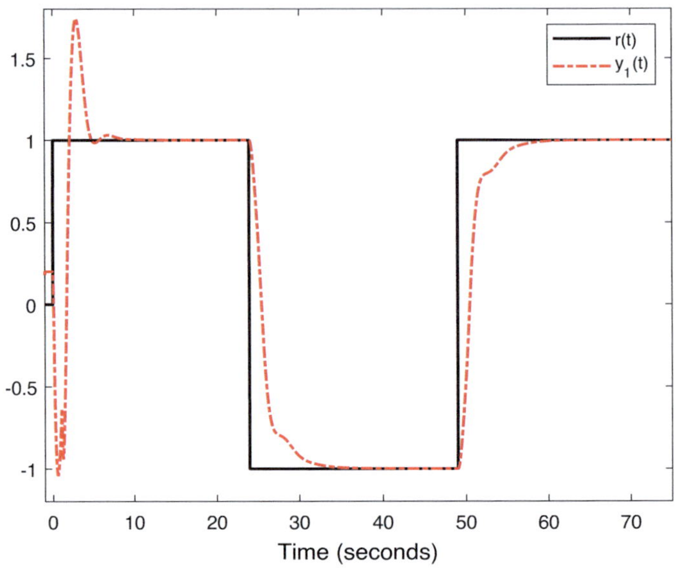

Fig. 14.11 Trajectories of $y_1(t)$ and $r(t)$ (Example 2, a second-order functional observer without internal delay, Chap. 14)

$$
\begin{aligned}
z_a(t) &= F^1 x(t) + F_d^1 x(t-\tau) \\
&= \begin{bmatrix} 1 & 1 \end{bmatrix} x(t) + \begin{bmatrix} 1 & -1 \end{bmatrix} x(t-\tau). \qquad (14.46)
\end{aligned}
$$

Note that with $z_a(t)$ defined above, $u_a(t)$ is independent of $y(t)$ since $\begin{bmatrix} C & C_d \\ F & F_d \\ F^1 & F_d^1 \end{bmatrix}$ is a full-row rank matrix. Based on the design method presented in Chap. 13, we obtain the following second-order observer

$$\hat{u}_p(t) = w_1(t) + My(t) + M_d y(t-\tau), \qquad (14.47)$$

$$\dot{w}(t) = \begin{bmatrix} \dot{w}_1(t) \\ \dot{w}_2(t) \end{bmatrix} = Nw(t) + Gy(t) + G_d y(t-\tau), \qquad (14.48)$$

where $M = [-13.1759 \quad 8.6759]$, $M_d = [11.6482 \quad -5.8241]$, $N = \begin{bmatrix} -1.5316 & 8.5798 \\ 0.2842 & -7.4684 \end{bmatrix}$, $G = \begin{bmatrix} 38.4316 & -7.4039 \\ -22.6042 & 14.7051 \end{bmatrix}$ and $G_d = \begin{bmatrix} -6.6645 & 12.0684 \\ 7.6907 & -4.3958 \end{bmatrix}$. Matrix N is stable with $\lambda(N) = \{-1.1459, -7.8541\}$.

Figure 14.11 shows responses of the output $y_1(t)$ to step changes in $r(t)$. After some initial and transient estimation errors (due to the observer), the output $y_1(t)$ asymptotically tracks step changes in $r(t)$ as required.

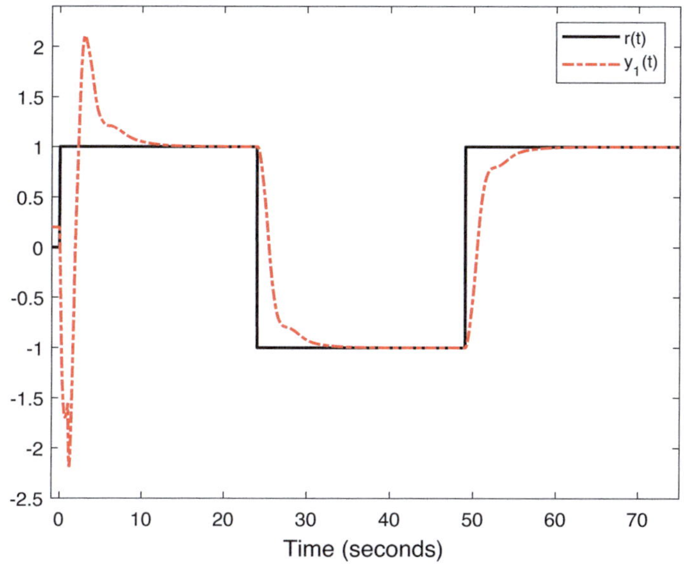

Fig. 14.12 Trajectories of $y_1(t)$ and $r(t)$ (Example 2, a first-order functional observer with internal delay, Chap. 14)

On the other hand, if a first-order functional observer with an internal delay is used to estimate $u_p(t)$, then the following observer is obtained

$$\hat{u}_p(t) = w(t) + My(t) + M_d y(t-\tau), \tag{14.49}$$

$$\dot{w}(t) = Nw(t) + N_d w(t-\tau) + Gy(t) + G_d y(t-\tau) + G_{dd} y(t-2\tau), \tag{14.50}$$

where $M = [-31.04 \quad 26.54]$, $M_d = [-24.08 \quad 12.04]$, $N = -6$, $N_d = 3$, $G = [169.24 \quad -84.62]$, $G_d = [165.24 \quad -82.62]$ and $G_{dd} = [-72.24 \;\; 36.12]$.

Figure 14.12 shows the responses of the output $y_1(t)$ to step changes in $r(t)$. Output $y_1(t)$ asymptotically tracks step changes in $r(t)$. Finally, Fig. 14.13 shows responses of $y_1(t)$ to step changes in $r(t)$ for three cases: (i) without an observer, (ii) with a first-order observer, and (iii) with a second-order observer. All three cases experienced the same initial condition where $x(\theta) = [0.2 \quad 0]^T$. Observe that the second-order functional observer exhibits better performance than the first-order functional observer during the first 10 seconds of the transient response.

14.4 Concluding Remark

In this chapter, we have applied the unified framework of estimating the general functional $z(t) = Fx(t) + F_d x(t-\tau)$ to solve the problem of designing dynamic output feedback controllers such that one of the selected outputs of the system asymptot-

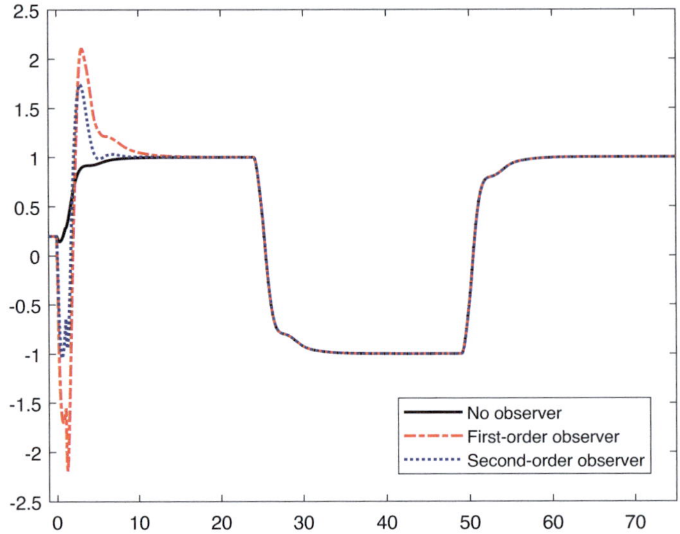

Fig. 14.13 Trajectories of $y_1(t)$ (Example 2, Chap. 14)

ically tracks any step (or ramp) reference command input, $r(t)$. By estimating $\hat{z}(t)$ directly, we have shown via a numerical example, the advantages of our approach in getting a simpler observer-based control scheme with fewer computation blocks and better transient responses.

We will continue our exploratory research into the design of generalized functional observers. Thus, in the next chapter, Chap. 15, we will look at time-delay system (5.28) and (5.29) and design an observer to directly estimate the following generalized functional

$$\tilde{z}(t) = F\bar{x}_1(t) + F_d\bar{x}_1(t-\tau) + F_{dd}\bar{x}_1(t-2\tau) + F_{3d}\bar{x}_1(t-3\tau),$$

where $\tilde{z}(t) \in \mathbb{R}^m$, F, F_d, F_{dd} and F_{3d} are known constant matrices of appropriate dimensions. This problem is motivated by the unknown input estimation problem for time-delay systems (5.1) and (5.2) where decomposition (5.12) is realized.

Reference

1. Wu, Z., Michiels, W.: Reliably computing all characteristic roots of delay differential equations in a given right half plane using a spectral method. J. Comput. Appl. Math. **236**(9), 2499–2514 (2012)

Chapter 15
Generalized Observers–Multiple Delays in State and Output Vectors

15.1 Background

We continue our exploratory journey, and for completeness, let us consider time-delay systems (5.28) and (5.29), i.e.,

$$\dot{\bar{x}}_1(t) = \bar{A}_{11}\bar{x}_1(t) + \tilde{A}_{d11}\bar{x}_1(t-\tau) + \tilde{A}_{dd11}\bar{x}_1(t-2\tau) + \bar{A}_{12}V_1 y(t) + \bar{A}_{d12}V_1 y(t-\tau),$$
$$\tilde{y}(t) = \tilde{C}_1\bar{x}_1(t) + \tilde{C}_{d1}\bar{x}_1(t-\tau) + \tilde{C}_{dd1}\bar{x}_1(t-2\tau),$$

and our aim is to design observers with various structures to estimate the following generalized functional

$$\tilde{z}(t) = F\bar{x}_1(t) + F_d\bar{x}_1(t-\tau) + F_{dd}\bar{x}_1(t-2\tau) + F_{3d}\bar{x}_1(t-3\tau), \tag{15.1}$$

where $\tilde{z}(t) \in \mathbb{R}^m$, F, F_d, F_{dd} and F_{3d} are known constant matrices of appropriate dimensions. This estimation problem is motivated by the unknown input estimation problem for time-delay systems (5.1) and (5.2) where decomposition (5.12) is realized. Thus, for the case where

$$F = -\bar{A}_{21}, \tag{15.2}$$
$$F_d = -\tilde{A}_{d21}, \tag{15.3}$$
$$F_{dd} = -\tilde{A}_{d22}, \tag{15.4}$$
$$F_{3d} = -V_1 C_d E \tilde{A}_{dd11}, \tag{15.5}$$

then the designed functional observer, $\hat{\tilde{z}}(t)$, is used to directly estimate the unknown input, $d(t)$, according to (5.39).

H. Trinh et al., *Unknown Inputs Estimation in Linear Time-Delay Systems Using Generalized Functional Observers*, Studies in Systems, Decision and Control 648,
https://doi.org/10.1007/978-3-032-11870-7_15

15.2 Main Results

To estimate the generalized functional $\tilde{z}(t)$ defined by (15.1) for system (5.28) and (5.29), we consider the following generalized functional observer

$$\begin{aligned}\hat{\tilde{z}}(t) &= w(t) + M\tilde{y}(t) + M_d\tilde{y}(t-\tau), \qquad (15.6)\\ \dot{w}(t) &= Nw(t) + N_dw(t-\tau) + N_{dd}w(t-2\tau) + G\tilde{y}(t) + G_d\tilde{y}(t-\tau)\\ &\quad + G_{dd}\tilde{y}(t-2\tau) + G_{3d}\tilde{y}(t-3\tau) + Jy(t) + J_dy(t-\tau) + J_{dd}y(t-2\tau)\\ &\quad + J_{3d}y(t-3\tau) + J_{4d}y(t-4\tau), \qquad (15.7)\end{aligned}$$

where $\tilde{y}(t) = V_2y(t) - V_2C_dDV_1y(t-\tau)$, $w(\theta) = \rho(\theta)$, $\forall\theta \in [0, 2\tau]$, $w(t) \in \mathbb{R}^m$ and $\hat{\tilde{z}}(t) \in \mathbb{R}^m$ is the estimate of $\tilde{z}(t)$. Matrices $M, M_d, N, N_d, N_{dd}, G, G_d, G_{dd}, G_{3d}, J, J_d, J_{dd}, J_{3d}$ and J_{4d} are unknown and to be determined such that $\hat{\tilde{z}}(t)$ converges asymptotically to $\tilde{z}(t)$.

Note that the estimation problem (15.1) is more general than the estimation problem (12.62). As such, when compared to (12.1) and (12.2), observer (15.6) and (15.7) contains extra delayed terms $M_d\tilde{y}(t-\tau)$, $G_{3d}\tilde{y}(t-3\tau)$ and $J_{4d}y(t-4\tau)$, which are necessary in order to relax their observer existence conditions.

For completeness, in the following, we derive their existence conditions by following similar lines of reasoning as we have done in the previous chapters. Once their existence conditions are derived, a procedure for determining their parameters will be presented. Now, let us define the following error vector

$$e(t) = \hat{\tilde{z}}(t) - \tilde{z}(t), \qquad (15.8)$$

and by substituting (15.6) and (15.1) into (15.8), we obtain

$$\begin{aligned}w(t) &= e(t) + F\bar{x}_1(t) + F_d\bar{x}_1(t-\tau) + F_{dd}\bar{x}_1(t-2\tau)\\ &\quad + F_{3d}\bar{x}_1(t-3\tau) - M\tilde{y}(t) - M_d\tilde{y}(t-\tau). \qquad (15.9)\end{aligned}$$

Taking derivative of both sides of (15.8) with respect to time, and after some substitutions and rearranging, the following error dynamic equation is obtained

$$\begin{aligned}\dot{e}(t) &= Ne(t) + N_de(t-\tau) + N_{dd}e(t-2\tau) + (J-\mathscr{A})y(t) + (J_d-\mathscr{B})y(t-\tau)\\ &\quad + (J_{dd}-\mathscr{C})y(t-2\tau) + (J_{3d}-\mathscr{D})y(t-3\tau) + (J_{4d}-\mathscr{E})y(t-4\tau) + \mathscr{F}\bar{x}_1(t)\\ &\quad + \mathscr{G}\bar{x}_1(t-\tau) + \mathscr{H}\bar{x}_1(t-2\tau) + \mathscr{I}\bar{x}_1(t-3\tau) + \mathscr{J}\bar{x}_1(t-4\tau) + \mathscr{K}\bar{x}_1(t-5\tau),\end{aligned}$$
(15.10)

where

$$\mathscr{A} = F\bar{A}_{12}V_1 - M\tilde{C}_1\bar{A}_{12}V_1, \tag{15.11}$$

$$\mathscr{B} = F\bar{A}_{d12}V_1 + F_d\bar{A}_{12}V_1 - M\tilde{C}_1\bar{A}_{d12}V_1 - M\tilde{C}_{d1}\bar{A}_{12}V_1 - M_d\tilde{C}_1\bar{A}_{12}V_1, \tag{15.12}$$

$$\begin{aligned}\mathscr{C} = {} & F_d\bar{A}_{d12}V_1 + F_{dd}\bar{A}_{12}V_1 - M\tilde{C}_{d1}\bar{A}_{d12}V_1 - M\tilde{C}_{dd1}\bar{A}_{12}V_1 - M_d\tilde{C}_1\bar{A}_{d12}V_1 \\ & - M_d\tilde{C}_{d1}\bar{A}_{12}V_1,\end{aligned} \tag{15.13}$$

$$\mathscr{D} = F_{dd}\bar{A}_{d12}V_1 + F_{3d}\bar{A}_{12}V_1 - M\tilde{C}_{dd1}\bar{A}_{d12}V_1 - M_d\tilde{C}_{d1}\bar{A}_{d12}V_1 - M_d\tilde{C}_{dd1}\bar{A}_{12}V_1, \tag{15.14}$$

$$\mathscr{E} = F_{3d}\bar{A}_{d12}V_1 - M_d\tilde{C}_{dd1}\bar{A}_{d12}V_1, \tag{15.15}$$

$$\mathscr{F} = NF + \tilde{T}\tilde{C}_1 + M\tilde{C}_1\bar{A}_{11} - \tilde{\Upsilon}_{a1}, \tag{15.16}$$

$$\mathscr{G} = NF_d + N_dF + \hat{W}\tilde{C}_1 + \tilde{T}\tilde{C}_{d1} + M\bar{C}_1 + M_d\tilde{C}_1\bar{A}_{11} - \tilde{\Upsilon}_{a2}, \tag{15.17}$$

$$\begin{aligned}\mathscr{H} = {} & NF_{dd} + N_dF_d + N_{dd}F + \hat{U}\tilde{C}_1 + \hat{W}\tilde{C}_{d1} + \tilde{T}\tilde{C}_{dd1} + M\bar{C}_{1a} + M_d\bar{C}_1 \\ & - \tilde{\Upsilon}_{a3},\end{aligned} \tag{15.18}$$

$$\begin{aligned}\mathscr{I} = {} & NF_{3d} + N_dF_{dd} + N_{dd}F_d + \hat{Y}\tilde{C}_1 + \hat{U}\tilde{C}_{d1} + \hat{W}\tilde{C}_{dd1} + M\hat{C}_1 + M_d\bar{C}_{1a} \\ & - \hat{\Upsilon}_{a4},\end{aligned} \tag{15.19}$$

$$\mathscr{J} = N_dF_{3d} + N_{dd}F_{dd} + \hat{Y}\tilde{C}_{d1} + \hat{U}\tilde{C}_{dd1} + M\hat{C}_{1a} + M_d\hat{C}_1 - \hat{\Upsilon}_{a5}, \tag{15.20}$$

$$\mathscr{K} = N_{dd}F_{3d} + \hat{Y}\tilde{C}_{dd1} + M_d\hat{C}_{1a} - \hat{\Upsilon}_{a6}, \tag{15.21}$$

where

$$\hat{W} = G_d - NM_d - N_dM, \tag{15.22}$$

$$\hat{U} = G_{dd} - N_dM_d - N_{dd}M, \tag{15.23}$$

$$\hat{Y} = G_{3d} - N_{dd}M_d, \tag{15.24}$$

$$\hat{\Upsilon}_{a4} = F_d\tilde{A}_{dd11} + F_{dd}\tilde{A}_{d11} + F_{3d}\bar{A}_{11}, \tag{15.25}$$

$$\hat{\Upsilon}_{a5} = F_{dd}\tilde{A}_{dd11} + F_{3d}\tilde{A}_{d11}, \tag{15.26}$$

$$\hat{\Upsilon}_{a6} = F_{3d}\tilde{A}_{dd11}, \tag{15.27}$$

$$\hat{C}_{1a} = \tilde{C}_{dd1}\tilde{A}_{dd11}, \tag{15.28}$$

$\tilde{T}$, $\bar{C}_1$, $\bar{C}_{1a}$, $\hat{C}_1$, and $\tilde{\Upsilon}_{a1}$, $\tilde{\Upsilon}_{a2}$, $\tilde{\Upsilon}_{a3}$ are as defined by (10.84), (12.4), (12.6), (12.7) and (12.75), respectively.

From (15.10), by letting

$$J = \mathscr{A},\ J_d = \mathscr{B},\ J_{dd} = \mathscr{C},\ J_{3d} = \mathscr{D},\ J_{4d} = \mathscr{E}, \tag{15.29}$$

and

$$\mathscr{F} = 0, \mathscr{G} = 0, \mathscr{H} = 0, \mathscr{I} = 0, \mathscr{J} = 0, \mathscr{K} = 0, \tag{15.30}$$

the error dynamic equation (15.10) is reduced to the following equation

$$\dot{e}(t) = Ne(t) + N_d e(t-\tau) + N_{dd} e(t-2\tau), \tag{15.31}$$

in which for a given τ and any initial condition $e(\theta)$, $\forall\theta \in [0, 2\tau]$, it is necessary to determine matrices N, N_d and N_{dd} such that the error $e(t)$ converges to zeros.

Thanks to our unified framework for solving coupled matrix equations, we can now systematically solve for the nine unknown matrices in (15.16)–(15.21) to satisfy the constraint (15.30). By letting

$$X = [N \;\; N_d \;\; N_{dd} \;\; \tilde{T} \;\; \hat{W} \;\; \hat{U} \;\; \hat{Y} \;\; M \;\; M_d], \tag{15.32}$$

$$\hat{\Theta} = \begin{bmatrix} F & F_d & F_{dd} & F_{3d} & 0 & 0 \\ 0 & F & F_d & F_{dd} & F_{3d} & 0 \\ 0 & 0 & F & F_d & F_{dd} & F_{3d} \\ \tilde{C}_1 & \tilde{C}_{d1} & \tilde{C}_{dd1} & 0 & 0 & 0 \\ 0 & \tilde{C}_1 & \tilde{C}_{d1} & \tilde{C}_{dd1} & 0 & 0 \\ 0 & 0 & \tilde{C}_1 & \tilde{C}_{d1} & \tilde{C}_{dd1} & 0 \\ 0 & 0 & 0 & \tilde{C}_1 & \tilde{C}_{d1} & \tilde{C}_{dd1} \\ \tilde{C}_1\bar{A}_{11} & \bar{C}_1 & \bar{C}_{1a} & \hat{C}_1 & \hat{C}_{1a} & 0 \\ 0 & \tilde{C}_1\bar{A}_{11} & \bar{C}_1 & \bar{C}_{1a} & \hat{C}_1 & \hat{C}_{1a} \end{bmatrix}, \tag{15.33}$$

$$\hat{\Upsilon} = [\tilde{\Upsilon}_{a1} \;\; \tilde{\Upsilon}_{a2} \;\; \tilde{\Upsilon}_{a3} \;\; \hat{\Upsilon}_{a4} \;\; \hat{\Upsilon}_{a5} \;\; \hat{\Upsilon}_{a6}], \tag{15.34}$$

and subject to the constraint (15.30), Eqs. (15.16)–(15.21) can be expressed as follows

$$X\hat{\Theta} = \hat{\Upsilon}. \tag{15.35}$$

A solution for the nine unknown matrices, i.e., X, exists if and only if the following condition holds

$$\text{rank}\begin{bmatrix} \hat{\Theta} \\ \hat{\Upsilon} \end{bmatrix} = \text{rank}\big[\hat{\Theta}\big]. \tag{15.36}$$

Similar to the methodology in Chap. 12 (see, from (12.25)–(12.32)) and provided that the solution to (15.35) leads to an asymptotic stable error system (15.31), the rest of the observer parameters can be obtained. Accordingly, once X is found, from (15.29), J, J_d, J_{dd}, J_{3d} and J_{4d} are obtained. Finally, from (10.84), (15.22)–(15.24), we obtain the following observer matrices

$$G = \tilde{T} + NM, \tag{15.37}$$

$$G_d = \hat{W} + NM_d + N_d M, \tag{15.38}$$

$$G_{dd} = \hat{U} + N_d M_d + N_{dd} M, \tag{15.39}$$

$$G_{3d} = \hat{Y} + N_{dd} M_d. \tag{15.40}$$

15.3 Numerical Examples

To demonstrate the effectiveness of the generalized functional observers designed in this chapter, we consider the following numerical examples.

Numerical Example 1: In this example, we undertake a comparative analysis on the estimation of $\tilde{z}(t)$ (as defined by (15.1)) by using two approaches. In the first approach, we design an observer, $\hat{\tilde{z}}(t)$, to directly estimate $\tilde{z}(t)$ based on the development presented in this chapter. In the second approach, instead, we design a state observer, $\hat{\bar{x}}_1(t)$, to estimate $\bar{x}_1(t)$. Then from (15.1), $\hat{\tilde{z}}(t)$ is realized as follows

$$\hat{\tilde{z}}(t) = F\hat{\bar{x}}_1(t) + F_d\hat{\bar{x}}_1(t-\tau) + F_{dd}\hat{\bar{x}}_1(t-2\tau) + F_{3d}\hat{\bar{x}}_1(t-3\tau). \tag{15.41}$$

Note that the second approach described above is the same as Option 1 (see, Sect. 4.2, Chap. 4). From this study, we highlight the salient features of estimating $\tilde{z}(t)$ directly by the first approach. The system under consideration is described by (12.47) and (12.48). For ease of simulation purpose, as in Numerical Example 1 (Chap. 12), the term $Bu(t)$ in (12.47) is used instead of $\bar{A}_{12}V_1y(t)$ and $\bar{A}_{d12}V_1y(t-\tau)$ as in (5.28).

For this numerical example, we let $B = \begin{bmatrix} 1 \\ 2 \end{bmatrix}$, matrices $\bar{A}_{11}$, $\tilde{A}_{d11}$, $\tilde{A}_{dd11}$, $\tilde{C}_1$, $\tilde{C}_{d1}$ and $\tilde{C}_{dd1}$ are as given in Numerical Example 3, Chap. 5. Here, we note that the system is a second-order system, i.e., $n = 2$. The function to be estimated, $\tilde{z}(t)$, is defined by (15.1) where F, F_d, F_{dd} and F_{3d} are given in Numerical Example 3, Chap. 5, i.e.,

$$\begin{aligned}\tilde{z}(t) = {} & [0 \ \ -1]\bar{x}_1(t) + [-4 \ \ -3]\bar{x}_1(t-\tau) \\ & + [-27 \ \ 2]\bar{x}_1(t-2\tau) + [-10 \ \ 0]\bar{x}_1(t-3\tau).\end{aligned} \tag{15.42}$$

For the purpose of comparison, let us impose an additional requirement on the designed observers such that they do not contain any internal delays. Thus, in the first approach, the designed functional observer, $\hat{\tilde{z}}(t)$, takes the following form

$$\hat{\tilde{z}}(t) = w(t) + M\tilde{y}(t) + M_d\tilde{y}(t-\tau), \tag{15.43}$$

$$\begin{aligned}\dot{w}(t) &= Nw(t) + G\tilde{y}(t) + G_d\tilde{y}(t-\tau) + G_{dd}\tilde{y}(t-2\tau)\\ &\quad + G_{3d}\tilde{y}(t-3\tau) + Ju(t) + J_d u(t-\tau) + J_{dd}u(t-2\tau)\\ &\quad + J_{3d}u(t-3\tau),\end{aligned} \tag{15.44}$$

where $N_d = 0$ and $N_{dd} = 0$. Likewise, a state observer for $\hat{\bar{x}}_1(t)$ can be designed.

First Approach: In the following, we show how to design an observer (15.43)–(15.44) for this numerical example.

First-order Observer: We consider a first-order functional observer as described by (15.43) and (15.44), i.e., $w(t) \in \mathbb{R}$. Since we require $N_d = 0$ and $N_{dd} = 0$, Eq. (15.35) is reduced to

$$X_{rr}\hat{\Theta}_{rr} = \hat{\Upsilon}, \tag{15.45}$$

where

$$X_{rr} = [N \;\; \tilde{T} \;\; \hat{W} \;\; \hat{U} \;\; \hat{Y} \;\; M \;\; M_d], \tag{15.46}$$

and condition (15.36) is therefore replaced with the following condition

$$\text{rank}\begin{bmatrix}\hat{\Theta}_{rr}\\ \hat{\Upsilon}\end{bmatrix} = \text{rank}\big[\hat{\Theta}_{rr}\big], \tag{15.47}$$

where

$$\hat{\Theta}_{rr} = \begin{bmatrix} F & F_d & F_{dd} & F_{3d} & 0 & 0\\ \tilde{C}_1 & \tilde{C}_{d1} & \tilde{C}_{dd1} & 0 & 0 & 0\\ 0 & \tilde{C}_1 & \tilde{C}_{d1} & \tilde{C}_{dd1} & 0 & 0\\ 0 & 0 & \tilde{C}_1 & \tilde{C}_{d1} & \tilde{C}_{dd1} & 0\\ 0 & 0 & 0 & \tilde{C}_1 & \tilde{C}_{d1} & \tilde{C}_{dd1}\\ \tilde{C}_1\bar{A}_{11} & \bar{C}_1 & \bar{C}_{1a} & \hat{C}_1 & \hat{C}_{1a} & 0\\ 0 & \tilde{C}_1\bar{A}_{11} & \bar{C}_1 & \bar{C}_{1a} & \hat{C}_1 & \hat{C}_{1a} \end{bmatrix}. \tag{15.48}$$

Accordingly, we first obtain matrices $\hat{\Theta}_{rr}$ and $\hat{\Upsilon}$ as defined by (15.48) and (15.34), respectively, with $\hat{\Theta}_{rr} \in \mathbb{R}^{7\times 12}$ and $\hat{\Upsilon} \in \mathbb{R}^{1\times 12}$. We then check condition (15.47), but it is found that rank $\begin{bmatrix}\hat{\Theta}_{rr}\\ \hat{\Upsilon}\end{bmatrix} = 8$, whereas, $\text{rank}\big[\hat{\Theta}_{rr}\big] = 7$. Therefore, condition (15.47) is not satisfied, and as a result, a first-order functional observer (15.43) and (15.44) does not exist.

Higher-order Observers: As first-order functional observers do not exist, we have an option of increasing the order of observer (15.43) and (15.44). In a similar way (see, for example, (13.41)), we introduce ν additional functions, $\tilde{z}^{\nu}(t)$, where

$$\tilde{z}^v(t) = F^v\bar{x}_1(t) + F_d^v\bar{x}_1(t-\tau) + F_{dd}^v\bar{x}_1(t-2\tau) + F_{3d}^v\bar{x}_1(t-3\tau), \tag{15.49}$$

and F^v, F_d^v, F_{dd}^v and F_{3d}^v are row matrices to be found such that condition (15.47) is satisfied and an asymptotic observer is obtained.

Let $\tilde{z}_a^v(t) = \begin{bmatrix} \tilde{z}(t) \\ \tilde{z}^v(t) \end{bmatrix} \in \mathbb{R}^{v+1}$. Then our task is to design an $(v+1)$-th order functional observer to estimate $\tilde{z}_a^v(t)$. Notice that if an $(v+1)$-th order observer is found (i.e., $\hat{\tilde{z}}_a^v(t)$), then $\hat{\tilde{z}}(t)$ is obtained from $\hat{\tilde{z}}_a^v(t)$, where $\hat{\tilde{z}}(t) = [1\ 0\ \cdots\ 0]\hat{\tilde{z}}_a^v(t)$.

As we have discussed previously, how to systematically find F^v, F_d^v, F_{dd}^v and F_{3d}^v with the minimum dimension, i.e., v, to construct an asymptotic observer is still an open research problem. Notice that the addition of F^v, F_d^v, F_{dd}^v and F_{3d}^v brings extra degrees of freedom to the design of generalized observers. Below is a simple way to satisfy condition (15.47) using a higher order observer.

To proceed with the design of a higher-order functional observer, it is necessary to increase the order of the observer so that condition (15.47) is satisfied. In this example, we notice that the last column of $\hat{\Theta}_{rr}$ is zeros since $\tilde{C}_{dd1} = \begin{bmatrix} 6 & 0 \end{bmatrix}$ and $\hat{C}_{1a} = \begin{bmatrix} -60 & 0 \end{bmatrix}$. Therefore, the maximum value of the right-hand side of condition (15.47) is 11, i.e., $\text{rank}\begin{bmatrix}\hat{\Theta}_{rr}\end{bmatrix} = 11$. On the other hand, we notice that the last column of $\hat{\Upsilon}_{a6}$ is also zeros since $\tilde{A}_{dd11} = \begin{bmatrix} -10 & 0 \\ -34 & 0 \end{bmatrix}$. Thus, the maximum value of the left-hand side of condition (15.47) is also 11, i.e., $\text{rank}\begin{bmatrix} \hat{\Theta}_{rr} \\ \hat{\Upsilon} \end{bmatrix} = 11$. Clearly, to increase the rank of $\hat{\Theta}_{rr}$ from 7 to 11, it is necessary to add at least four additional rows, or $\tilde{z}^v(t) \in \mathbb{R}^4$. We therefore design a fifth-order observer. Note that the four additional rows must be linearly independent from $\hat{\Theta}_{rr}$ in order to increase its rank from 7 to 11. We can simply select them as follows

$$\begin{bmatrix} F & F_d & F_{dd} & F_{3d} \\ F^v & F_d^v & F_{dd}^v & F_{3d}^v \end{bmatrix} = \begin{bmatrix} 0 & -1 & -4 & -3 & -27 & 2 & -10 & 0 \\ 1 & 0 & 0 & 0 & 0 & 0 & 0 & 0 \\ 0 & 1 & 0 & 0 & 0 & 0 & 0 & 0 \\ 0 & 0 & 1 & 0 & 0 & 0 & 0 & 0 \\ 0 & 0 & 0 & 1 & 0 & 0 & 0 & 0 \end{bmatrix}, \tag{15.50}$$

where the above matrix has full row rank.

Then from (15.48) and (15.34), we obtain matrices $\hat{\Theta}_{rr}$ and $\hat{\Upsilon}$, respectively, where F, F_d, F_{dd} and F_{3d} are replaced with $\begin{bmatrix} F \\ F^v \end{bmatrix}$, $\begin{bmatrix} F_d \\ F_d^v \end{bmatrix}$, $\begin{bmatrix} F_{dd} \\ F_{dd}^v \end{bmatrix}$ and $\begin{bmatrix} F_{3d} \\ F_{3d}^v \end{bmatrix}$, respectively. It is found that condition (15.47) is satisfied where $\text{rank}\begin{bmatrix} \hat{\Theta}_{rr} \\ \hat{\Upsilon} \end{bmatrix} = \text{rank}\begin{bmatrix}\hat{\Theta}_{rr}\end{bmatrix} = 11$. Thus, we can now proceed to solve for the unknown matrix X_{rr} in (15.45). Its solution is given as follows

$$X_{rr} = \hat{\Upsilon}\hat{\Theta}_{rr}^+ + Z_r(I - \hat{\Theta}_{rr}\hat{\Theta}_{rr}^+), \tag{15.51}$$

where $\hat{\Theta}_{rr}^{+}$ is the Moore-Penrose inverse of $\hat{\Theta}_{rr}$, and Z_r is an arbitrary matrix of appropriate dimension.

In this example, it is found that $\hat{\Theta}_{rr}\hat{\Theta}_{rr}^{+} = I_{11}$ and hence $X_{rr} = \hat{\Upsilon}\hat{\Theta}_{rr}^{+}$. In this case, the eigenvalues of matrix N are fixed. Solving for X_{rr} and from (15.37)–(15.40), we obtain the following observer gain matrices

$$N = \begin{bmatrix} 3.2980 & -17.1481 & 2.8535 & -81.9899 & 20.6902 \\ 0 & -8.3333 & -2 & -25 & 3.6667 \\ 0 & 5.6667 & -7 & -8 & -0.3333 \\ 1 & 0.5 & 1 & -4.5 & 1.5 \\ -0.0909 & 1.6667 & -0.0909 & 10.4545 & -5.6061 \end{bmatrix}, \quad M = \begin{bmatrix} -3.7222 \\ 0 \\ 0 \\ 0 \\ 0 \end{bmatrix},$$

$$M_d = \begin{bmatrix} -1.6667 \\ 0 \\ 0 \\ 0 \\ 0 \end{bmatrix}, \quad G = \begin{bmatrix} -32.3499 \\ -1.6667 \\ -5.6667 \\ -4.2222 \\ -1.3283 \end{bmatrix}, \quad G_d = \begin{bmatrix} -25.0926 \\ 0 \\ 0 \\ -1.6667 \\ -5.6667 \end{bmatrix},$$

$$G_{dd} = \begin{bmatrix} -21.3333 \\ 0 \\ 0 \\ 0 \\ 0 \end{bmatrix} \text{ and } G_{3d} = 0_{5\times 1}.$$

Matrix N is stable with eigenvalues at $\{-4.8048 \pm j7.7996, -7.5076, -2.5121 \pm j0.3554\}$. Since $\bar{A}_{12}V_1 y(t)$ and $\bar{A}_{d12}V_1 y(t-\tau)$ are replaced with $Bu(t)$, and thus from (15.29), J, J_d, J_{2d} and J_{3d} of observer (15.44) are obtained as follows

$$J = FB - M\tilde{C}_1 B, \tag{15.52}$$

$$J_d = F_d B - M\tilde{C}_{d1} B - M_d\tilde{C}_1 B, \tag{15.53}$$

$$J_{dd} = F_{dd} B - M\tilde{C}_{dd1} B - M_d\tilde{C}_{d1} B, \tag{15.54}$$

$$J_{3d} = F_{3d} B - M_d\tilde{C}_{dd1} B. \tag{15.55}$$

Finally, from (15.52)–(15.55), we obtain

$$J = \begin{bmatrix} 1.7222 \\ 1 \\ 2 \\ 0 \\ 0 \end{bmatrix}, \; J_d = \begin{bmatrix} 10.2778 \\ 0 \\ 0 \\ 1 \\ 2 \end{bmatrix}, \; J_{dd} = \begin{bmatrix} 7.6667 \\ 0 \\ 0 \\ 0 \\ 0 \end{bmatrix} \text{ and } J_{3d} = 0_{5\times 1}.$$

Thus, the design of a fifth-order observer to estimate the vector $\tilde{z}_a^v(t)$ is now completed, and the estimation of the required function $\tilde{z}(t)$ defined by (15.42) is obtained as $\hat{\tilde{z}}(t) = [1\ 0\ 0\ 0\ 0]\hat{\tilde{z}}_a^v(t)$, i.e.,

$$\hat{\tilde{z}}(t) = w_1(t) - 3.7222\tilde{y}(t) - 1.6667\tilde{y}(t-\tau), \tag{15.56}$$

$w(t) = [w_1(t)\ w_2(t)\ w_3(t)\ w_4(t)\ w_5(t)]^T$ is the observer state vector and $\dot{w}(t)$ is as defined by (15.44).

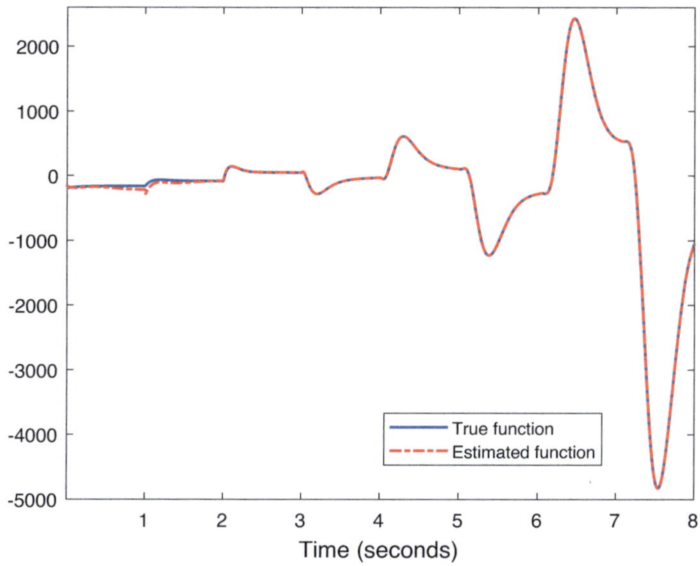

Fig. 15.1 Trajectories of $\tilde{z}(t)$ and $\hat{\tilde{z}}(t)$ (Example 1, a fifth-order functional observer without internal delays using the First Approach, Chap. 15)

To show the performance of the designed generalized observer, Fig. 15.1 shows the trajectories of $\hat{\tilde{z}}(t)$ and $\tilde{z}(t)$ to a step change in $u(t)$. Figure 15.2 shows the error $e(t) = \tilde{z}(t) - \hat{\tilde{z}}(t)$. Observe that the error converges to zero within two seconds, as expected.

It is worthwhile to mention that this is the first time such a generalized observer of order much higher than n is designed to estimate the function $\tilde{z}(t)$ defined by (15.1) for time-delay system (5.28) and (5.29). Thus, the results presented in this chapter are significant, as from now on we can expand the search for a more complex generalized functional observer to meet our requirement.

Second Approach: Next, we design a state observer to estimate $\bar{x}_1(t)$ and then $\hat{\tilde{z}}(t)$ is realized according to (15.41). From (12.49), if we let $z(t) = \bar{x}_1(t)$, i.e., $F = I_2$, then estimating $z(t)$ is the same as estimating the vector $\bar{x}_1(t)$. However, it is found that condition (12.43) is not satisfied. This is also reaffirmed by checking condition (15.47), where we let $F = I_2$ and $F_d = F_{dd} = F_{3d} = 0_{2\times 2}$. What this implies is that it is not possible to estimate $\bar{x}_1(t)$ (i.e., $z(t)$) by using a second-order observer of either the form (12.39) and (12.40) or a more general form (15.43) and (15.44). Note that in checking condition (15.47), it was found that $\text{rank}\big[\hat{\Theta}_{rr}\big] = 8$, but $\text{rank}\begin{bmatrix}\hat{\Theta}_{rr}\\ \hat{\Upsilon}\end{bmatrix} = 10$. Thus, the only way that we can design an observer to estimate $\bar{x}_1(t)$ is to increase its order. Thanks to our results developed in this chapter, it is now possible to do so.

Similar to the First Approach and as discussed previously, the simplest way to satisfy condition (15.47) is to increase the order of the designed observer so that

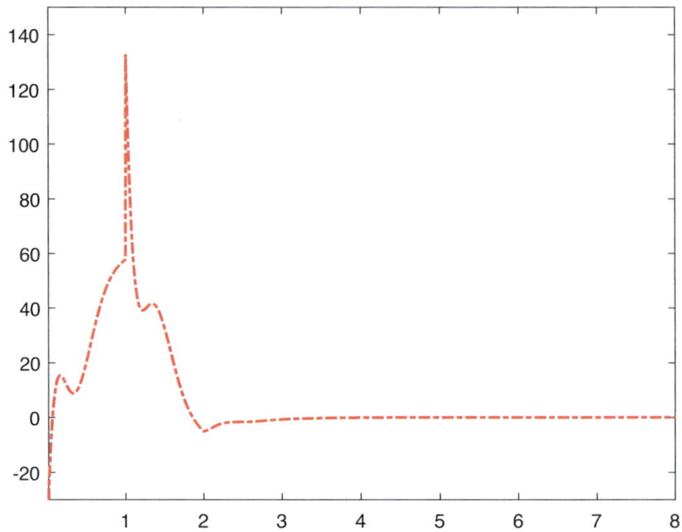

Fig. 15.2 Trajectory of $e(t) = \tilde{z}(t) - \hat{\tilde{z}}(t)$ (Example 1, using the First Approach, Chap. 15)

rank$\left[\hat{\Theta}_{rr}\right] = 11$. To achieve this, we need to add at least three more rows. Again, the three additional rows must be linearly independent from $\hat{\Theta}_{rr}$. There exist many possibilities to select them. Obviously, if an added row is the same as the one given in (15.42) (i.e., the first row in (15.50)), then $\hat{\tilde{z}}(t)$ can be directly estimated. However, in this analysis, to show the difference between the two approaches, we specifically design an observer to estimate $\bar{x}_1(t)$ and then $\hat{\tilde{z}}(t)$ is realized according to (15.41).

Note that we can use the observer that we designed in the First Approach to obtain $\hat{\bar{x}}_1(t)$. From (15.50), $\hat{\bar{x}}_1(t)$ can be obtained from $\hat{\tilde{z}}_a^v(t)$ as follows

$$\hat{\bar{x}}_1(t) = \begin{bmatrix} 0\,1\,0\,0\,0 \\ 0\,0\,1\,0\,0 \end{bmatrix} \hat{\tilde{z}}_a^v(t) = \begin{bmatrix} w_2(t) \\ w_3(t) \end{bmatrix}. \tag{15.57}$$

Hence, from (15.41), $\hat{\tilde{z}}(t)$ is realized as follows

$$\begin{aligned} \hat{\tilde{z}}(t) = [0\ -1]\begin{bmatrix} w_2(t) \\ w_3(t) \end{bmatrix} + [-4\ -3]\begin{bmatrix} w_2(t-\tau) \\ w_3(t-\tau) \end{bmatrix} \\ + [-27\ \ 2]\begin{bmatrix} w_2(t-2\tau) \\ w_3(t-2\tau) \end{bmatrix} + [-10\ \ 0]\begin{bmatrix} w_2(t-3\tau) \\ w_3(t-3\tau) \end{bmatrix}. \end{aligned} \tag{15.58}$$

Notice that (15.56) is different from (15.58), where $\hat{\tilde{z}}(t)$ is directly obtained from $w_1(t)$ and both outputs $\tilde{y}(t)$ and $\tilde{y}(t-\tau)$. On the other hand, in (15.58), $\hat{\tilde{z}}(t)$ comprises multiple delays of $w(t)$. It is thus expected that realizing (15.58) is more

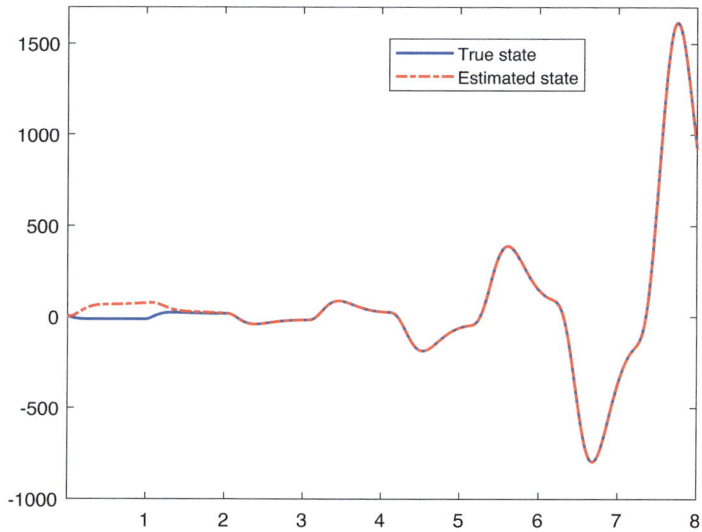

Fig. 15.3 Trajectories of $\hat{\bar{x}}_{11}(t)$ and $\bar{x}_{11}(t)$ (Example 1, a fifth-order functional observer without internal delays using the Second Approach, Chap. 15)

complicated than (15.56) which involves more computation blocks, such as delay units and proportional gains (see, Chap. 14).

Before we compare the performance of both realizations, Figs. 15.3 and 15.4 show the responses of $\hat{\bar{x}}_{11}(t)$, $\bar{x}_{11}(t)$ and $\hat{\bar{x}}_{12}(t)$, $\bar{x}_{12}(t)$, respectively, to a step change in $u(t)$. It is clear from the figures that asymptotic tracking is achieved.

To show the performance of two realizations (15.56) and (15.58), Fig. 15.5 shows the responses of $\hat{\tilde{z}}(t)$ and $\tilde{z}(t)$ to a step change in $u(t)$ using two approaches. It is clear that the First Approach gives better transient performance than the Second Approach (due to (15.58) having multiple delays which results in more transient errors and hence a longer settling time). Note that this fact was highlighted in Remark 1 (Chap. 14). Again, via this numerical example, we can see clearly the advantages of employing a generalized functional observer to directly realize $\tilde{z}(t)$.

Finally, observer (15.56) is used to estimate the unknown input $d(t)$ in Numerical Example 3, Sect. 5.4.1 of Chap. 5.

Numerical Example 2: We reconsider Numerical Example 1 but we change matrices $\tilde{C}_1$ and $\tilde{C}_{d1}$, where $\tilde{C}_1 = [0\ 0]$ and $\tilde{C}_{d1} = [1\ 1]$. The designed functional observer is then used to estimate the unknown input $d(t)$ for system (5.1) and (5.2), where C is not a full row matrix (see Numerical Example 4 in Sect. 5.4.2, Chap. 5). Notice that in this example, matrix $\tilde{C}_1 = [0\ 0]$ and thus the output vector $\tilde{y}(t)$ as defined by (12.48) does not contain the instantaneous vector $\bar{x}_1(t)$. Thanks to our unified observer design method, this does not cause any additional difficulties in the observer design process. Our objective here is to design a generalized observer of form (15.43) and (15.44) to estimate $\tilde{z}(t)$ defined by (15.42).

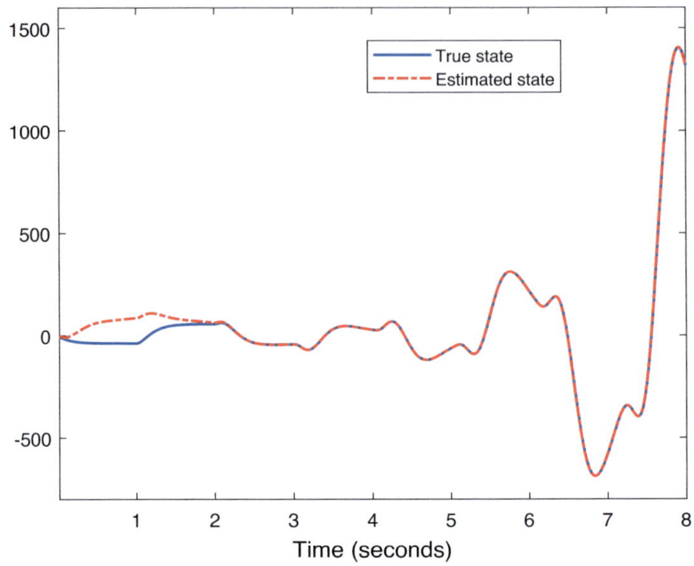

Fig. 15.4 Trajectories of $\hat{\bar{x}}_{12}(t)$ and $\bar{x}_{12}(t)$ (Example 1, a fifth-order functional observer without internal delays using the Second Approach, Chap. 15)

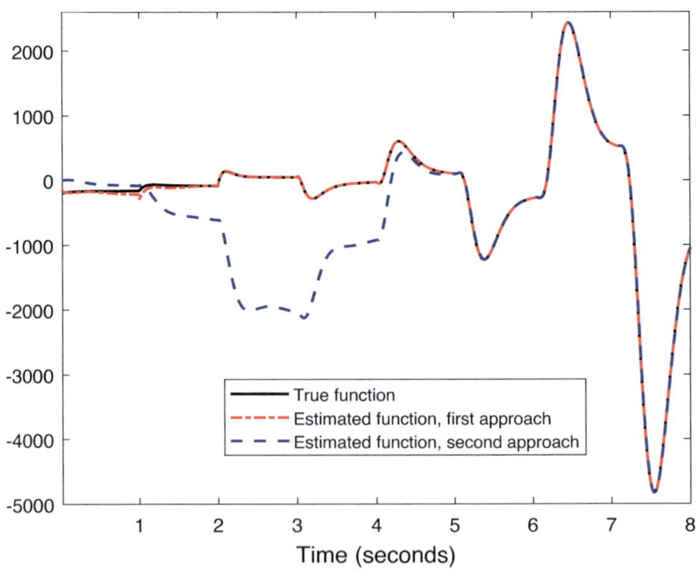

Fig. 15.5 Trajectories of $\tilde{z}(t)$ and $\hat{\tilde{z}}(t)$ (Example 1, First approach versus Second approach, Chap. 15)

By following the same process as in Numerical Example 1 and with F^v, F^v_d, F^v_{dd} and F^v_{3d} chosen as in (15.50), we obtain the following fifth-order generalized observer

$$\hat{\tilde{z}}(t) = w_1(t) - 4.2778\tilde{y}(t) - 1.6667\tilde{y}(t-\tau), \tag{15.59}$$

$$\begin{aligned}\dot{w}(t) &= Nw(t) + G\tilde{y}(t) + G_d\tilde{y}(t-\tau) + G_{dd}\tilde{y}(t-2\tau)\\ &\quad + Ju(t) + J_d u(t-\tau) + J_{dd}u(t-2\tau),\end{aligned} \tag{15.60}$$

where

$$N = \begin{bmatrix} -3.3687 & 0 & 3.6313 & 21.0572 & -1.2963 \\ 0 & -10 & -2 & -28.3333 & 3.6667 \\ 0 & 0 & -7 & -19.3333 & -0.3333 \\ 1 & 0 & 1 & -5.5 & 1.5 \\ -0.0909 & 0 & -0.0909 & 3.2424 & -3.6667 \end{bmatrix}, G = \begin{bmatrix} 2.1007 \\ -1.6667 \\ -5.6667 \\ -4.7778 \\ -3.2172 \end{bmatrix},$$

$$G_d = \begin{bmatrix} -21.0926 \\ 0 \\ 0 \\ -1.6667 \\ -5.6667 \end{bmatrix}, G_{dd} = \begin{bmatrix} -21.3333 \\ 0 \\ 0 \\ 0 \\ 0 \end{bmatrix}, J = \begin{bmatrix} -2 \\ 1 \\ 2 \\ 0 \\ 0 \end{bmatrix}, J_d = \begin{bmatrix} 2.8333 \\ 0 \\ 0 \\ 1 \\ 2 \end{bmatrix},$$

$$J_{dd} = \begin{bmatrix} 7.6667 \\ 0 \\ 0 \\ 0 \\ 0 \end{bmatrix}.$$

Note that the eigenvalues of N are fixed and stable, where

$$\lambda\{N\} = \{-10, -7.5333, -7, -2.5010 \pm j0.6110\}.$$

Figure 15.6 shows the trajectories of $\hat{\tilde{z}}(t)$ and $\tilde{z}(t)$ to a step change in $u(t)$. It is clear from the figure that asymptotic tracking is achieved. Observer (15.59) and (15.60) is used to estimate the unknown input $d(t)$ in Numerical Example 4, Sect. 5.4.2 in Chap. 5.

15.4 Concluding Remark

In this chapter, we have considered the problem of designing observers to estimate a generalized functional of time-delay system (5.28) and (5.29) that contains multiple delays in both state and output vectors. Via addressing the estimation of a generalized functional containing multiple state-delayed terms, namely $\tilde{z}(t) = F\bar{x}_1(t) + F_d\bar{x}_1(t-\tau) + F_{dd}\bar{x}_1(t-2\tau) + F_{3d}\bar{x}_1(t-3\tau)$, we have shown the advantages and generality of our results. It is clear from Numerical Examples 1 and 2, the addition of F^v, F^v_d, F^v_{dd} and F^v_{3d} brings extra degrees of freedom to the design of generalized functional observers. With our results, we can expand the dimensions of the

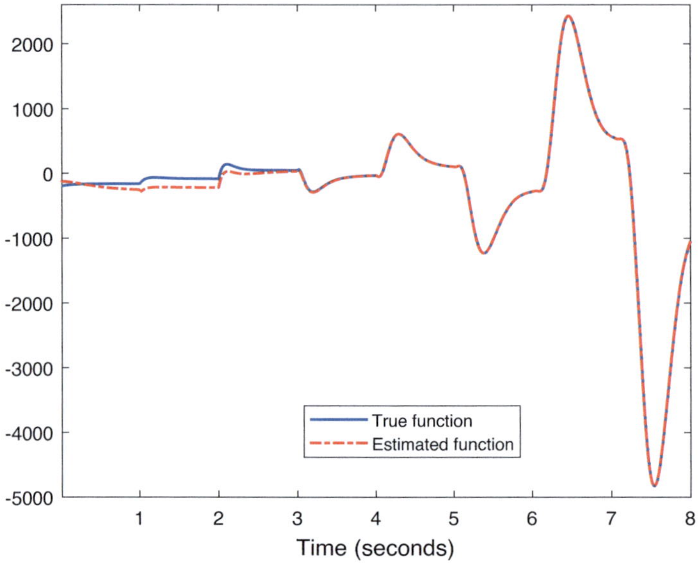

Fig. 15.6 Trajectories of $\hat{\tilde{z}}(t)$ and $\tilde{z}(t)$ (Example 2, a fifth-order functional observer without internal delays, Chap. 15)

designed observers to meet their existence conditions. Thus, the design of an observer for time-delay system (5.28) and (5.29) can be extended beyond order n.

As we continue our exploratory journey, in Chap. 16, we will further demonstrate the generality and advantages of estimating an even more general functional, $z_{2\tau}(t) = Fx(t) + F_d x(t-\tau) + F_{dd} x(t-2\tau)$ for time-delay system (16.1) and (16.2).

Chapter 16
Further Discussions on Generalized Functional Observers

16.1 Motivations: The Case for $z_{2\tau}(t) = Fx(t) + F_d x(t-\tau) + F_{dd} x(t-2\tau)$

So far in this book, we have given a plethora of numerical examples to show the advantages of estimating a generalized functional that includes multiple state-delayed terms. Particularly, in Chap. 13, we considered time-delay systems of the forms (4.3)–(4.4) and (5.13)–(5.14) and demonstrated the advantages of estimating the functional $\bar{z}(t) = F\bar{x}_1(t) + F_d\bar{x}_1(t-\tau)$. In this chapter, we further demonstrate the generality and advantages of estimating an even more general functional, $z_{2\tau}(t) = Fx(t) + F_d x(t-\tau) + F_{dd} x(t-2\tau)$ for the following time-delay system (16.1)–(16.2) by using a more general observer structure. The results of this chapter can be used for the estimation of various unknown inputs discussed earlier in this book, particularly in Chap. 6.

16.2 System Description, Problem Statement, General Observer Structure

We consider the following linear system with a time delay in both the state and output vectors

$$\dot{x}(t) = Ax(t) + A_d x(t-\tau) + Bu(t), \quad (16.1)$$

$$y(t) = Cx(t) + C_d x(t-\tau), \quad (16.2)$$

where $x(t) \in \mathbb{R}^n$, $y(t) \in \mathbb{R}^p$ and $u(t) \in \mathbb{R}^r$ are, respectively, the state vector, output vector and control input vector. Matrices $A, A_d \in \mathbb{R}^{n \times n}$, $B \in \mathbb{R}^{n \times r}$ and $C, C_d \in \mathbb{R}^{p \times n}$ are known constant matrices. Term τ is a known constant time delay. Our aim is to design an observer to estimate the following vector with multiple delayed terms

H. Trinh et al., *Unknown Inputs Estimation in Linear Time-Delay Systems Using Generalized Functional Observers*, Studies in Systems, Decision and Control 648,
https://doi.org/10.1007/978-3-032-11870-7_16

$$z_{2\tau}(t) = Fx(t) + F_d x(t-\tau) + F_{dd} x(t-2\tau), \tag{16.3}$$

where $z_{2\tau}(t) \in \mathbb{R}^m$, and F, F_d, $F_{dd} \in \mathbb{R}^{m\times n}$ are known constant matrices. Note that when $C_d \neq 0_{p\times n}$, and F_d, $F_{dd} \neq 0_{m\times n}$, the problem of estimating $z_{2\tau}(t)$ is a more general problem than just estimating a linear function of the state vector, $z(t) = Fx(t)$, which was considered in the literature (see, [1–4]). For the case where $m = 1$, $C = 0_{p\times n}$ and $F = 0_{1\times n}$, this problem is reduced to estimating vector (6.13) for time-delay system (6.15)–(6.16), which is used to estimate the delayed unknown input $d(t-\tau)$ as defined by (6.14).

To estimate the required vector (16.3) for system (16.1)–(16.2), we consider the following generalized observer with multiple delayed terms

$$\hat{z}_{2\tau}(t) = w(t) + My(t) + M_d y(t-\tau), \tag{16.4}$$

$$\begin{aligned} \dot{w}(t) &= Nw(t) + N_d w(t-\tau) + N_{dd} w(t-2\tau) + Gy(t) + G_d y(t-\tau) + G_{dd} y(t-2\tau) \\ &+ G_{3d} y(t-3\tau) + Ju(t) + J_d u(t-\tau) + J_{dd} u(t-2\tau), \end{aligned} \tag{16.5}$$

where $w(\theta) = \rho(\theta)$, $\forall\theta \in [0, 2\tau]$, $w(t) \in \mathbb{R}^m$, and $\hat{z}_{2\tau}(t)$ is the estimate of $z_{2\tau}(t)$. Matrices M, M_d, N, N_d, N_{dd}, G, G_d, G_{dd}, G_{3d}, J, J_d and J_{dd} are to be determined such that $\hat{z}_{2\tau}(t)$ converges asymptotically to $z_{2\tau}(t)$.

16.2.1 Observer Existence Conditions

By following a similar analytical approach, as presented in Chap. 13 (here, the detail is omitted), we present existence conditions for observer (16.4)–(16.5).

Existence Conditions: The estimate $\hat{z}_{2\tau}(t)$ converges asymptotically to $z_{2\tau}(t)$ for any initial condition $w(\theta)$ and any $u(t)$ provided that the following conditions hold

$$\dot{e}(t) = Ne(t) + N_d e(t-\tau) + N_{dd} e(t-2\tau) \text{ is asymptotically stable,} \tag{16.6}$$

$$J - FB + MCB = 0, \tag{16.7}$$

$$J_d - F_d B + MC_d B + M_d CB = 0, \tag{16.8}$$

$$J_{dd} - F_{dd} B + M_d C_d B = 0, \tag{16.9}$$

$$NF + \tilde{T}C + MCA - FA = 0, \tag{16.10}$$

$$NF_d + N_d F + \tilde{T}C_d + \hat{W}C + M(CA_d + C_d A) + M_d CA - FA_d - F_d A = 0, \tag{16.11}$$

$$\begin{aligned} &NF_{dd} + N_d F_d + N_{dd} F + \hat{U}C + \hat{W}C_d + M_d(CA_d + C_d A) + MC_d A_d - F_d A_d \\ &- F_{dd} A = 0, \end{aligned} \tag{16.12}$$

$$N_d F_{dd} + N_{dd} F_d + \hat{Y} C + \hat{U} C_d + M_d C_d A_d - F_{dd} A_d = 0, \tag{16.13}$$

$$N_{dd} F_{dd} + \hat{Y} C_d = 0, \tag{16.14}$$

where $\tilde{T}$, $\hat{W}$, $\hat{U}$ and $\hat{Y}$ are defined by (10.84), (15.22), (15.23) and (15.24), respectively.

Thus, to find the observer parameters to satisfy conditions (16.6)–(16.14), we need to solve matrix equations (16.10)–(16.14), while ensuring the time-delay error system (16.6) is asymptotically stable. We augment (16.10)–(16.14) into the following form

$$X_\tau \Theta_\tau = \Upsilon_\tau, \tag{16.15}$$

where

$$X_\tau = [N \;\; N_d \;\; N_{dd} \;\; \tilde{T} \;\; \hat{W} \;\; \hat{U} \;\; \hat{Y} \;\; M \;\; M_d], \tag{16.16}$$

$$\Theta_\tau = \begin{bmatrix} F & F_d & F_{dd} & 0 & 0 \\ 0 & F & F_d & F_{dd} & 0 \\ 0 & 0 & F & F_d & F_{dd} \\ C & C_d & 0 & 0 & 0 \\ 0 & C & C_d & 0 & 0 \\ 0 & 0 & C & C_d & 0 \\ 0 & 0 & 0 & C & C_d \\ CA & C_\tau & C_d A_d & 0 & 0 \\ 0 & CA & C_\tau & C_d A_d & 0 \end{bmatrix}, \tag{16.17}$$

$$C_\tau = C A_d + C_d A, \tag{16.18}$$

and

$$\Upsilon_\tau = [FA \;\; FA_d + F_d A \;\; F_d A_d + F_{dd} A \;\; F_{dd} A_d \;\; 0]. \tag{16.19}$$

From (16.15), X_τ exists if and only if the following condition holds

$$\operatorname{rank} \begin{bmatrix} \Theta_\tau \\ \Upsilon_\tau \end{bmatrix} = \operatorname{rank}\big[\Theta_\tau\big]. \tag{16.20}$$

Upon satisfying condition (16.20), Eq. (16.15) is solved for the unknown matrix X_τ. In the same way as was done previously (see, Chap. 12, (12.25)–(12.33)), the solution, X_τ, is expressed in terms of an arbitrary matrix, Z_τ, and the time-delay error system (16.6) can be expressed as follows

$$\dot{e}(t) = (N_1 + Z_\tau N_2)e(t) + (N_{d1} + Z_\tau N_{d2})e(t-\tau) + (N_{dd1} + Z_\tau N_{dd2})e(t-2\tau), \tag{16.21}$$

where N_1, N_2, N_{d1}, N_{d2}, N_{dd1} and N_{dd2} are known constant matrices. The reader can refer to Chap. 20 for stability and stabilizability conditions of (16.21). Once asymptotic stability of the time-delay error system (16.21) is established, the rest of the observer parameters can be obtained.

16.2.2 Numerical Examples

In this section, we provide numerical examples to extensively discuss the design of various types of observers for time-delay system (16.1)–(16.2). Through the examples, we show the salient features of our design method and the generality of our results over those reported in [1, 2].

Numerical Example 1

This example is used to show the generality of our results. We consider a numerical example taken from Example 4.3 in [2], where $\tau = 1s$, and matrices A, A_d, B, C, C_d, F, F_d and F_{dd} are given below

$$A = \begin{bmatrix} -1 & 2 & 1 \\ 1 & 0 & 2 \\ 1 & 3 & -2 \end{bmatrix}, \quad A_d = \begin{bmatrix} -1 & 0 & 0 \\ 2 & 1 & 0 \\ 1 & 2 & -1 \end{bmatrix}, \quad B = \begin{bmatrix} 1 \\ 1 \\ 0 \end{bmatrix}, \quad C = \begin{bmatrix} 1 & 0 & 0 \end{bmatrix},$$

$C_d = \begin{bmatrix} 0 & 0 & 0 \end{bmatrix}$, $F = \begin{bmatrix} 0 & 0 & 1 \end{bmatrix}$ and $F_d = F_{dd} = \begin{bmatrix} 0 & 0 & 0 \end{bmatrix}$.

Note that in this example, we have $C_d = F_d = F_{dd} = 0_{1\times 3}$, and thus the considered time-delay system belongs to a special case of system (16.1)–(16.2) (i.e., the output vector $y(t)$ does not contain any delayed-state terms). Moreover, the vector to be estimated is $z(t) = Fx(t) = x_3(t)$ instead of a more general vector $z_{2\tau}(t)$ as defined by (16.3).

Firstly, in [1], the author proposed the following obsever to estimate $z(t)$ (the reader can also refer to Chap. 10, Sect. 10.2 for an alternative design approach to [1])

$$\hat{z}(t) = w(t) + My(t), \tag{16.22}$$
$$\dot{w}(t) = Nw(t) + N_d w(t-\tau) + Gy(t) + G_d y(t-\tau) + Ju(t), \tag{16.23}$$

and presented their necessary existence condition [1] as follows

$$\operatorname{rank}\begin{bmatrix} F & 0 \\ 0 & F \\ C & 0 \\ 0 & C \\ CA & CA_d \\ FA & FA_d \end{bmatrix} = \operatorname{rank}\begin{bmatrix} F & 0 \\ 0 & F \\ C & 0 \\ 0 & C \\ CA & CA_d \end{bmatrix}. \tag{16.24}$$

In this example, condition (16.24) is, however, not satisfied. Thus, to estimate $z(t)$, a first-order observer of the form (16.22)–(16.23) does not exist.

We note here that condition (16.24) can be derived from our general condition (16.20). This is achieved by letting C_d, F_d, F_{dd}, M_d, N_{dd}, G_{dd}, G_{3d}, J_d and J_{dd} be zero matrices. Accordingly, matrix equations (16.10)–(16.14) are reduced to the following two equations

$$NF + \tilde{T}C + MCA - FA = 0, \tag{16.25}$$

$$N_d F + \bar{W}C + MCA_d - FA_d = 0, \tag{16.26}$$

where $\bar{W}$ is defined by (11.34). Hence from (16.25)–(16.26), we can deduce that the necessary and sufficient for their solvability is indeed condition (16.24).

On the other hand, in [2], the authors proposed a more general obsever than the one considered in [1] to estimate $z(t)$ (refer to Chap. 10, Sect. 10.3), where

$$\hat{z}(t) = w(t) + My(t) + M_d y(t-\tau), \tag{16.27}$$

$$\begin{aligned} \dot{w}(t) = {} & Nw(t) + N_d w(t-\tau) + Gy(t) + G_d y(t-\tau) + G_{dd} y(t-2\tau) \\ & + Ju(t) + J_d u(t-\tau). \end{aligned} \tag{16.28}$$

The necessary existence condition for observer (16.27)–(16.28) is as follows [2]

$$\operatorname{rank}\begin{bmatrix} F & 0 & 0 \\ 0 & F & 0 \\ C & 0 & 0 \\ 0 & C & 0 \\ 0 & 0 & C \\ CA & CA_d & 0 \\ 0 & CA & CA_d \\ FA & FA_d & 0 \end{bmatrix} = \operatorname{rank}\begin{bmatrix} F & 0 & 0 \\ 0 & F & 0 \\ C & 0 & 0 \\ 0 & C & 0 \\ 0 & 0 & C \\ CA & CA_d & 0 \\ 0 & CA & CA_d \end{bmatrix}. \tag{16.29}$$

In this numerical example, condition (16.29) is satisfied and the authors in [2] designed a first-order observer of the form (16.27)–(16.28) to asymptotically estimate $z(t)$.

Note also that condition (16.29) can be derived from our condition (16.20). This is done by letting C_d, F_d, F_{dd}, N_{dd}, G_{3d} and J_{dd} to be zero matrices. Accordingly, matrix equations (16.10)–(16.14) are reduced to the following three equations

$$NF + \tilde{T}C + MCA - FA = 0, \tag{16.30}$$

$$N_d F + \hat{W}C + MCA_d + M_d CA - FA_d = 0, \tag{16.31}$$

$$\tilde{U}C + M_d CA_d = 0, \tag{16.32}$$

where $\tilde{U}$ is as defined by (10.86). Again, from (16.30)–(16.32), we can deduce the necessary and sufficient condition for their solvability, which is indeed condition (16.29).

The above analysis shows the generality of observer (16.4)–(16.5) and its existence condition (16.20). To provide further analysis of our results, we now let A and A_d be as follows

$$A = \begin{bmatrix} -1 & 2 & 1 \\ 1 & -4 & 2 \\ 1 & 3 & -15 \end{bmatrix} \text{ and } A_d = \begin{bmatrix} 0 & 0 & 1 \\ 2 & -2 & 0 \\ 1 & 2 & -5 \end{bmatrix}.$$

Now, with A and A_d given above, we test conditions (16.24) and (16.29). However, it is found that neither of the conditions are satisfied. Thus, first-order observers (16.22)–(16.23) and (16.27)–(16.28) do not exist. We note here that a second-order observer (16.22)–(16.23) exists and can be designed to estimate $z(t)$.

We now turn our attention to observer (16.4)–(16.5). First, let us deduce some resuts for a specific case where $C_d = 0$, $F_d = 0$ and $F_{dd} = 0$. In this case, we have from (16.9), $J_{dd} = 0$, and matrix equations (16.10)–(16.14) are reduced to the following four equations

$$NF + \tilde{T}C + MCA - FA = 0, \tag{16.33}$$

$$N_d F + \hat{W}C + MCA_d + M_d CA - FA_d = 0, \tag{16.34}$$

$$N_{dd} F + \hat{U}C + M_d CA_d = 0, \tag{16.35}$$

$$\hat{Y}C = 0. \tag{16.36}$$

Since C is a full row rank matrix, from (16.36), we have $\hat{Y} = 0$ as the necessary and sufficient condition. Hence from (15.24), G_{3d} is obtained as follows

$$G_{3d} = N_{dd} M_d. \tag{16.37}$$

From (16.33)–(16.35), we obtain the following necessary and sufficient condition for their solvability

$$\text{rank}\begin{bmatrix} F & 0 & 0 \\ 0 & F & 0 \\ 0 & 0 & F \\ C & 0 & 0 \\ 0 & C & 0 \\ 0 & 0 & C \\ CA & CA_d & 0 \\ 0 & CA & CA_d \\ FA & FA_d & 0 \end{bmatrix} = \text{rank}\begin{bmatrix} F & 0 & 0 \\ 0 & F & 0 \\ 0 & 0 & F \\ C & 0 & 0 \\ 0 & C & 0 \\ 0 & 0 & C \\ CA & CA_d & 0 \\ 0 & CA & CA_d \end{bmatrix}. \tag{16.38}$$

Next, condition (16.38) is tested for this example and found to be satisfied. We can then design a first-order observer to asymptotically estimate $z(t) = x_3(t)$, where

$$\hat{x}_3(t) = w(t) + 1.5y(t) + y(t-1), \tag{16.39}$$

$$\begin{aligned}\dot{w}(t) = &-16.5w(t) - 7.5w(t-1) - w(t-2) - 22.25y(t) - 25.75y(t-1) \\ &- 9y(t-2) - y(t-3) - 1.5u(t) - u(t-1).\end{aligned} \tag{16.40}$$

Note that in this example, the error dynamic system (16.6) (i.e., $\dot{e}(t) = -16.5e(t) - 7.5e(t-1) - e(t-2)$ is 0.5472-exponentially stable according to LMI (20.7) in Chap. 20. Thus, even though a first-order observer of either the form (16.22)–(16.23) or (16.27)–(16.28) does not exist in this example, we can still design a first-order observer (16.39)–(16.40), which needs a few more extra delayed terms to asymptotically estimate $z(t) = x_3(t)$. Compared with observer (16.27)–(16.28), there are two extra delayed terms in observer (16.39)–(16.40), namely, $-w(t-2)$ and $-y(t-3)$. This is a trade-off for employing a first-order observer to estimate $z(t)$ where more delay units are needed. Otherwise a second-order observer of the form (16.22)–(16.23) can be designed to estimate $z(t)$. Thus, the designer can have more options to design observers. Figure 16.1 shows the trajectories of $x_3(t)$ and $\hat{x}_3(t)$. It is clear that asymptotic estimation has been achieved.

For the sake of completeness and comparison, we also design a second-order observer of the form (16.22)–(16.23) to estimate $z(t)$. Here, even though condition (16.24) is not satisfied for the case where $F = [0\ \ 0\ \ 1]$, we can increase the order of the observer to a second-order observer by inserting an additional row into F. So far, this approach has been extensively discussed in this book. Now, we design a second-order observer to estimate the following vector

$$z_a(t) = F_a x(t) = \begin{bmatrix} 0\,0\,1 \\ 0\,1\,0 \end{bmatrix} x(t) = \begin{bmatrix} x_3(t) \\ x_2(t) \end{bmatrix}. \tag{16.41}$$

Hence, $\hat{z}(t)$ can be extracted from $\hat{z}_a(t)$ as $\hat{z}(t) = [1\ \ 0]\hat{z}_a(t)$.

Next, we check condition (16.24) with F being replaced with F_a and it is found that the condition is satisfied. For the purpose of comparison, we select $\beta = 0.5472$ (which is the same value as the one obtained for the first-order observer (16.39)–(16.40)) and we solve the 0.5472-exponential stabilization of system (20.1) by using

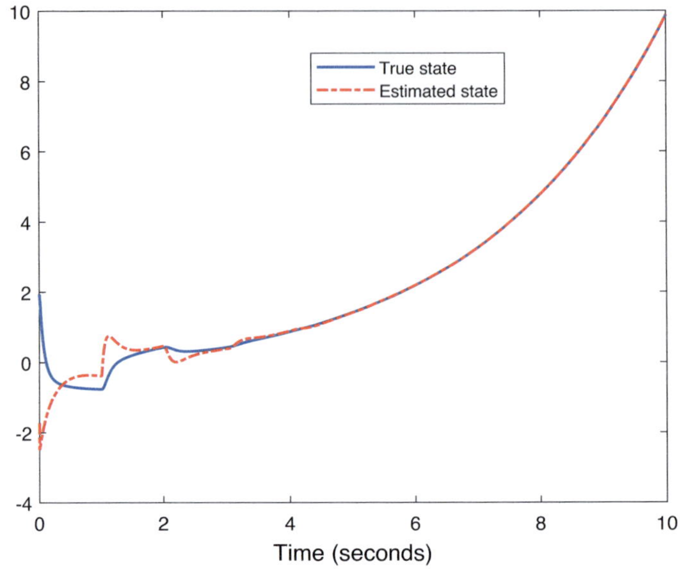

Fig. 16.1 Trajectories of $x_3(t)$ and $\hat{x}_3(t)$ (Numerical Example 1, in this chapter)

LMI (20.5) in Chap. 20. We then obtain the following second-order observer to estimate $z(t) = x_3(t)$

$$\hat{x}_3(t) = w_1(t) - 1.7962y(t), \tag{16.42}$$

$$\begin{aligned}\dot{w}_1(t) &= -13.2038w_1(t) + 6.5924w_2(t) - 3.2038w_1(t-1) + 2w_2(t-1)\\ &+ 34.6993y(t) + 10.3282y(t-1) + 1.7962u(t),\\ \dot{w}_2(t) &= 0.2132w_1(t) - 7.5735w_2(t) - 1.7868w_1(t-1) - 2w_2(t-1)\\ &- 11.1284y(t) + 1.6358y(t-1) - 0.7868u(t).\end{aligned} \tag{16.43}$$

Observer (16.39)–(16.40) requires six delay units (i.e., $y(t-1)$ $y(t-2)$, $y(t-3)$, $w(t-1)$, $w(t-2)$ and $u(t-1)$), seven proportional gains and one integrator for its implementation. On the other hand, compared with observer (16.39)–(16.40), we see that observer (16.42)–(16.43) requires three fewer delay units (i.e., it requires $w_1(t-1)$, $w_2(t-1)$ and $y(t-1)$) but with eight more proportional gains and one more integrator for its implementation. We also note that better error convergent rate can be achieved by employing a second-order observer (16.42)–(16.43) since the β-exponential stabilization of the error system (20.1) (see, LMI (20.5) in Chap. 20) results in a larger β value than for the β-exponential stability of system (20.2) (see, LMI (20.7) in Chap. 20).

Numerical Example 2

In this example, we design an observer to estimate $z_{2\tau}(t)$ as defined by (16.3) for system (16.1)–(16.2), where $\tau = 1s$, matrices A, A_d, B, C, C_d, F, F_d and F_{dd} are given below

$$A = \begin{bmatrix} -3 & 2 \\ 1 & -7 \end{bmatrix}, \quad A_d = \begin{bmatrix} -2 & 2 \\ 0 & 3 \end{bmatrix}, \quad B = \begin{bmatrix} 1 \\ 2 \end{bmatrix}, \quad C = \begin{bmatrix} 1 & 0 \end{bmatrix}, \quad C_d = \begin{bmatrix} 1 & 1 \end{bmatrix},$$

$F = \begin{bmatrix} 1 & 2 \end{bmatrix}$, $F_d = \begin{bmatrix} 3 & 4 \end{bmatrix}$ and $F_{dd} = \begin{bmatrix} 5 & 6 \end{bmatrix}$.

Note that the output contains delayed state variables, i.e., $y(t) = x_1(t) + x_1(t-1) + x_2(t-1)$, and the problem of estimating $z_{2\tau}(t) = x_1(t) + 2x_2(t) + 3x_1(t-1) + 4x_2(t-1) + 5x_1(t-2) + 6x_2(t-2)$ is a more general estimation problem than the ones discussed in [1, 2].

To design an observer to estimate $z_{2\tau}(t)$, we first consider a first-order observer of form (16.4)–(16.5). For this, we test condition (16.20), but it is found that the condition is not satisfied. Thus, a first-order observer of form (16.4)–(16.5) does not exist in this numerical example.

As discussed extensively in this book, we increase the order of the observer and design a second-order observer to estimate $z_{2\tau}(t)$ and an extra function, $x_1(t)$. Thus, let us design a second-order observer to estimate the following vector

$$z_a(t) = \begin{bmatrix} z_{2\tau}(t) \\ x_1(t) \end{bmatrix} = F_a x(t) + F_{ad} x(t-1) + F_{add} x(t-2), \tag{16.44}$$

where $F_a = \begin{bmatrix} 1 & 2 \\ 1 & 0 \end{bmatrix}$, $F_{ad} = \begin{bmatrix} 3 & 4 \\ 0 & 0 \end{bmatrix}$ and $F_{add} = \begin{bmatrix} 5 & 6 \\ 0 & 0 \end{bmatrix}$.

Here, we have rank $\begin{bmatrix} F_a & F_{ad} & F_{add} \\ C & C_d & 0_{1\times 2} \end{bmatrix} = 3$, and when F, F_d and F_{dd} are replaced with F_a, F_{ad} and F_{add}, respectively, it is found that condition (16.20) is satisfied. Provided that the error system (16.6) is asymptotically stable, we can complete the design of a second-order observer (16.4)–(16.5) to asymptotically estimate $z_a(t)$. Hence, $\hat{z}_{2\tau}(t)$ can be obtained from $\hat{z}_a(t)$ as $\hat{z}_{2\tau}(t) = [1 \ \ 0]\hat{z}_a(t)$. Furthermore, $\hat{x}_1(t)$ is also obtained as $\hat{x}_1(t) = [0 \ \ 1]\hat{z}_a(t)$.

Since condition (16.20) is satisfied, we can further test to see whether an observer (16.4)–(16.5) without any internal delay, i.e., $N_d = N_{dd} = 0_{2\times 2}$, exists or not. In this case, condition (16.20) is replaced with the following condition

$$\text{rank}\begin{bmatrix} \bar{\Theta}_\tau \\ \Upsilon_{\tau a} \end{bmatrix} = \text{rank}\big[\bar{\Theta}_\tau\big], \tag{16.45}$$

where

$$\bar{\Theta}_\tau = \begin{bmatrix} F_a & F_{ad} & F_{add} & 0 & 0 \\ C & C_d & 0 & 0 & 0 \\ 0 & C & C_d & 0 & 0 \\ 0 & 0 & C & C_d & 0 \\ 0 & 0 & 0 & C & C_d \\ CA & C_\tau & C_d A_d & 0 & 0 \\ 0 & CA & C_\tau & C_d A_d & 0 \end{bmatrix}, \tag{16.46}$$

and

$$\Upsilon_{\tau a} = [F_a A \;\; F_a A_d + F_{ad} A \;\; F_{ad} A_d + F_{add} A \;\; F_{add} A_d \;\; 0]. \tag{16.47}$$

Thus, condition (16.46) is tested; however, it is not satisfied. This implies a second-order observer (16.4)–(16.5) without any internal delay, i.e., $N_d = N_{dd} = 0_{2\times 2}$, does not exist.

Next, we further test condition (16.48) to see whether an observer (16.4)–(16.5) with $N_{dd} = 0_{2\times 2}$ exists or not. In this case, condition (16.48) is satisfied and thus we can proceed to design a second-order observer with a simpler structure (i.e., $N_{dd} = 0_{2\times 2}$) to estimate $z_a(t)$.

$$\text{rank}\begin{bmatrix} \tilde{\Theta}_\tau \\ \Upsilon_{\tau a} \end{bmatrix} = \text{rank}\big[\tilde{\Theta}_\tau\big], \tag{16.48}$$

where

$$\tilde{\Theta}_\tau = \begin{bmatrix} F_a & F_{ad} & F_{add} & 0 & 0 \\ 0 & F_a & F_{ad} & F_{add} & 0 \\ C & C_d & 0 & 0 & 0 \\ 0 & C & C_d & 0 & 0 \\ 0 & 0 & C & C_d & 0 \\ 0 & 0 & 0 & C & C_d \\ CA & C_\tau & C_d A_d & 0 & 0 \\ 0 & CA & C_\tau & C_d A_d & 0 \end{bmatrix}. \tag{16.49}$$

We then select $\beta = 1$ and solve the β-exponential stabilization of system (16.21) by using LMI (20.5) in Chap. 20. Accordingly, we obtain the following second-order observer to estimate both $z_{2\tau}(t)$ and $x_1(t)$

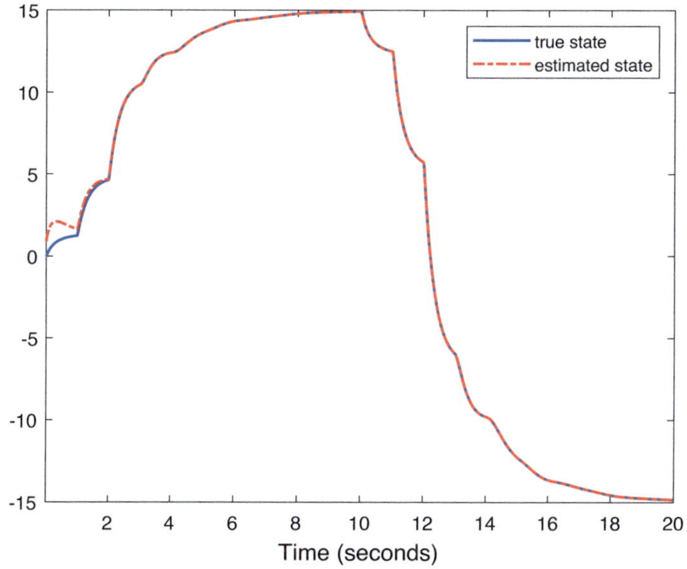

Fig. 16.2 Trajectories of $z_{2\tau}(t)$ and $\hat{z}_{2\tau}(t)$ (Numerical Example 2, second-order observer with an internal delay, in this chapter)

$$\begin{bmatrix} \hat{z}_{2\tau}(t) \\ \hat{x}_1(t) \end{bmatrix} = \begin{bmatrix} w_1(t) - 1.0719y(t) + 5.2225y(t-1) \\ w_2(t) + 0.3046y(t) + 0.0658y(t-1) \end{bmatrix}, \tag{16.50}$$

$$\begin{aligned} \begin{bmatrix} \dot{w}_1(t) \\ \dot{w}_2(t) \end{bmatrix} &= \begin{bmatrix} -4.9281 & 11.5452 \\ 0.6954 & -3.7029 \end{bmatrix} \begin{bmatrix} w_1(t) \\ w_2(t) \end{bmatrix} \\ &+ \begin{bmatrix} 1.4423 & -2.2716 \\ -0.4603 & 0.5253 \end{bmatrix} \begin{bmatrix} w_1(t-1) \\ w_2(t-1) \end{bmatrix} + \begin{bmatrix} -2.0336 \\ -0.9519 \end{bmatrix} y(t) \\ &+ \begin{bmatrix} 3.6106 \\ 0.3848 \end{bmatrix} y(t-1) + \begin{bmatrix} 0.6166 \\ 0.0636 \end{bmatrix} y(t-2) + \begin{bmatrix} 6.0719 \\ 0.6954 \end{bmatrix} u(t) \\ &+ \begin{bmatrix} 8.9931 \\ -0.9796 \end{bmatrix} u(t-1) + \begin{bmatrix} 1.3324 \\ -0.1973 \end{bmatrix} u(t-2). \end{aligned} \tag{16.51}$$

Figures 16.2 and 16.3 illustrate the show trajectories of $z_{2\tau}(t)$, $\hat{z}_{2\tau}(t)$ and $x_1(t)$, $\hat{x}_1(t)$, respectively. From these figures, it is clear that asymptotic estimation has been achieved.

Numerical Example 3

To further show the attractiveness of our result, we next show that we can design a third-order observer (16.4)–(16.5) with $N_d = N_{dd} = 0_{3\times 3}$ to estimate the unmeasur-

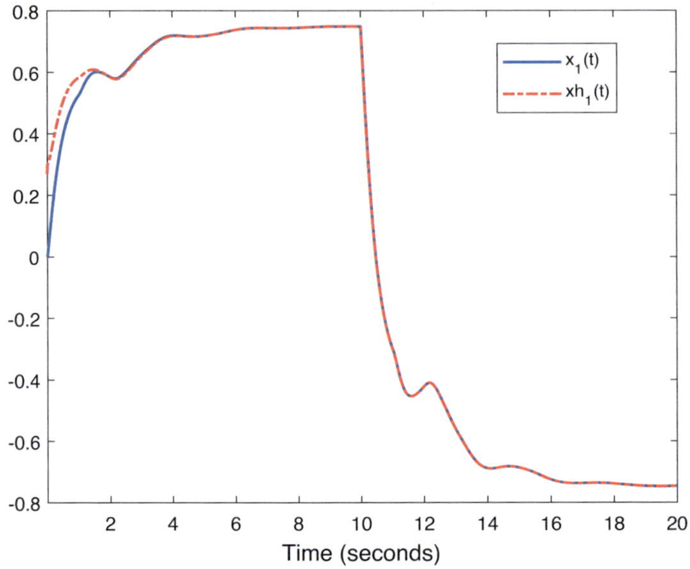

Fig. 16.3 Trajectories of $x_1(t)$ and $\hat{x}_1(t)$ (Numerical Example 2, second-order observer with an internal delay, in this chapter)

able vector $x(t) = \begin{bmatrix} x_1(t) \\ x_2(t) \end{bmatrix}$ (i.e., $z(t) = Fx(t)$ where $F = I_2$) in Example 2. Thanks to our general framework to the estimation problem of the vector $z_{2\tau}(t)$ (instead of $z(t)$), it is now possible to do so.

Firstly, with a second-order observer, we have $F = \begin{bmatrix} 1 & 0 \\ 0 & 1 \end{bmatrix}$ and $F_d = F_{dd} = \begin{bmatrix} 0 & 0 \\ 0 & 0 \end{bmatrix}$. We test condition (16.45) and find it unsatisfied. This implies that a second-order observer (16.4)–(16.5) with no internal delays does not exist. Since F is already a full rank matrix, and given the standard problem of estimating the vector $z(t) = Fx(t)$, the designer has to stop.

On the other hand, with our general approach to the estimation problem of the vector $z_{2\tau}(t)$ as defined by (16.3), it is possible to satisfy condition (16.45) by increasing the order of the observer. Now, we design a third-order observer to estimate the following vector

$$\begin{aligned} z_{2\tau}(t) &= \begin{bmatrix} z(t) \\ x_1(t-1) \end{bmatrix} \\ &= F_a x(t) + F_{ad} x(t-1) + F_{add} x(t-2), \end{aligned} \tag{16.52}$$

where $F_a = \begin{bmatrix} 1 & 0 \\ 0 & 1 \\ 0 & 0 \end{bmatrix}$, $F_{ad} = \begin{bmatrix} 0 & 0 \\ 0 & 0 \\ 1 & 0 \end{bmatrix}$ and $F_{add} = \begin{bmatrix} 0 & 0 \\ 0 & 0 \\ 0 & 0 \end{bmatrix}$.

Therefore, in the above, in addition to estimating the required vector $z(t) = x(t)$, we estimate an extra function, namely, $x_1(t-1)$. With F_a, F_{ad} and F_{add} as given above, it is easy to check that rank $\begin{bmatrix} F_a & F_{ad} & F_{add} \\ C & C_d & 0_{1\times 2} \end{bmatrix} = 4$, and condition (16.45) is now satisfied. We then carry out the rest of the observer design and obtain the following third-order observer with no internal delays

$$\begin{bmatrix} \hat{x}_1(t) \\ \hat{x}_2(t) \\ \hat{x}_1(t-1) \end{bmatrix} = \begin{bmatrix} w_1(t) \\ w_2(t) \\ w_3(t) + 0.5714y(t) \end{bmatrix}, \tag{16.53}$$

$$\begin{bmatrix} \dot{w}_1(t) \\ \dot{w}_2(t) \\ \dot{w}_3(t) \end{bmatrix} = \begin{bmatrix} -5 & 2 & -4 \\ -2 & -7 & -3 \\ -2 & -1.1429 & -3.5714 \end{bmatrix} \begin{bmatrix} w_1(t) \\ w_2(t) \\ w_3(t) \end{bmatrix} + \begin{bmatrix} -0.2857 \\ 1.2857 \\ 1.6735 \end{bmatrix} y(t)$$
$$+ \begin{bmatrix} 0 \\ 0 \\ -0.8571 \end{bmatrix} y(t-1) + \begin{bmatrix} 1 \\ 2 \\ -0.5714 \end{bmatrix} u(t) + \begin{bmatrix} 0 \\ 0 \\ -0.7143 \end{bmatrix} u(t-1), \tag{16.54}$$

and the error vector $\dot{e}(t) = Ne(t)$ is now independent to the time delay, τ, and stable with $\lambda(N) = \{-1.1858, -7.1928 \pm j1.2781\}$. Figures 16.4 and 16.5 show the trajectories of $x_1(t)$, $\hat{x}_1(t)$ and $x_2(t)$, $\hat{x}_2(t)$, respectively. From the figures, it is clear that asymptotic estimation has been achieved.

Numerical Example 4

In this example, we further demonstrate the generality of our result which can deal with the case where $C = 0_{p\times n}$. Let us reconsider Numerical Example 3 with $C = 0_{1\times 2}$, $F = I_2$ and $F_d = F_{dd} = 0_{2\times 2}$. In this case, we consider the problem of estimating the vector $x(t)$ where the output vector contains only delayed state variables, i.e., $y(t) = C_d x(t-\tau)$. To make this problem more interesting, our objective is to design an observer (16.4)–(16.5) with no internal delays.

By following the same approach in Numerical Example 3, we expand the dimension of the designed observer to satisfy condition (16.45). Accordingly, we construct the following third-order observer with no internal delays

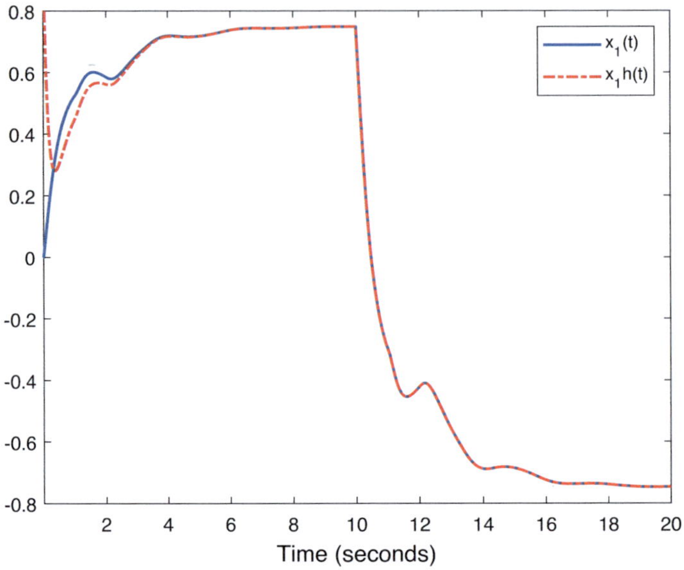

Fig. 16.4 Trajectories of $x_1(t)$ and $\hat{x}_1(t)$ (Numerical Example 3, third-order observer with no internal delays, in this chapter)

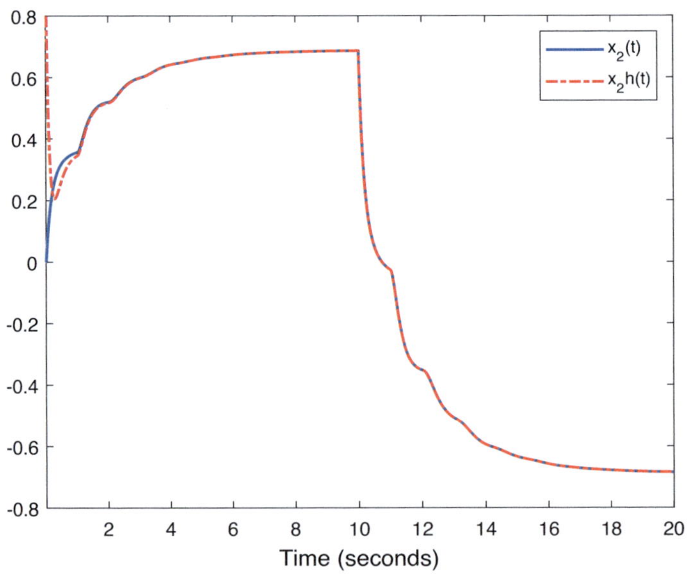

Fig. 16.5 Trajectories of $x_2(t)$ and $\hat{x}_2(t)$ (Numerical Example 3, third-order observer with no internal delays, in this chapter)

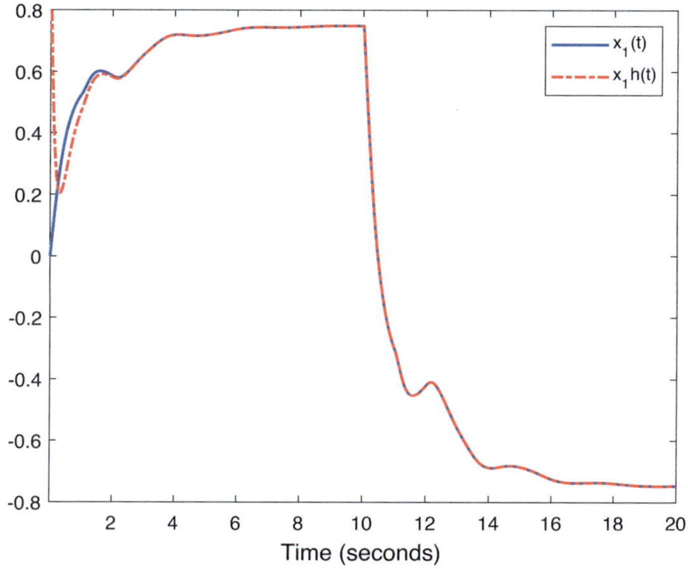

Fig. 16.6 Trajectories of $x_1(t)$ and $\hat{x}_1(t)$ (Numerical Example 4, $C = 0_{1\times 2}$, third-order observer with no internal delays, in this chapter)

$$\begin{bmatrix} \hat{x}_1(t) \\ \hat{x}_2(t) \\ \hat{x}_1(t-1) \end{bmatrix} = \begin{bmatrix} w_1(t) \\ w_2(t) \\ w_3(t) + 0.5714y(t) \end{bmatrix}, \tag{16.55}$$

$$\begin{bmatrix} \dot{w}_1(t) \\ \dot{w}_2(t) \\ \dot{w}_3(t) \end{bmatrix} = \begin{bmatrix} -3 & 2 & -4 \\ 1 & -7 & -3 \\ 0 & 0 & -6.7143 \end{bmatrix} \begin{bmatrix} w_1(t) \\ w_2(t) \\ w_3(t) \end{bmatrix} + \begin{bmatrix} -0.2857 \\ 1.2857 \\ 1.0204 \end{bmatrix} y(t)$$
$$+ \begin{bmatrix} 0 \\ 0 \\ -0.8571 \end{bmatrix} y(t-1) + \begin{bmatrix} 1 \\ 2 \\ 0 \end{bmatrix} u(t) + \begin{bmatrix} 0 \\ 0 \\ -0.7143 \end{bmatrix} u(t-1), \tag{16.56}$$

and N is stable with $\lambda(N) = \{-2.5505, -7.4495, -6.7143\}$. This example shows the significance and generality of our result. Figures 16.6 and 16.7 show the trajectories of $x_1(t)$, $\hat{x}_1(t)$ and $x_2(t)$, $\hat{x}_2(t)$, respectively. It is clear that asymptotic estimation has been achieved.

Numerical Example 5

Finally, in this example, we reconsider Numerical Example 2 with $C = 0_{1\times 2}$, $C_d = [1 \;\; 0]$ and $F = [0 \;\; 0]$. In this case, the problem is reduced to estimating vector (6.13) (i.e., $z_\tau(t) = F_d x(t-\tau) + F_{dd} x(t-2\tau)$) for time-delay system (6.15)–

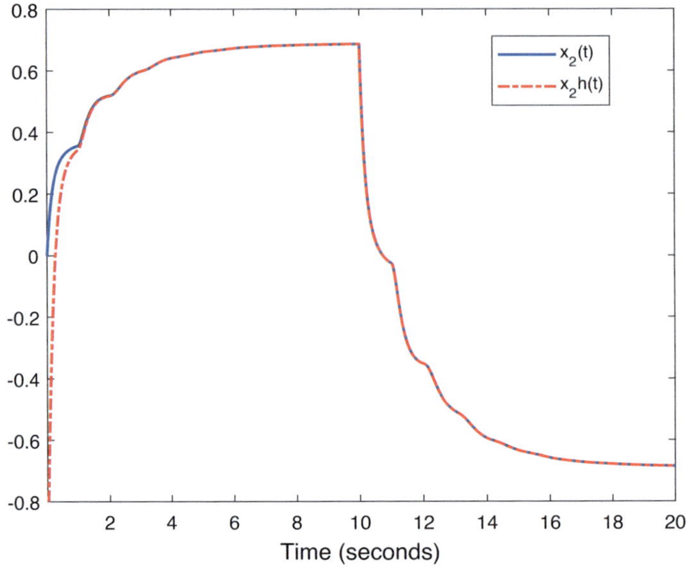

Fig. 16.7 Trajectories of $x_2(t)$ and $\hat{x}_2(t)$ (Numerical Example 4, $C = 0_{1\times 2}$, third-order observer with no internal delays, in this chapter)

(6.16), which is used to obtain the estimation of the delayed unknown input, $\hat{d}(t-\tau)$, as defined by (6.14).

With $C = 0_{1\times 2}$, $C_d = [1 \;\; 0]$, $F = [0 \;\; 0]$, $F_d = [3 \;\; 4]$ and $F_{dd} = [5 \;\; 6]$, we first test condition (16.45) and find it to be satisfied. We then can design the following first-order observer with no internal delays to estimate $z_\tau(t)$

$$\hat{z}_\tau(t) = w(t) + 9y(t) + 14y(t-1) \tag{16.57}$$

$$\dot{w}(t) = -10w(t) - 38y(t) - 45y(t-1) + 18y(t-2) + 2u(t-1) + 3u(t-2). \tag{16.58}$$

It is clear from Fig. 16.8 that $\hat{z}_\tau(t)$ converges to $z_\tau(t)$ asymptotically.

16.3 Concluding Remark

In this chapter, we have further demonstrated the generality and advantages of estimating a more general functional, $z_{2\tau}(t) = Fx(t) + F_d x(t-\tau) + F_{dd} x(t-2\tau)$ for time-delay system (16.1)–(16.2). We have provided five numerical examples to discuss the design of various types of observers for time-delay system (16.1)–(16.2). Through the examples, we have shown the salient features of our design method. We have also shown the generality of our results over those reported in [1, 2].

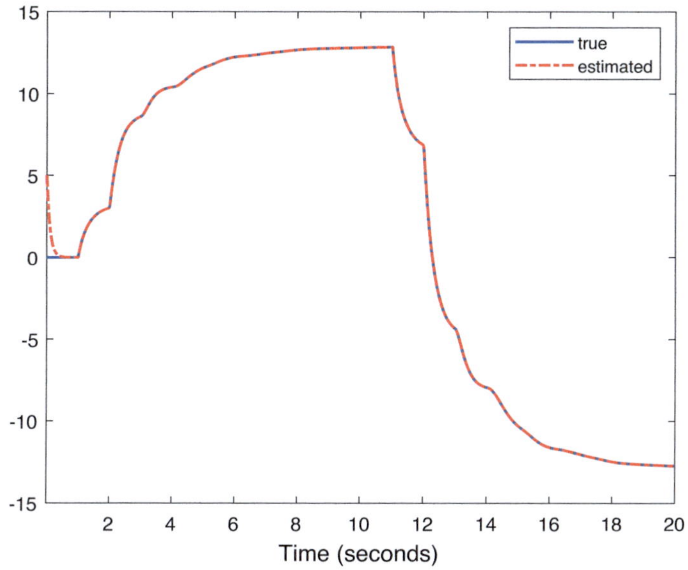

Fig. 16.8 Trajectories of $z_\tau(t)$ and $\hat{z}_\tau(t)$, (Numerical Example 5, first-order observer with no internal delays, in this chapter)

We will continue our exploratory journey in the following chapters, and in Chap. 17, we will look at time-delay system (17.1)–(17.2) with two non-commensurate delays time delays, τ and h, in state and output vectors, i.e.,

$$\dot{x}(t) = Ax(t) + A_d x(t-\tau) + A_h x(t-h) + Bu(t),$$
$$y(t) = Cx(t) + C_d x(t-\tau) + C_h x(t-h).$$

For this system, we design generalized observers to estimate the following functional

$$z_{\tau h}(t) = Fx(t) + F_d x(t-\tau) + F_h x(t-h).$$

The outcome of this chapter will be used for the estimation of the unknown input, $d(t)$, as defined by (8.19) in Chap. 8.

References

1. Darouach, M.: Linear functional observers for systems with delays in state variables. IEEE Trans. Automat. Contr. **46**(3), 491–496 (2001)
2. Nam, P.T., Pathirana, P.N., Trinh, H.: ϵ-bounded state estimation for time-delay systems with bounded disturbances. Int. J. Control **87**(9), 1747–1756 (2014)

3. Trinh, H.: Linear functional state observers for time-delay systems. Int. J. Control **72**(18), 1642–1658 (1999)
4. Trinh, H., Fernando, T.: Functional Observers for Dynamical Systems. Springer, Berlin, Heidelberg (2012)

Chapter 17
Generalized Observers: Two Non-Commensurate Time Delays

17.1 Background and Motivations

In this chapter, we consider the following linear system with two different time delays in the state and output vectors

$$\dot{x}(t) = Ax(t) + A_d x(t-\tau) + A_h x(t-h) + Bu(t), \tag{17.1}$$

$$y(t) = Cx(t) + C_d x(t-\tau) + C_h x(t-h), \tag{17.2}$$

where $x(t) \in \mathbb{R}^n$, $y(t) \in \mathbb{R}^p$ and $u(t) \in \mathbb{R}^r$ are, respectively, the state vector, output vector and control input vector. Matrices $A, A_d, A_h \in \mathbb{R}^{n \times n}$, $B \in \mathbb{R}^{n \times r}$ and $C, C_d, C_h \in \mathbb{R}^{p \times n}$ are known constant matrices. Also, $0 < \tau < h$, where τ and h are known constant non-commensurate delays. Our aim is to design generalized observers to estimate the following functional

$$z_{\tau h}(t) = Fx(t) + F_d x(t-\tau) + F_h x(t-h), \tag{17.3}$$

where $z_{\tau h}(t) \in \mathbb{R}^m$, and F, F_d and F_h are known constant matrices of approriate dimensions. This problem is more general than the problem of designing functional observers to estimate $z(t) = Fx(t)$, i.e., the particular case where $F_d = 0$ and $F_h = 0$. As discussed in Case 1 of Sect. 8.2.1, $\hat{z}_{\tau h}(t)$ is used to estimate the unknown input, $d(t)$, as defined by (8.19).

H. Trinh et al., *Unknown Inputs Estimation in Linear Time-Delay Systems Using Generalized Functional Observers*, Studies in Systems, Decision and Control 648,
https://doi.org/10.1007/978-3-032-11870-7_17

17.2 Main Results

17.2.1 General Observer Structure

To estimate functional (17.3) for system (17.1)–(17.2), we consider the following generalized observer with multiple delayed terms

$$\hat{z}_{\tau h}(t) = w(t) + My(t) + M_d y(t-\tau) + M_h y(t-h) + M_{dh} y(t-\tau-h), \tag{17.4}$$

$$\begin{aligned}\dot{w}(t) &= Nw(t) + N_d w(t-\tau) + N_h w(t-h) + Gy(t) + G_d y(t-\tau) + G_h y(t-h)\\ &+ G_{hh} y(t-2h) + G_{dd} y(t-2\tau) + G_{dh} y(t-\tau-h) + G_{d2h} y(t-\tau-2h)\\ &+ G_{2dh} y(t-2\tau-h) + Ju(t) + J_d u(t-\tau) + J_h u(t-h) + J_{dd} u(t-2\tau)\\ &+ J_{hh} u(t-2h) + J_{dh} u(t-\tau-h) + J_{2dh} u(t-2\tau-h) + J_{d2h} u(t-\tau-2h),\end{aligned} \tag{17.5}$$

where $w(\theta) = \rho(\theta)$, $\forall\theta \in [0, h]$, $w(t) \in \mathbb{R}^m$, and $\hat{z}_{\tau h}(t)$ is the estimate of $z_{\tau h}(t)$. Matrices M, M_d, M_h, M_{dh}, N, N_d, N_h, G, G_d, G_h, G_{hh}, G_{dd}, G_{dh}, G_{d2h}, G_{2dh}, J, J_d, J_h, J_{dd}, J_{hh}, J_{dh}, J_{2dh} and J_{d2h} are to be determined such that $\hat{z}_{\tau h}(t)$ converges asymptotically to $z_{\tau h}(t)$.

17.2.2 Observer Existence Conditions

Following the same approach in Chap. 12, we derive conditions for the existence of observer (17.4)–(17.5). The conditions are given below.

Existence Conditions: The estimate $\hat{z}_{\tau h}(t)$ converges asymptotically to $z_{\tau h}(t)$ for any initial condition $w(\theta)$ and any $u(t)$ provided that the following conditions hold

$$\dot{e}(t) = Ne(t) + N_d e(t-\tau) + N_h e(t-h) \text{ is asymptotically stable,} \tag{17.6}$$

$$J - FB + MCB = 0, \tag{17.7}$$

$$J_d - F_d B + MC_d B + M_d CB = 0, \tag{17.8}$$

$$J_h - F_h B + MC_h B + M_h CB = 0, \tag{17.9}$$

$$J_{dd} + M_d C_d B = 0, \tag{17.10}$$

$$J_{hh} + M_h C_h B = 0, \tag{17.11}$$

$$J_{dh} + M_d C_h B + M_h C_d B + M_{dh} CB = 0, \tag{17.12}$$

$$J_{2dh} + M_{dh} C_d B = 0, \tag{17.13}$$

$$J_{d2h} + M_{dh} C_h B = 0, \tag{17.14}$$

$$NF + \tilde{T}C + MCA - FA = 0, \tag{17.15}$$

$$NF_d + N_dF + \tilde{T}C_d + \hat{W}C + M(CA_d + C_dA) + M_dCA - FA_d - F_dA = 0, \tag{17.16}$$

$$NF_h + N_hF + \tilde{T}C_h + \hat{W}_hC + M(CA_h + C_hA) + M_hCA - FA_h - F_hA = 0, \tag{17.17}$$

$$N_dF_d + \tilde{U}C + \hat{W}C_d + MC_dA_d + M_d(CA_d + C_dA) - F_dA_d = 0, \tag{17.18}$$

$$N_hF_h + \tilde{U}_{hh}C + \hat{W}_hC_h + MC_hA_h + M_h(CA_h + C_hA) - F_hA_h = 0, \tag{17.19}$$

$$N_dF_h + N_hF_d + \tilde{U}_{dh}C + \hat{W}C_h + \hat{W}_hC_d + M(C_dA_h + C_hA_d) + M_d(CA_h + C_hA) + M_h(CA_d + C_dA) + M_{dh}CA - F_dA_h - F_hA_d = 0, \tag{17.20}$$

$$\tilde{U}C_h + \tilde{U}_{dh}C_d + \tilde{U}_{2dh}C + M_d(C_dA_h + C_hA_d) + M_hC_dA_d + M_{dh}(CA_d + C_dA) = 0, \tag{17.21}$$

$$\tilde{U}_{hh}C_d + \tilde{U}_{dh}C_h + \tilde{U}_{d2h}C + M_dC_hA_h + M_h(C_dA_h + C_hA_d) + M_{dh}(CA_h + C_hA) = 0, \tag{17.22}$$

$$\tilde{U}_{d2h}C_d + \tilde{U}_{2dh}C_h + M_{dh}(C_dA_h + C_hA_d) = 0, \tag{17.23}$$

$$\tilde{U}C_d + M_dC_dA_d = 0, \tag{17.24}$$

$$\tilde{U}_{hh}C_h + M_hC_hA_h = 0, \tag{17.25}$$

$$\tilde{U}_{2dh}C_d + M_{dh}C_dA_d = 0, \tag{17.26}$$

$$\tilde{U}_{d2h}C_h + M_{dh}C_hA_h = 0, \tag{17.27}$$

where $\tilde{T}$, $\hat{W}$, $\hat{W}_h$, $\tilde{U}$, $\tilde{U}_{hh}$, $\tilde{U}_{dh}$, $\tilde{U}_{2dh}$ and $\tilde{U}_{d2h}$ are defined as follows

$$\tilde{T} = G - NM, \tag{17.28}$$

$$\hat{W} = G_d - NM_d - N_dM, \tag{17.29}$$

$$\hat{W}_h = G_h - NM_h - N_hM, \tag{17.30}$$

$$\tilde{U} = G_{dd} - N_dM_d, \tag{17.31}$$

$$\tilde{U}_{hh} = G_{hh} - N_hM_h, \tag{17.32}$$

$$\tilde{U}_{dh} = G_{dh} - N_dM_h - N_hM_d - NM_{dh}, \tag{17.33}$$

$$\tilde{U}_{2dh} = G_{2dh} - N_dM_{dh}, \tag{17.34}$$

$$\tilde{U}_{d2h} = G_{d2h} - N_hM_{dh}. \tag{17.35}$$

The above existence conditions (17.6)–(17.27) are derived based on the approach presented in Chap. 12. For completeness, an outline of the process is described as follows. Firstly, denoting $e(t)$ as the error vector, where

$$\begin{aligned} e(t) &= \hat{z}_{\tau h}(t) - z_{\tau h}(t) \\ &= w(t) + My(t) + M_d y(t-\tau) + M_h y(t-h) + M_{dh} y(t-\tau-h) \\ &- Fx(t) - F_d x(t-\tau) - F_h x(t-h). \end{aligned} \tag{17.36}$$

From (17.36), we obtain

$$\begin{aligned} w(t) &= e(t) + Fx(t) + F_d x(t-\tau) + F_h x(t-h) - My(t) - M_d y(t-\tau) \\ &- M_h y(t-h) - M_{dh} y(t-\tau-h). \end{aligned} \tag{17.37}$$

Secondly, we take the derivative of both sides of (17.36) with respect to time, i.e.,

$$\begin{aligned} \dot{e}(t) &= \dot{w}(t) + M\dot{y}(t) + M_d \dot{y}(t-\tau) + M_h \dot{y}(t-h) + M_{dh} \dot{y}(t-\tau-h) \\ &- F\dot{x}(t) - F_d \dot{x}(t-\tau) - F_h \dot{x}(t-h), \end{aligned} \tag{17.38}$$

and after some substitutions using (17.1), (17.2), (17.5) and (17.37), the following error dynamic equation is obtained

$$\begin{aligned} \dot{e}(t) &= Ne(t) + N_d e(t-\tau) + N_h e(t-h) + \mathscr{L}_{(17.7)} u(t) + \mathscr{L}_{((17.8))} u(t-\tau) \\ &+ \mathscr{L}_{(17.9)} u(t-h) + \mathscr{L}_{(17.10)} u(t-2\tau) + \mathscr{L}_{(17.11)} u(t-2h) + \mathscr{L}_{(17.12)} u(t-\tau-h) \\ &+ \mathscr{L}_{(17.13)} u(t-2\tau-h) + \mathscr{L}_{(17.14)} u(t-\tau-2h) + \mathscr{L}_{(17.15)} x(t) + \mathscr{L}_{(17.16)} x(t-\tau) \\ &+ \mathscr{L}_{(17.17)} x(t-h) + \mathscr{L}_{(17.18)} x(t-2\tau) + \mathscr{L}_{(17.19)} x(t-2h) + \mathscr{L}_{(17.20)} x(t-\tau-h) \\ &+ \mathscr{L}_{(17.21)} x(t-2\tau-h) + \mathscr{L}_{(17.22)} x(t-\tau-2h) + \mathscr{L}_{(17.23)} x(t-2\tau-2h) \\ &+ \mathscr{L}_{(17.24)} x(t-3\tau) + \mathscr{L}_{(17.25)} x(t-3h) + \mathscr{L}_{(17.26)} x(t-3\tau-h) \\ &+ \mathscr{L}_{(17.27)} x(t-\tau-3h), \end{aligned} \tag{17.39}$$

where for brevity, $\mathscr{L}_{(.)}$ is used to denote all the terms on the left-hand side of the corresponding equation specified in the subscript. For example, from equation (17.7), we have $\mathscr{L}_{(17.7)} = J - FB + MCB$.

Thirdly, from (17.39), conditions for error vector $e(t)$ converging asymptotically to zero are derived, which are as given in (17.6)–(17.27).

17.2.3 Observer Design Procedure

The next task is to solve for the unknown matrices such that existence conditions (17.6)–(17.27) are satisfied. From (17.7)–(17.14), we can obtain J, J_d, J_h, J_{dd}, J_{hh}, J_{dh}, J_{2dh} and J_{d2h} as follows

$$J = FB - MCB, \tag{17.40}$$
$$J_d = F_d B - MC_d B - M_d CB, \tag{17.41}$$
$$J_h = F_h B - MC_h B - M_h CB, \tag{17.42}$$
$$J_{dd} = -M_d C_d B, \tag{17.43}$$
$$J_{hh} = -M_h C_h B, \tag{17.44}$$
$$J_{dh} = -M_d C_h B - M_h C_d B - M_{dh} CB, \tag{17.45}$$
$$J_{2dh} = -M_{dh} C_d B, \tag{17.46}$$
$$J_{d2h} = -M_{dh} C_h B. \tag{17.47}$$

Also, from (17.28)–(17.35), we can obtain G, G_d, G_h, G_{dd}, G_{hh}, G_{dh}, G_{2dh} and G_{d2h} as follows

$$G = \tilde{T} + NM, \tag{17.48}$$
$$G_d = \hat{W} + NM_d + N_d M, \tag{17.49}$$
$$G_h = \hat{W}_h + NM_h + N_h M, \tag{17.50}$$
$$G_{dd} = \tilde{U} + N_d M_d, \tag{17.51}$$
$$G_{hh} = \tilde{U}_{hh} + N_h M_h, \tag{17.52}$$
$$G_{dh} = \tilde{U}_{dh} + N_d M_h + N_h M_d + NM_{dh}, \tag{17.53}$$
$$G_{2dh} = \tilde{U}_{2dh} + N_d M_{dh}, \tag{17.54}$$
$$G_{d2h} = \tilde{U}_{d2h} + N_h M_{dh}. \tag{17.55}$$

Now, examination of (17.6) and (17.15)–(17.27) shows that there are 15 unknown matrices (they are augmented and denoted by $X \in \mathbb{R}^{m\times(3m+12p)}$ in (17.56)) that we need to solve for, such that the coupled matrix equations (17.15)–(17.27) are satisfied. At the same time, we need to ensure the time-delay error system (17.6) is asymptotically stable.

$$X = [N \ \ N_d \ \ N_h \ \ \tilde{T} \ \ \hat{W} \ \ \hat{W}_h \ \ \tilde{U} \ \ \tilde{U}_{hh} \ \ \tilde{U}_{dh} \ \ \tilde{U}_{2dh} \ \ \tilde{U}_{d2h} \ \ M \ \ M_d \ \ M_h \ \ M_{dh}]. \tag{17.56}$$

We first derive the necessary and sufficient condition for the solvability of (17.15)–(17.27). For this purpose, we augment them as follows

$$X\Theta_{dh} = \Upsilon_{dh}, \tag{17.57}$$

where X is as defined by (17.56), and $\Upsilon_{dh} \in \mathbb{R}^{m\times 13n}$ and $\Theta_{dh} \in \mathbb{R}^{(3m+12p)\times 13n}$ are given as follows

$$\Upsilon_{dh} = [\Upsilon_{dh_1} \ \ \Upsilon_{dh_2} \ \ \Upsilon_{dh_3} \ \ \Upsilon_{dh_4}], \tag{17.58}$$
$$\Theta_{dh} = [\Theta_{dh_1} \ \ \Theta_{dh_2} \ \ \Theta_{dh_3} \ \ \Theta_{dh_4}], \tag{17.59}$$

$$\Upsilon_{dh_1} = [FA \quad FA_d + F_d A \quad FA_h + F_h A \quad F_d A_d], \tag{17.60}$$

$$\Upsilon_{dh_2} = [F_h A_h \quad F_d A_h + F_h A_d \quad 0], \tag{17.61}$$

$$\Upsilon_{dh_3} = \Upsilon_{dh_4} = [0 \quad 0 \quad 0], \tag{17.62}$$

and

$$\Theta_{dh_1} = \begin{bmatrix} F & F_d & F_h & 0 \\ 0 & F & 0 & F_d \\ 0 & 0 & F & 0 \\ C & C_d & C_h & 0 \\ 0 & C & 0 & C_d \\ 0 & 0 & C & 0 \\ 0 & 0 & 0 & C \\ 0 & 0 & 0 & 0 \\ 0 & 0 & 0 & 0 \\ 0 & 0 & 0 & 0 \\ 0 & 0 & 0 & 0 \\ CA & CA_d + C_d A & CA_h + C_h A & C_d A_d \\ 0 & CA & 0 & CA_d + C_d A \\ 0 & 0 & CA & 0 \\ 0 & 0 & 0 & 0 \end{bmatrix}, \tag{17.63}$$

$$\Theta_{dh_2} = \begin{bmatrix} 0 & 0 & 0 \\ 0 & F_h & 0 \\ F_h & F_d & 0 \\ 0 & 0 & 0 \\ 0 & C_h & 0 \\ C_h & C_d & 0 \\ 0 & 0 & C_h \\ C & 0 & 0 \\ 0 & C & C_d \\ 0 & 0 & C \\ 0 & 0 & 0 \\ C_h A_h & C_d A_h + C_h A_d & 0 \\ 0 & CA_h + C_h A & C_d A_h + C_h A_d \\ CA_h + C_h A & CA_d + C_d A & C_d A_d \\ 0 & CA & CA_d + C_d A \end{bmatrix}, \tag{17.64}$$

$$\Theta_{dh_3} = \begin{bmatrix} 0 & 0 & 0 \\ 0 & 0 & 0 \\ 0 & 0 & 0 \\ 0 & 0 & 0 \\ 0 & 0 & 0 \\ 0 & 0 & 0 \\ 0 & 0 & C_d \\ C_d & 0 & 0 \\ C_h & 0 & 0 \\ 0 & C_h & 0 \\ C & C_d & 0 \\ 0 & 0 & 0 \\ C_h A_h & 0 & C_d A_d \\ C_d A_h + C_h A_d & 0 & 0 \\ C A_h + C_h A & C_d A_h + C_h A_d & 0 \end{bmatrix}, \tag{17.65}$$

$$\Theta_{dh_4} = \begin{bmatrix} 0 & 0 & 0 \\ 0 & 0 & 0 \\ 0 & 0 & 0 \\ 0 & 0 & 0 \\ 0 & 0 & 0 \\ 0 & 0 & 0 \\ 0 & 0 & 0 \\ C_h & 0 & 0 \\ 0 & 0 & 0 \\ 0 & C_d & 0 \\ 0 & 0 & C_h \\ 0 & 0 & 0 \\ 0 & 0 & 0 \\ C_h A_h & 0 & 0 \\ 0 & C_d A_d & C_h A_h \end{bmatrix}. \tag{17.66}$$

From (17.57), X exists if and only if the following condition holds

$$\text{rank}\begin{bmatrix} \Theta_{dh} \\ \Upsilon_{dh} \end{bmatrix} = \text{rank}\big[\Theta_{dh}\big]. \tag{17.67}$$

Since Θ_{dh} and Υ_{dh} are known matrices, checking condition (17.67) is an easy task. Upon its satisfaction, equation (17.57) is solved for the unknown matrix X. In the same approach in Chap. 12 (see, from (12.25)–(12.32)), the solution, X, is expressed in terms of an arbitrary matrix, Z_h, and the time-delay error system (17.6) can be expressed as follows

$$\dot{e}(t) = (N_1 + Z_h N_2)e(t) + (N_{d1} + Z_h N_{d2})e(t-\tau) + (N_{h1} + Z_h N_{h2})e(t-h), \tag{17.68}$$

where $N_1, N_2, N_{d1}, N_{d2}, N_{h1}$ and N_{h2} are known constant matrices. It is now required that the above time-delay error system is asymptotically stable. To meet this requirement, stability analysis (i.e., for the case where $N_2 = N_{d2} = N_{h2} = 0$) or the stabilization problem (i.e., finding Z_h) can be carried out to establish asymptotic stability of (17.68). The reader can refer to Chap. 20 for some stability and stabilizability conditions of (17.68). Once asymptotic stability of the time-delay error system (17.68) is established, the rest of the observer parameters can be obtained. Please refer to Numerical Example 1 for an illustrative example.

Remark 1 Condition (17.67) still valid for the particular case where $F_d = F_h = 0$, i.e., $z_{\tau h}(t) = z(t) = Fx(t)$. In such case, we simply let $F_d = 0$ and $F_h = 0$ in both Υ_{dh} and Θ_{dh} where they are as defined by (17.58) and (17.59), respectively.

As we mentioned earlier in this chapter, the problem of estimating the functional $z_{\tau h}(t)$ as defined by (17.3) is more general than the problem of estimating $z(t) = Fx(t)$. We refer the reader to Numerical Example 1, where we show that $F_d, F_h \neq 0$ provides more flexibility in the design of generalized observers.

Remark 2 We can also consider generalized observers with a simpler structure than (17.4)–(17.5) at the expense of more conservative existence conditions. In particular, generalized observers without any internal delay, i.e., by letting $N_d = N_h = 0$ in (17.5). In this case, from (17.56), there are now only 13 unknown matrices in X, with $N_d = 0$ and $N_h = 0$, and equation (17.57) is reduced to the following

$$X\tilde{\Theta}_{dh} = \Upsilon_{dh}, \tag{17.69}$$

where $\tilde{\Theta}_{dh} \in \mathbb{R}^{(m+12p)\times 13n}$ is a sub-matrix of Θ_{dh} obtained by removing the second and third row-block of Θ_{dh}, i.e.,

$$\begin{bmatrix} 0 & F & 0 & F_d & 0 & F_h & 0 & 0 & 0 & 0 & 0 & 0 & 0 \\ 0 & 0 & F & 0 & F_h & F_d & 0 & 0 & 0 & 0 & 0 & 0 & 0 \end{bmatrix}.$$

Accordingly, the condition for the solvability of (17.69) is as follows

$$\text{rank}\begin{bmatrix} \tilde{\Theta}_{dh} \\ \Upsilon_{dh} \end{bmatrix} = \text{rank}\big[\tilde{\Theta}_{dh}\big]. \tag{17.70}$$

While condition (17.70) is generally more conservative than condition (17.67), the error system (17.6), i.e., $\dot{e}(t) = Ne(t)$, is now independent of both time delays τ and h. Thus, we only need N to be a stable matrix, to ensure that the error vector $e(t)$ converges asymptotically to zeros.

In a similar way, we can also consider various types of functional observers with for example $N_h = 0$ or $N_d = 0$. Thus, for example, if $N_d \neq 0$ and $N_h = 0$, the observer (17.4)–(17.5) now has only one internal delay, and there is only τ in the time-delay error system (17.6). In this case, $\tilde{\Theta}_{dh} \in \mathbb{R}^{(2m+12p)\times 13n}$ is a sub-matrix of Θ_{dh} obtained by removing the third row-block of Θ_{dh}.

17.2.4 Numerical Examples

In this section, we provide three numerical examples to discuss the design of various types of observers for time-delay system (17.1)–(17.2). Through the examples, we will demonstrate the salient features of our design method and results.

Numerical Example 1

This example is given to illustrate the design of various types of functional observers to estimate the function $z_{\tau h}(t) \in \mathbb{R}^m$ defined by (17.3). We provide a detailed explanation on how to go about designing observers with the extra degrees of freedom in having F_d and F_h.

We consider system (17.1)–(17.2), where $\tau = 0.6s,\ h = 1s,\ A = \begin{bmatrix} -3 & -3 \\ 1 & -5 \end{bmatrix}$, $A_d = \begin{bmatrix} -4 & 0 \\ 0 & -2 \end{bmatrix}$, $A_h = \begin{bmatrix} -3 & 1 \\ -1 & -2 \end{bmatrix}$, $C = \begin{bmatrix} 1 & 0 \\ 0 & 0 \end{bmatrix}$, $C_d = \begin{bmatrix} 0 & 1 \\ 0 & 1 \end{bmatrix}$, $C_h = \begin{bmatrix} 1 & 1 \\ 0 & 1 \end{bmatrix}$ and $B = \begin{bmatrix} 1 \\ 2 \end{bmatrix}$.

Note that there are two outputs for this system, namely, $y_1(t)$ and $y_2(t)$, where

$$y_1(t) = x_1(t) + x_2(t-\tau) + x_1(t-h) + x_2(t-h), \tag{17.71}$$
$$y_2(t) = x_2(t-\tau) + x_2(t-h), \tag{17.72}$$

but it is not possible to explicitly derive $x_1(t)$ and $x_2(t)$ in terms of the outputs $y_1(t)$ and $y_2(t)$ due to the presence of multiple delayed terms in (17.71)–(17.72). Thus, even though the number of outputs is the same as the number of the state variables, it is still necessary to design a dynamical observer to estimate the state variables or any linear combination of the state variables that is linearly independent to the outputs.

Let us now design a functional observer to estimate the following function

$$z_{\tau h}(t) = [1\ \ -1]x(t) + [0\ \ 1]x(t-\tau) + [-1\ \ 1]x(t-h), \tag{17.73}$$

i.e., $F = [1\ \ -1]$, $F_d = [0\ \ 1]$ and $F_h = [-1\ \ 1]$. Note that $z_{\tau h}(t)$ is linearly independent to $y(t)$ since it is easy to check that

$$\text{rank}\begin{bmatrix} C & C_d & C_h \\ F & F_d & F_h \end{bmatrix} = 3, \tag{17.74}$$

which implies that a dynamical observer is required to estimate $z_{\tau h}(t)$, as it is not possible to algebraically express $z_{\tau h}(t)$ in terms of $y(t)$.

Case 1: First-order Observer: To estimate $z_{\tau h}(t)$, we first consider a first-order generalized observer of the form (17.4)–(17.5). To proceed with the design, we obtain matrices Υ_{dh} and Θ_{dh}, as defined by (17.58) and (17.59), respectively. We then check condition (17.67) and find it unsatisfied since rank $\begin{bmatrix} \Theta_{dh} \\ \Upsilon_{dh} \end{bmatrix} = 24$, but rank$\begin{bmatrix} \Theta_{dh} \end{bmatrix} = 23$, i.e., rank $\begin{bmatrix} \Theta_{dh} \\ \Upsilon_{dh} \end{bmatrix} \neq$ rank$\begin{bmatrix} \Theta_{dh} \end{bmatrix}$. Note that condition (17.67) is the necessary and sufficient condition for the solvability of the coupled matrix equations (17.15)–(17.27). Thus, a first-order generalized observer (17.4)–(17.5) does not exist.

We now have an option of increasing the order of the designed generalized observer by estimating a new augmented vector, $\bar{z}_{\tau h}(t) \in \mathbb{R}^{(m+\upsilon)}$, where

$$\bar{z}_{\tau h}(t) = \begin{bmatrix} z_{\tau h}(t) \\ z_{\tau h_a}(t) \end{bmatrix}, \tag{17.75}$$

and $z_{\tau h_a}(t) \in \mathbb{R}^{\upsilon}$ is defined as follows

$$z_{\tau h_a}(t) = F_a x(t) + F_{d_a} x(t-\tau) + F_{h_a} x(t-h), \tag{17.76}$$

$F_a, F_{d_a}, F_{h_a} \in \mathbb{R}^{\upsilon \times n}$ are row matrices to be found such that condition (17.67) is satisfied. It should be noted that the newly added vector, $z_{\tau h_a}(t)$, must be linearly independent to the vectors $z_{\tau h}(t)$ and $y(t)$, i.e.,

$$\text{rank} \begin{bmatrix} C & C_d & C_h \\ F & F_d & F_h \\ F_a & F_{d_a} & F_{h_a} \end{bmatrix} = p + m + \upsilon. \tag{17.77}$$

When testing condition (17.67), we need to replace F, F_d and F_h with $\bar{F} = \begin{bmatrix} F \\ F_a \end{bmatrix}$, $\bar{F}_d = \begin{bmatrix} F_d \\ F_{d_a} \end{bmatrix}$ and $\bar{F}_h = \begin{bmatrix} F_h \\ F_{h_a} \end{bmatrix}$, respectively, in both matrices Θ_{dh} and Υ_{dh}. Thus, if F_a, F_{d_a} and F_{h_a} are found such that condition (17.67) is satisfied, we solve (17.57) for the unknown matrix X and then we establish the asymptotic stability of the observer error vector $e(t)$. Upon its satisfaction, the rest of the observer matrices are obtained according to (17.40)–(17.47) and (17.48)–(17.55).

Case 2: Second-order observer without internal Delays ($N_d = N_h = 0_{2\times 2}$): Let us first increase the order of the observer by one, i.e., $\upsilon = 1$, and we select F_a, F_{d_a} and F_{h_a} as follows

$$\begin{bmatrix} F_a & F_{d_a} & F_{h_a} \end{bmatrix} = [1\ \ 0\ \ 0\ \ 0\ \ 0\ \ 0]. \tag{17.78}$$

Here, the extra function to be estimated is $z_{\tau h_a}(t) = x_1(t)$ and it is also clear that $\text{rank}\begin{bmatrix} C & C_d & C_h \\ \bar{F} & \bar{F}_d & \bar{F}_h \end{bmatrix} = 4$.

With F_a, F_{d_a} and F_{h_a} selected above, we proceed to test condition (17.67). For this, as we discussed in Remark 2, we should first check by letting $N_d = N_h = 0_{2\times 2}$ whether condition (17.70) is satisfied or not, since this would give a simpler observer without internal delays. However, it is found that condition (17.70) is still not satisfied since $\text{rank}\begin{bmatrix} \tilde{\Theta}_{dh} \\ \Upsilon_{dh} \end{bmatrix} = 24$, but $\text{rank}\big[\tilde{\Theta}_{dh}\big] = 23$. This implies that to estimate the augmented vector, $\bar{z}_{\tau h}(t) \in \mathbb{R}^2$, a second-order generalized observer without internal delays does not exist.

Case 3: Second-order observer with an internal delay: Next, we consider the case where $N_d \neq 0_{2\times 2}$ and $N_h = 0_{2\times 2}$. Note that this type of observer has a simpler structure than the one with both $N_d \neq 0_{2\times 2}$ and $N_h \neq 0_{2\times 2}$. It is also preferable to the one with $N_h \neq 0_{2\times 2}$ and $N_d = 0_{2\times 2}$, since we deal with the stability/stabilization of a time-delay error system with a smaller time delay, i.e., $\tau < h$.

As per Remark 2, condition (17.70) is tested and found to be satisfied where $\text{rank}\begin{bmatrix} \tilde{\Theta}_{dh} \\ \Upsilon_{dh} \end{bmatrix} = \text{rank}\big[\tilde{\Theta}_{dh}\big] = 24$. Accordingly, we solve (17.69) for the unknown matrix X, where

$$X = \Upsilon_{dh}\tilde{\Theta}_{dh}^{+} + Z\Omega, \tag{17.79}$$

$\tilde{\Theta}_{dh}^{+} \in \mathbb{R}^{13n\times(2m+12p)}$ is the Moore-Penrose inverse of $\tilde{\Theta}_{dh}$, $\Omega \triangleq (I_{2m+12p} - \tilde{\Theta}_{dh}\tilde{\Theta}_{dh}^{+})$, and $Z \in \mathbb{R}^{m\times(2m+12p)}$ is an arbitrary matrix.

In this example, we have $\Omega \neq 0$ and thus we can exploit the arbitrariness of Z to solve the stabilization problem to ensure that the following time-delay error system is β-exponentially stable

$$\dot{e}(t) = (N_1 + ZN_2)e(t) + (N_{d1} + ZN_{d2})e(t - 0.6), \tag{17.80}$$

where $N_1 = \begin{bmatrix} 1 & -6.7888 \\ 3 & -5.6601 \end{bmatrix}$, $N_{d1} = \begin{bmatrix} 2 & -9.2112 \\ 1 & -6.3399 \end{bmatrix}$, $\tau = 0.6s$. Here, $N_2 \in \mathbb{R}^{28\times 2}$ and $N_{d2} \in \mathbb{R}^{28\times 2}$ are sub-matrices of Ω obtained by picking the first two columns of Ω and the next two columns of Ω, respectively, i.e.,

$$N_2 = \Omega\mathscr{I}^1, \tag{17.81}$$

$$N_{d2} = \Omega\mathscr{I}^2, \tag{17.82}$$

where $\mathscr{I}^1 \triangleq \begin{bmatrix} I_m \\ 0_{(m+12n)\times m} \end{bmatrix}$ and $\mathscr{I}^2 \triangleq \begin{bmatrix} 0_{m\times m} \\ I_m \\ 0_{12n\times m} \end{bmatrix}$.

It is worth mentioning that by letting $Z = 0$, the nominal time-delay system (17.80), i.e.,

$$\dot{e}(t) = \begin{bmatrix} 1 & -6.7888 \\ 3 & -5.6601 \end{bmatrix} e(t) + \begin{bmatrix} 2 & -9.2112 \\ 1 & -6.3399 \end{bmatrix} e(t-0.6),$$

is an unstable system. This can be verified by using the MATLAB package developed by Wu and Michiels [2], where it is found that there is a pair of unstable eigenvalues at $\mathbf{0.4185} \pm j3.9556$.

Therefore, it is necessary to find a $Z \neq 0$ such that the time-delay system (17.80) is β-exponentially stable. Notice that for this example, we have $\text{rank}[N_2] = \text{rank}[N_{d2}] = 1$, and instead of solving the stabilization problem of system (17.80), we solve the stabilization problem of the following time-delay system

$$\dot{e}(t) = (N_1 + Z_d \bar{N}_2)e(t) + (N_{d1} + Z_d \bar{N}_{d2})e(t-0.6), \tag{17.83}$$

where we pick two non-zeros row matrices $\bar{N}_2 \in \mathbb{R}^{1\times 2}$ and $\bar{N}_{d2} \in \mathbb{R}^{1\times 2}$ from the ith-row of N_2 and N_{d2}, respectively, and $Z_d \in \mathbb{R}^{2\times 1}$ is an unknown matrix to be found so that system (17.83) is β-exponentially stable. Note that the stabilization problem (17.83) is simpler than the stabilization problem (17.80), since there are only two unknowns in Z_d.

To proceed with the problem of finding a Z_d to stabilize system (17.83). We pick the sixth-row of N_2 and N_{d2} and obtain $\bar{N}_2 = [0 \;\; 0.2058]$ and $\bar{N}_{d2} = [0 \;\; -0.2058]$, respectively. (There is no particular reason for selecting them to be the sixth-row, apart from the fact that with this selection, we have $\bar{N}_2 \neq 0$, $\bar{N}_{d2} \neq 0$). Next, we solve the stabilization problem (see, LMI (20.5) of Chap. 20). It is found that LMI (20.5) is feasible with $\tau = 0.6s$ and $\beta = 2.5$. This ensures the error time-delay system (17.83) converges to zero with a convergence rate of $\beta = 2.5$. We therefore obtain $Z_d = \begin{bmatrix} -29.4453 \\ -23.2568 \end{bmatrix}$, and by substituting N_1, N_{d1}, Z_d, $\bar{N}_2$ and $\bar{N}_{d2}$ into (17.83), the following time-delay error system is obtained

$$\dot{e}(t) = \begin{bmatrix} 1 & -12.8482 \\ 3 & -10.4460 \end{bmatrix} e(t) + \begin{bmatrix} 2 & -3.1518 \\ 1 & -1.5540 \end{bmatrix} e(t-0.6).$$

The eigenvalues of the above time-delay error system are computed [2], and it is found that they are all stable with a pair of dominant eigenvalues at $-2.8482 \pm j2.1659$.

With the obtained Z_d, we let all the columns of Z be zeros except its sixth-column to be Z_d, i.e.,

$$Z = [0_{2\times 5} \;\; Z_d \;\; 0_{2\times 22}]. \tag{17.84}$$

Substituting (17.84) into (17.79), we obtain X, and from (17.40)–(17.47) and (17.48)–(17.55), all the matrices of observer (17.4)–(17.5) are obtained, where $N = \begin{bmatrix} 1 & -12.8482 \\ 3 & -10.4460 \end{bmatrix}$, $N_d = \begin{bmatrix} 2 & -3.1518 \\ 1 & -1.5540 \end{bmatrix}$, $N_h = 0_{2\times 2}$, $M = \begin{bmatrix} -1 & -2.2347 \\ 0 & -2.2567 \end{bmatrix}$,

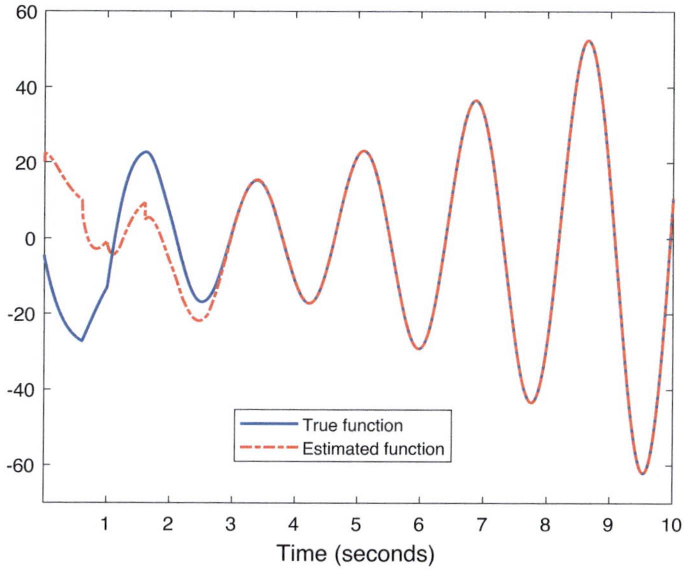

Fig. 17.1 Trajectories of $z_{\tau h}(t)$ and $\hat{z}_{\tau h}(t)$ (Numerical Example 1, second-order observer with an internal delay, in this chapter)

$$M_d = \begin{bmatrix} 0.0932 & 0.3484 \\ -0.0216 & 0.2084 \end{bmatrix}, M_h = \begin{bmatrix} 0.0932 & 0.0962 \\ -0.0216 & -0.0122 \end{bmatrix}, M_{dh} = \begin{bmatrix} 0 & 0.4415 \\ 0 & 0.1869 \end{bmatrix},$$
$$G = \begin{bmatrix} 3.8482 & 4.0180 \\ 1.4460 & -0.9246 \end{bmatrix}, G_d = \begin{bmatrix} -3.9639 & -5.6139 \\ -1.7493 & -3.1850 \end{bmatrix}, G_h = \begin{bmatrix} -3.9639 & -2.7495 \\ -1.7493 & -2.0554 \end{bmatrix},$$
$$G_{hh} = \begin{bmatrix} 0.4689 & -0.1832 \\ -0.0984 & 0.0525 \end{bmatrix}, G_{dd} = \begin{bmatrix} 0.1853 & 0.9918 \\ -0.1464 & 0.6713 \end{bmatrix}, G_{dh} = \begin{bmatrix} 0.6542 & 1.3410 \\ -0.2449 & 1.0998 \end{bmatrix},$$
$$G_{d2h} = \begin{bmatrix} 0.4415 & 0.4415 \\ 0.1869 & 0.1869 \end{bmatrix}, G_{2dh} = \begin{bmatrix} 0.4415 & 0.7356 \\ 0.1869 & 0.3380 \end{bmatrix}, J = \begin{bmatrix} 0 \\ 1 \end{bmatrix}, J_d = \begin{bmatrix} 8.3763 \\ 4.5350 \end{bmatrix},$$
$$J_h = \begin{bmatrix} 8.3763 \\ 4.5350 \end{bmatrix}, \quad J_{dd} = \begin{bmatrix} -0.8831 \\ -0.3738 \end{bmatrix}, \quad J_{hh} = \begin{bmatrix} -0.4719 \\ 0.0890 \end{bmatrix}, \quad J_{dh} = \begin{bmatrix} -1.3550 \\ -0.2847 \end{bmatrix},$$
$$J_{2dh} = \begin{bmatrix} -0.8831 \\ -0.3738 \end{bmatrix} \text{ and } J_{d2h} = \begin{bmatrix} -0.8831 \\ -0.3738 \end{bmatrix}.$$

For $\tau = 0.6s$ and $h = 1s$, $u(t) = 1, \forall t \geq 0$, $x(\theta) = \begin{bmatrix} 5 \\ -5 \end{bmatrix}$, $\rho(\theta) = \begin{bmatrix} 0 \\ 0 \end{bmatrix}$ $\forall \theta \in [-1, 0]$, Fig. 17.1 shows the trajectories of $z_{\tau h}(t)$ and $\hat{z}_{\tau h}(t)$. It is clear that $\hat{z}_{\tau h}(t)$ converges to $z_{\tau h}(t)$.

Furthermore, Fig. 17.2 shows the trajectories of $z_{\tau h_a}(t) = x_1(t)$ and $\hat{z}_{\tau h_a}(t)$, i.e., $\hat{x}_1(t)$ (see, (17.75)–(17.76) where we increase the order of the designed generalized observer by estimating an extra function $z_{\tau h_a}(t)$). It is clear from Fig. 17.2 that $\hat{z}_{\tau h_a}(t)$ converges to $z_{\tau h_a}(t)$.

Case 4: Third-order observer without internal delays ($N_d = N_h = 0_{3\times3}$): As mentioned earlier, $F_d \neq 0$ and $F_h \neq 0$ provide more flexibility in designing generalized

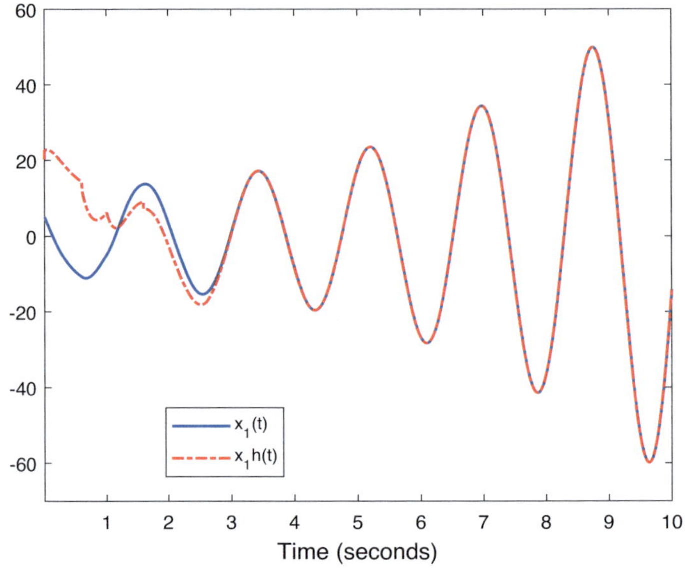

Fig. 17.2 Trajectories of $x_1(t)$ and $\hat{x}_1(t)$ (Numerical Example 1, second-order observer with an internal delay, in this chapter)

observers. To demonstrate this point, we design a third-order generalized observer without internal delays to estimate function (17.73). As discussed in Case 2, it was not possible to design a second-order generalized observer without internal delays; instead, we design a second-order generalized observer with an internal delay, i.e., $N_d \neq 0$, as discussed in Case 3.

To proceed, we increase the order of the designed observer by two, i.e., $v = 2$, and the extra vector, $z_{\tau h_a}(t) \in \mathbb{R}^2$, to be estimated is defined by (17.76). Furthermore, to show the advantage of having $F_{d_a} \neq 0$ and/or $F_{h_a} \neq 0$ as opposed to the case where $F_{d_a} = F_{h_a} = 0$, we first choose $z_{\tau h_a}(t)$ to be as follows

$$z_{\tau h_a}(t) = F_a x(t), \tag{17.85}$$

i.e., we let $F_{d_a} = F_{h_a} = 0_{2\times2}$. Here, $F_a \in \mathbb{R}^{2\times2}$ is a full rank matrix, to be found such that condition (17.70) is satisfied. However, $F_a \in \mathbb{R}^{2\times2}$ does not exist to satisfy condition (17.70).

Next, we select F_a, F_{d_a} and F_{h_a} as follows

$$\begin{bmatrix} F_a \; F_{d_a} \; F_{h_a} \end{bmatrix} = \begin{bmatrix} 1\,0\,0\,0\,0\,0 \\ 0\,1\,1\,1\,0\,0 \end{bmatrix}. \tag{17.86}$$

Thus, we have $F_{d_a} = \begin{bmatrix} 0\,0 \\ 1\,1 \end{bmatrix} \neq 0_{2\times2}$ and the extra vector to be estimated is

$$z_{\tau h_a}(t) = \begin{bmatrix} x_1(t) \\ x_2(t) + x_1(t-\tau) + x_2(t-\tau) \end{bmatrix}, \tag{17.87}$$

which contains delayed state variables $x_1(t-\tau)$ and $x_2(t-\tau)$. It is also clear that

$$\text{rank}\begin{bmatrix} C & C_d & C_h \\ \bar{F} & \bar{F}_d & \bar{F}_h \end{bmatrix} = 5.$$

With F_a, F_{d_a} and F_{h_a} given by (17.86), we proceed to test condition (17.70) and find it satisfied, where rank $\begin{bmatrix} \tilde{\Theta}_{dh} \\ \Upsilon_{dh} \end{bmatrix} = \text{rank}\big[\tilde{\Theta}_{dh}\big] = 24$.

Accordingly, we solve (17.69) for the unknown matrix X, where

$$X = \Upsilon_{dh}\tilde{\Theta}_{dh}^{+} + Z\Omega, \tag{17.88}$$

$\tilde{\Theta}_{dh}^{+} \in \mathbb{R}^{13n\times(m+12p)}$ is the Moore-Penrose inverse of $\tilde{\Theta}_{dh}$, $\Omega \triangleq (I_{m+12p} - \tilde{\Theta}_{dh}\tilde{\Theta}_{dh}^{+})$, and $Z \in \mathbb{R}^{m\times(m+12p)}$ is an arbitrary matrix. Note that in this example, the first three columns of Ω, i.e., N_2, is zeros. Thus, Z does not play any roles in the stabilization of the error system, $\dot{e}(t) = (N_1 + ZN_2)e(t)$, and so we let $Z = 0_{3\times 27}$, and the solution given by (17.88), i.e., $X = \Upsilon_{dh}\tilde{\Theta}_{dh}^{+}$, results in a matrix $N = N_1$ being stable, where

$$N = \begin{bmatrix} -1 & -6 & -2 \\ 2 & -7 & -1 \\ 2 & 2 & -6 \end{bmatrix}.$$

As matrix N is stable with eigenvalues at $\{-5, -4.5 \pm j2.7839\}$, the error vector $e(t)$ converges to zero for any non-zero initial conditions. Furthermore, $\dot{e}(t) = Ne(t)$ is now independent of both time delays τ and h. Accordingly, a third-order observer without internal delays to estimate the vector $\bar{z}_{\tau h}(t) = \begin{bmatrix} z_{\tau h}(t) \\ z_{\tau h_a}(t) \end{bmatrix}$ is obtained, where the rest of the observer matrices are given below

$$N_d = N_h = 0_{3\times 3},\ M = \begin{bmatrix} 1 & -0.2986 \\ 0 & -0.8803 \\ -1 & 2.0730 \end{bmatrix},\ M_d = \begin{bmatrix} 0.3845 & -0.0440 \\ 0.3175 & -0.0908 \\ 0.1698 & -1.3535 \end{bmatrix},$$

$$M_h = \begin{bmatrix} 0.3845 & 0.3169 \\ 0.3175 & 0.3022 \\ 0.1698 & -1.0967 \end{bmatrix},\ M_{dh} = \begin{bmatrix} 0 & 0.3405 \\ 0 & 0.2267 \\ 0 & 0.8162 \end{bmatrix},\ G = \begin{bmatrix} 3 & -3.9049 \\ 1 & -2.9570 \\ -2 & -8.9215 \end{bmatrix},$$

$$G_d = \begin{bmatrix} -5.1771 & 2.9494 \\ -2.7905 & 2.4413 \\ -0.1782 & 2.6425 \end{bmatrix},\ G_h = \begin{bmatrix} -5.1771 & 2.2906 \\ -2.7905 & 2.7554 \\ 0.8218 & 6.2330 \end{bmatrix},\ G_{hh} = \begin{bmatrix} 1.8550 & -0.8366 \\ 1.5723 & -0.6504 \\ -0.4177 & -1.6060 \end{bmatrix},$$

$$G_{dd} = \begin{bmatrix} 1.1975 & -0.5165 \\ 1.0434 & -0.5900 \\ -2.1372 & -0.2303 \end{bmatrix},\ G_{dh} = \begin{bmatrix} 3.0526 & -2.5991 \\ 2.6156 & -1.5114 \\ -2.5549 & -1.3483 \end{bmatrix},\ J_{d2h} = \begin{bmatrix} -0.6811 \\ -0.4534 \\ -1.6325 \end{bmatrix}$$

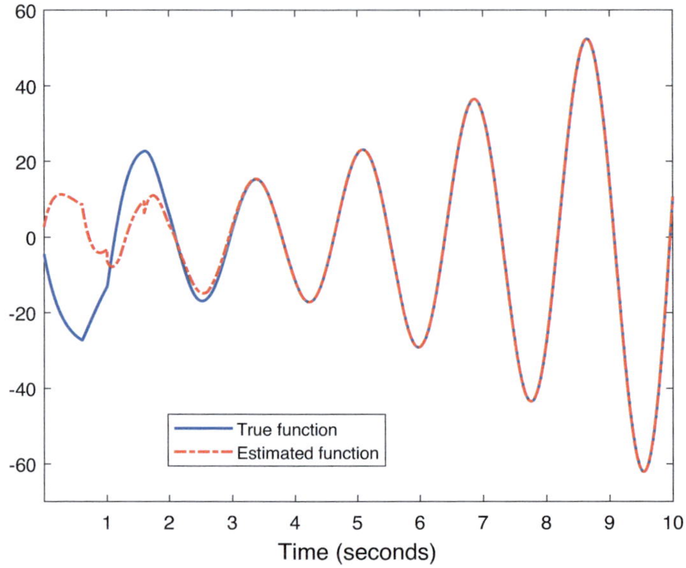

Fig. 17.3 Trajectories of $z_{\tau h}(t)$ and $\hat{z}_{\tau h}(t)$, $\tau = 0.6s$ and $h = 1s$ (Numerical Example 1, third-order observer without internal delays, in this chapter)

$$G_{d2h} = G_{2dh} = \begin{bmatrix} 0.3405 & 0.3405 \\ 0.2267 & 0.2267 \\ 0.8162 & 0.8162 \end{bmatrix}, J = \begin{bmatrix} 0 \\ 1 \\ 3 \end{bmatrix}, J_d = \begin{bmatrix} 4.2126 \\ 1.4431 \\ 0.6842 \end{bmatrix}, J_h = \begin{bmatrix} 4.2126 \\ 1.4431 \\ -1.3158 \end{bmatrix},$$

$$J_{dd} = \begin{bmatrix} -0.6811 \\ -0.4534 \\ 2.3675 \end{bmatrix}, J_{hh} = \begin{bmatrix} -1.7874 \\ -1.5569 \\ 1.6842 \end{bmatrix}, J_{dh} = \begin{bmatrix} -2.4685 \\ -2.0103 \\ 4.0517 \end{bmatrix} \text{ and } J_{2dh} = \begin{bmatrix} -0.6811 \\ -0.4534 \\ -1.6325 \end{bmatrix}.$$

For $\tau = 0.6s$ and $h = 1s$, $u(t) = 1, \forall t \geq 0$, $x(\theta) = \begin{bmatrix} 5 \\ -5 \end{bmatrix}$, $\forall \theta \in [-1, 0]$, $w(0) = \begin{bmatrix} 0 \\ 0 \\ 0 \end{bmatrix}$, Fig. 17.3 shows the trajectories of $z_{\tau h}(t)$ and $\hat{z}_{\tau h}(t)$. It is clear from Fig. 17.3 that $\hat{z}_{\tau h}(t)$ asymptotically converges to $z_{\tau h}(t)$.

Finally, notice that the error vector $e(t)$ is independent of both delays τ and h. Thus, the designed third-order generalized observer works for any given non-commensurate delays τ and h. Figure 17.4 shows the trajectories of $z_{\tau h}(t)$ and $\hat{z}_{\tau h}(t)$ for the case where $\tau = 0.8\,s$ and $h = 1.1\,s$, $u(t) = 1, \forall t \geq 0$, $x(\theta) = \begin{bmatrix} 5 \\ -5 \end{bmatrix}$, $\forall \theta \in [-1.1, 0]$, $w(0) = \begin{bmatrix} 0 \\ 0 \\ 0 \end{bmatrix}$. Again, it is clear from Fig. 17.4 that $\hat{z}_{\tau h}(t)$ converges to $z_{\tau h}(t)$.

Thus, via this numerical example, we can see the flexibility and attractive features of solving the generalized estimation problem (17.3).

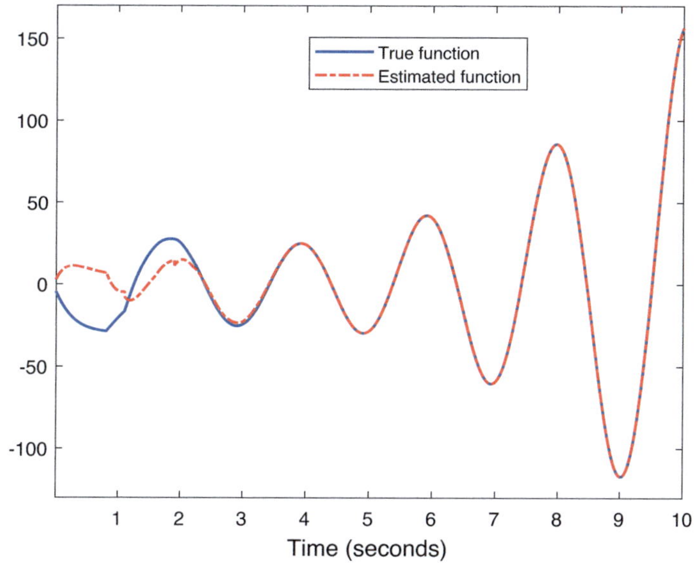

Fig. 17.4 Trajectories of $z_{\tau h}(t)$ and $\hat{z}_{\tau h}(t)$, $\tau = 0.8\,s$ and $h = 1.1\,s$ (Numerical Example 1, third-order observer without internal delays, in this chapter)

Numerical Example 2

In this example, we consider system (17.1)–(17.2), where $\tau = 0.8\,s$, $h = 1.1\,s$, $C = [1\ \ 0]$, $C_d = [0\ \ 0]$, $C_h = [0\ \ 1]$, $A = \begin{bmatrix} -3 & 2 \\ 1 & -7 \end{bmatrix}$, $A_d = \begin{bmatrix} -4 & 1 \\ 0 & -1 \end{bmatrix}$, $A_h = \begin{bmatrix} -2 & 2 \\ 0 & 3 \end{bmatrix}$ and $B = \begin{bmatrix} 1 \\ 2 \end{bmatrix}$.

In this example, we have $C_d = 0_{1\times 2}$ and there is only one output, namely, $y(t)$, where

$$y(t) = x_1(t) + x_2(t - 1.1). \tag{17.89}$$

Our objective is to design an observer to estimate the following function

$$z_{\tau h}(t) = [1\ 2]x(t) + [3\ \ 4]x(t - 0.8) + [5\ \ 6]x(t - 1.1), \tag{17.90}$$

i.e., $F = [1\ 2]$, $F_d = [3\ \ 4]$ and $F_h = [5\ \ 6]$. As for this function, $z_{\tau h}(t)$ is linearly independent from $y(t)$, i.e.,

$$\text{rank} \begin{bmatrix} C & C_d & C_h \\ F & F_d & F_h \end{bmatrix} = \text{rank} \begin{bmatrix} 1\,0\,0\,0\,0\,1 \\ 1\,2\,3\,4\,5\,6 \end{bmatrix} = 2,$$

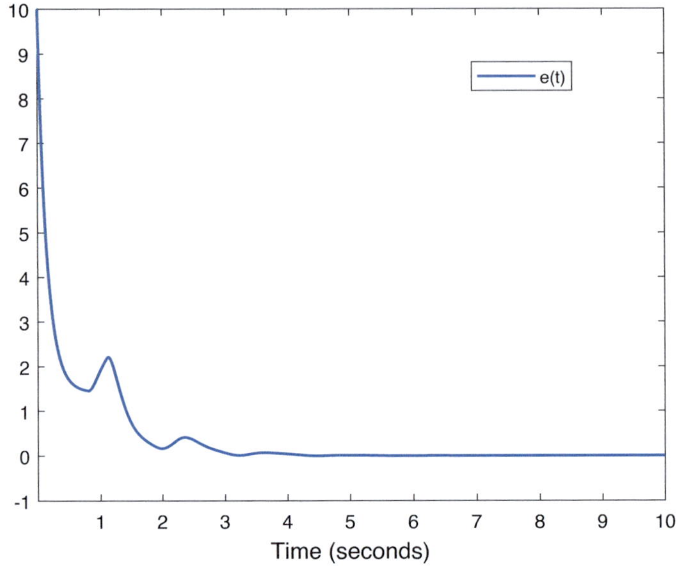

Fig. 17.5 Trajectory of $e(t)$ as defined by (17.91) where $e(\theta) = 10, \ \forall \theta \in [-1.1, 0]$ (Numerical Example 2, in this chapter)

and thus a dynamical observer is required to estimate $z_{\tau h}(t)$.

To estimate function (17.90), we consider a first-order functional observer (17.4)–(17.5). First, we check condition (17.67) and it is found that the condition is satisfied. Note that the solution to equation (17.57), i.e., $X = \Upsilon_{dh}\Theta_{dh}^{+} + Z\Omega$, results in $\Omega = 0_{15\times 15}$. Thus, Z does not play any roles in the stabilization of the time-delay error dynamic equation (17.68). With $X = \Upsilon_{dh}\Theta_{dh}^{+}$, we obtain the following time-delay error dynamic equation

$$\dot{e}(t) = -7e(t) - e(t - 0.8) + 2e(t - 1.1). \tag{17.91}$$

It is required that the above time-delay error system is asymptotically stable. To meet this requirement, stability analysis can be carried out to establish β-exponential stability of (17.91) (see, Chap. 20 or some LMI conditions as presented in [1]). For simplicity, Fig. 17.5 shows trajectory of $e(t)$ for a non-zero initial function. Thus, we can ensure that the error $e(t)$ converges to zero for any non-zero initial function. Accordingly, we obtain a first-order functional observer (17.4)–(17.5), where

$N = -7$, $N_d = -1$, $N_h = 2$, $M = 1$, $M_d = 3$, $M_h = 5$, $M_{dh} = 0$, $G = 2$, $G_d = 4$, $G_h = 5$, $G_{hh} = -5$, $G_{dd} = 0$, $G_{dh} = -3$, $G_{d2h} = G_{2dh} = 0$, $J = 4$, $J_d = 8$, $J_h = 10$, $J_{dd} = 0$, $J_{hh} = -10$, $J_{dh} = -6$, $J_{2dh} = J_{d2h} = 0$.

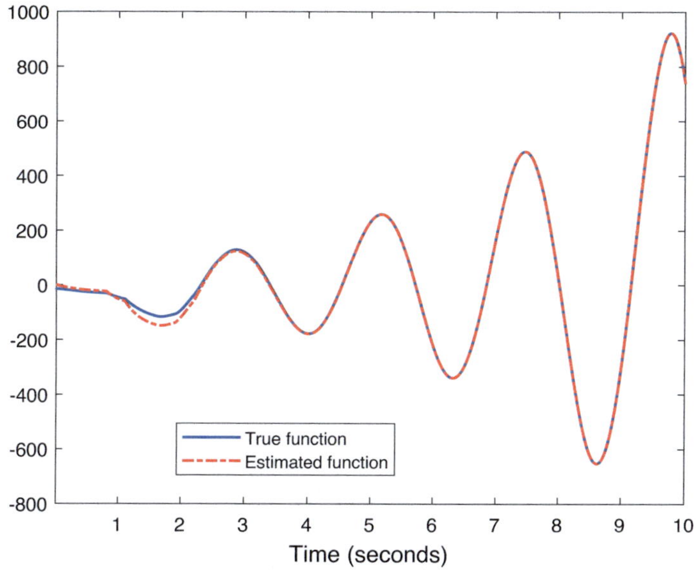

Fig. 17.6 Trajectories of $z_{\tau h}(t)$ and $\hat{z}_{\tau h}(t)$, $\tau = 0.8\,s$ and $h = 1.1\,s$ (Numerical Example 2, in this chapter

For $\tau = 0.8\,s$ and $h = 1.1\,s$, $u(t) = 1$, $\forall t \geq 0$, $x(\theta) = \begin{bmatrix} 5 \\ -5 \end{bmatrix}$, $\forall \theta \in [-1.1, 0]$, $w(0) = 0$. Figure 17.6 shows the trajectories of $z_{\tau h}(t)$ and $\hat{z}_{\tau h}(t)$. It is clear from Fig. 17.6 that $\hat{z}_{\tau h}(t)$ asymptotically converges to $z_{\tau h}(t)$.

Numerical Example 3

In this example, we design a functional observer to estimate function (17.3) for system (17.1)–(17.2) subject to the constraint $C = 0_{p\times n}$. This is an interesting case since now the output vector, $y(t)$, contains only delayed state variables. Thanks to our unified design framework, this case does not pose any additional difficulty in the design of functional observers. Here, we simply replace $C = 0_{p\times n}$ in Θ_{dh} (where Θ_{dh} is as defined by (17.59)) and the existence condition (17.67) remains unchanged.

Let us now reconsider Numerical Example 1, where we let $C = 0_{2\times 2}$, and our objective is to design a generalized observer to estimate the function $z_{\tau h}(t)$ defined by (17.73).

Notice that with $C = 0_{2\times 2}$, the two outputs of this system are as follows

$$y_1(t) = x_2(t - \tau) + x_1(t - h) + x_2(t - h), \tag{17.92}$$

$$y_2(t) = x_2(t - \tau) + x_2(t - h), \tag{17.93}$$

where both outputs do not contain any instantaneous state variables. It is also clear that $z_{\tau h}(t)$ is linearly independent to $y_1(t)$ and $y_2(t)$, i.e.,

$$\text{rank}\begin{bmatrix} C & C_d & C_h \\ F & F_d & F_h \end{bmatrix} = \text{rank}\begin{bmatrix} 0 & 0\,0\,1 & 1\,1 \\ 0 & 0\,0\,1 & 0\,1 \\ 1 & -1\,0\,1 & -1\,1 \end{bmatrix} = 3,$$

and thus it is necessary to design a dynamical observer to estimate the required function $z_{\tau h}(t)$. By following exactly the same steps as discussed in Case 4 of Numerical Example 1 (i.e., design of a third-order observer without internal delays), we obtain the following third-order observer for $z_{\tau h}(t)$,

$$\hat{z}_{\tau h}(t) = w_1(t) + M_1 y(t) + M_{d1} y(t-\tau) + M_{h1} y(t-h) + M_{dh1} y(t-\tau-h), \tag{17.94}$$

where $M_1 = \begin{bmatrix} -1 & -1 \end{bmatrix}$, $M_{d1} = \begin{bmatrix} 0.0946 & -0.0946 \end{bmatrix}$, $M_{h1} = \begin{bmatrix} 0.0946 & 0.2606 \end{bmatrix}$, $M_{dh1} = \begin{bmatrix} 0 & 0.0821 \end{bmatrix}$, $w(t) = \begin{bmatrix} w_1(t) \\ w_2(t) \\ w_3(t) \end{bmatrix}$, and $\dot{w}(t)$ is as defined by (17.5), where

$N = \begin{bmatrix} -3 & -1 & -1 \\ 2 & -5 & -1 \\ 0 & 1 & -5 \end{bmatrix}$, $\lambda(N) = \{-5, -4 \pm j1.4142\}$, $N_d = N_h = 0_{3\times 3}$, $G = \begin{bmatrix} 2 & -5 \\ 1 & -9 \\ -2 & -6 \end{bmatrix}$,

$G_d = \begin{bmatrix} -3.2491 & 1.2491 \\ -3.1546 & 1.1546 \\ 0.0296 & 3.9704 \end{bmatrix}$, $G_h = \begin{bmatrix} -3.2491 & -1.5782 \\ -3.1546 & -1.3176 \\ 1.0296 & 5.7237 \end{bmatrix}$, $G_{hh} = \begin{bmatrix} 0.6389 & -0.0231 \\ 0.6389 & -0.0231 \\ -1.1357 & 0.2689 \end{bmatrix}$,

$G_{dd} = \begin{bmatrix} 0.2961 & -0.2961 \\ 0.2961 & -0.2961 \\ -5.0579 & -2.9421 \end{bmatrix}$, $G_{dh} = \begin{bmatrix} 0.9350 & -0.3977 \\ 0.9350 & -0.3156 \\ -6.1935 & -2.7918 \end{bmatrix}$,

$G_{d2h} = G_{2dh} = \begin{bmatrix} 0.0821 & 0.0821 \\ 0.0821 & 0.0821 \\ 0.2553 & 0.2553 \end{bmatrix}$, $J = \begin{bmatrix} -1 \\ 1 \\ 2 \end{bmatrix}$, $J_d = \begin{bmatrix} 6 \\ 6 \\ -3 \end{bmatrix}$, $J_h = \begin{bmatrix} 6 \\ 6 \\ -5 \end{bmatrix}$,

$J_{dd} = \begin{bmatrix} 0 \\ 0 \\ 8 \end{bmatrix}$, $J_{hh} = \begin{bmatrix} -0.8050 \\ -0.8050 \\ 1.2681 \end{bmatrix}$, $J_{dh} = \begin{bmatrix} -0.8050 \\ -0.8050 \\ 9.2681 \end{bmatrix}$, $J_{2dh} = J_{d2h} = \begin{bmatrix} -0.1643 \\ -0.1643 \\ -0.5105 \end{bmatrix}$.

As the designed observer does not contain internal delays (i.e., $N_d = N_h = 0_{3\times 3}$), it works for any given non-commensurate delays τ and h. Figure 17.7 shows the trajectories of $z_{\tau h}(t)$ and $\hat{z}_{\tau h}(t)$ for the case where $\tau = 0.8\,s$ and $h = 1.1\,s$, $u(t) = 1, \forall t \geq 0$, $x(\theta) = \begin{bmatrix} 5 \\ -5 \end{bmatrix}$, $\forall\theta \in [-1.1, 0]$, $w(0) = \begin{bmatrix} 0 \\ 0 \\ 0 \end{bmatrix}$. It is clear from Fig. 17.7 that $\hat{z}_{\tau h}(t)$ converges to $z_{\tau h}(t)$.

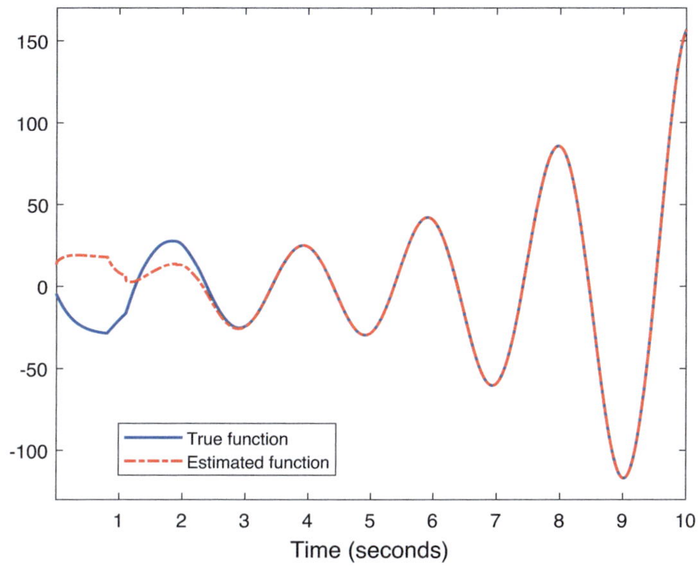

Fig. 17.7 Trajectories of $z_{\tau h}(t)$ and $\hat{z}_{\tau h}(t)$, $\tau = 0.8\,s$ and $h = 1.1\,s$ (Numerical Example 3, $C = 0_{2\times 2}$, third-order observer with no internal delays, in this chapter)

17.3 Concluding Remark

In this chapter, we have considered the problem of designing a generalized observer to estimate a generalized functional (17.3) of time-delay system (17.1)–(17.2) with two non-commensurate delays in state and output vectors. This is the first time the problem is reported and solved in this book. Thanks to our unified framework, we can readily obtain observer existence conditions, and via Numerical Examples 1-3, we have demonstrated in detail the design of various types of observers and their salient features.

References

1. Mondié, S., Kharitonov, V.L.: Exponential estimates for retarded time-delay systems: an lmi approach. IEEE Trans. Autom. Control **50**(2), 268–273 (2005)
2. Wu, Z., Michiels, W.: Reliably computing all characteristic roots of delay differential equations in a given right half plane using a spectral method. J. Comput. Appl. Math. **236**(9), 2499–2514 (2012)

Chapter 18
Generalized Observers for Time-Delay Systems with Unknown Inputs

18.1 System Description and Problem Statement

In this chapter, we consider the following time-delay system with unknown inputs

$$\dot{x}(t) = Ax(t) + A_d x(t-\tau) + A_{dd} x(t-2\tau) + Bu(t) + Dd(t), \tag{18.1}$$
$$y(t) = Cx(t) + C_d x(t-\tau) + C_{dd} x(t-2\tau), \tag{18.2}$$

where $x(t) \in \mathbb{R}^n$, $y(t) \in \mathbb{R}^p$, $u(t) \in \mathbb{R}^r$ and $d(t) \in \mathbb{R}^q$ are, respectively, the state vector, output vector, control input vector and unknown input vector. Matrices $A, A_d, A_{dd} \in \mathbb{R}^{n\times n}$, $B \in \mathbb{R}^{n\times r}$, $D \in \mathbb{R}^{n\times q}$ and $C, C_d, C_{dd} \in \mathbb{R}^{p\times n}$ are known constant matrices. Term $\tau > 0$ is a known constant time delay.

Our objective is to design an observer to estimate the following generalized functional

$$z_{3\tau}(t) = Fx(t) + F_d x(t-\tau) + F_{dd} x(t-2\tau) + F_{3d} x(t-3\tau), \tag{18.3}$$

where $z_{3\tau}(t) \in \mathbb{R}^m$, and F, F_d, F_{dd} and F_{3d} are known constant matrices of appropriate dimensions. This problem arises from the unknown input estimation problem of time-delay systems with more than one unknown input as discussed in Chap. 9. Also, note that in Chap. 15, we considered the problem of estimating the same vector (i.e., $z_{3\tau}(t)$) but for a class of time-delay systems (18.1)–(18.2) with no unknown input vector $d(t)$. Thus, in this chapter, we extend the work in Chap. 15 to include unknown inputs. Since this is the first time in this book that we look at generalized observers for time-delay systems with an unknown input vector, and thus for completeness, we provide their existence conditions and some numerical examples used in Chap. 9.

To estimate vector (18.3) for system (18.1)–(18.2), we consider the following generalized functional observer

H. Trinh et al., *Unknown Inputs Estimation in Linear Time-Delay Systems Using Generalized Functional Observers*, Studies in Systems, Decision and Control 648,
https://doi.org/10.1007/978-3-032-11870-7_18

$$\hat{z}_{3\tau}(t) = w(t) + My(t) + M_d y(t-\tau), \tag{18.4}$$

$$\begin{aligned}\dot{w}(t) &= Nw(t) + N_d w(t-\tau) + N_{dd} w(t-2\tau) + Gy(t) \\ &+ G_d y(t-\tau) + G_{dd} y(t-2\tau) + G_{3d} y(t-3\tau) + Ju(t) \\ &+ J_d u(t-\tau) + J_{dd} u(t-2\tau) + J_{3d} u(t-3\tau),\end{aligned} \tag{18.5}$$

where $w(\theta) = \rho(\theta)$, $\forall\theta \in [0, 2\tau]$, $w(t) \in \mathbb{R}^m$, and $\hat{z}_{3\tau}(t) \in \mathbb{R}^m$ is the estimate of $z_{3\tau}(t)$. Matrices M, M_d, N, N_d, N_{dd}, G, G_d, G_{dd}, G_{3d}, J, J_d, J_{dd} and J_{3d} are unknown and to be determined such that $\hat{z}_{3\tau}(t)$ converges asymptotically to $z_{3\tau}(t)$ in the presence of the unknown input vector $d(t)$.

For completeness, we derive their existence conditions by following a similar approach as in Chap. 15. Now, let us define the following error vector

$$\begin{aligned}e(t) &= \hat{z}_{3\tau}(t) - z_{3\tau}(t) \\ &= w(t) + My(t) + M_d y(t-\tau) - Fx(t) - F_d x(t-\tau) \\ &- F_{dd} x(t-2\tau) - F_{3d} x(t-3\tau).\end{aligned} \tag{18.6}$$

Taking derivative of both sides of (18.6) with respect to time and after some substitutions and rearranging, we obtain an error dynamic equation, in which we can establish the following existence conditions for observer (18.4)–(18.5).

18.2 Observer Existence Conditions

Existence Conditions: The estimate $\hat{z}_{3\tau}(t)$ converges asymptotically to $z_{3\tau}(t)$ for any initial condition $w(\theta)$, any $u(t)$ and $d(t)$, provided that the following conditions hold

$$\dot{e}(t) = Ne(t) + N_d e(t-\tau) + N_{dd} e(t-2\tau) \quad \text{is asymptotically stable,} \tag{18.7}$$

$$NF + \tilde{T}C + MCA - FA = 0, \tag{18.8}$$

$$NF_d + N_d F + \hat{W}C + \tilde{T}C_d + MC_\tau + M_d CA - FA_d - F_d A = 0, \tag{18.9}$$

$$\begin{aligned}&NF_{dd} + N_d F_d + N_{dd} F + \hat{U}C + \hat{W}C_d + \tilde{T}C_{dd} + MC_{2\tau} \\ &+ M_d C_\tau - FA_{dd} - F_d A_d - F_{dd} A = 0,\end{aligned} \tag{18.10}$$

$$NF_{3d} + N_d F_{dd} + N_{dd} F_d + \hat{Y} C + \hat{U} C_d + \hat{W} C_{dd} + M C_{3\tau} \\ + M_d C_{2\tau} - F_d A_{dd} - F_{dd} A_d - F_{3d} A = 0, \tag{18.11}$$

$$N_d F_{3d} + N_{dd} F_{dd} + \hat{Y} C_d + \hat{U} C_{dd} + M C_{dd} A_{dd} + M_d C_{3\tau} - F_{dd} A_{dd} - F_{3d} A_d = 0, \tag{18.12}$$

$$N_{dd} F_{3d} + \hat{Y} C_{dd} + M_d C_{dd} A_{dd} - F_{3d} A_{dd} = 0, \tag{18.13}$$

$$MCD - FD = 0, \tag{18.14}$$

$$MC_d D + M_d CD - F_d D = 0, \tag{18.15}$$

$$MC_{dd} D + M_d C_d D - F_{dd} D = 0, \tag{18.16}$$

$$M_d C_{dd} D - F_{3d} D = 0, \tag{18.17}$$

$$J = FB - MCB, \tag{18.18}$$

$$J_d = F_d B - MC_d B - M_d CB, \tag{18.19}$$

$$J_{dd} = F_{dd} B - MC_{dd} B - M_d C_d B, \tag{18.20}$$

$$J_{3d} = F_{3d} B - M_d C_{dd} B, \tag{18.21}$$

where

$$C_{2\tau} = CA_{dd} + C_d A_d + C_{dd} A, \tag{18.22}$$

$$C_{3\tau} = C_d A_{dd} + C_{dd} A_d, \tag{18.23}$$

and $\tilde{T}$, $\hat{W}$, $\hat{U}$, $\hat{Y}$ and C_τ are defined by (10.84), (15.22), (15.23), (15.24) and (16.18), respectively.

Remark 1 Compared with those existence conditions derived in Chap. 15, we can see that the presence of the unknown input vector $d(t)$ in (18.1) has resulted in four extra conditions, i.e., (18.14)–(18.17). These conditions are necessary to completely decouple the effect of $d(t)$ from the error dynamic equation, and hence ensuring that the designed observer is insensitive to $d(t)$. This approach has been well-documented in the literature to deal with unknown inputs (see, for example, [1–5]).

It is clear from (18.18)–(18.21) that we can explicitly obtain observer gains J, J_d, J_{dd} and J_{3d}, once M and M_d are found. Thus, there is a total of 9 unknown matrices, namely, X, where X is as defined by (15.32), need to be solved for, to meet conditions (18.7)–(18.17). Firstly, from (18.14)–(18.17), we can establish the following necessary and sufficient condition for their solvability

$$\operatorname{rank}\begin{bmatrix} CD & C_dD & C_{dd}D & 0_{p\times q} \\ 0_{p\times q} & CD & C_dD & C_{dd}D \\ FD & F_dD & F_{dd}D & F_{3d}D \end{bmatrix} = \operatorname{rank}\begin{bmatrix} CD & C_dD & C_{dd}D & 0_{p\times q} \\ 0_{p\times q} & CD & C_dD & C_{dd}D \end{bmatrix}. \tag{18.24}$$

Remark 2 Condition (18.24) can be regarded as a generalized observer matching condition. In general, due to the complexity in handling multiple time delays in the output vector and function $z_{3\tau}(t)$, condition (18.24) can be quite restrictive. For the simplest case where $C_d = C_{dd} = 0_{p\times n}$, and $F_d = F_{dd} = F_{3d} = 0_{m\times n}$, then condition (18.24) is reduced to the following well-known functional observer matching condition as reported in [5],

$$\operatorname{rank}\begin{bmatrix} CD \\ FD \end{bmatrix} = \operatorname{rank}\begin{bmatrix} CD \end{bmatrix}. \tag{18.25}$$

Generally, condition (18.25) is less conservative than the observer matching condition (see, for example, [1, 2], i.e., $\operatorname{rank}[CD] = \operatorname{rank} D$). Note that the observer matching condition is required in the design of state observers to estimate the full state vector of linear systems with unknown inputs.

Thus, to be able to solve for the 9 unknown matrices as defined by (15.32), it is first necessary to check whether condition (18.24) is satisfied or not. If (18.24) holds, we can proceed in the same way as discussed in Chap. 15 to solve for X. Provided that the solution leads to an asymptotic stable error system (18.7), the rest of the observer parameters can be obtained.

Note that the condition for the solvability of (18.8)–(18.17) is presented below

$$\operatorname{rank}\begin{bmatrix} \Theta_{3\tau} \\ \Upsilon_{3\tau} \end{bmatrix} = \operatorname{rank}\begin{bmatrix} \Theta_{3\tau} \end{bmatrix}, \tag{18.26}$$

where

$$\Theta_{3\tau} = \begin{bmatrix} F & F_d & F_{dd} & F_{3d} & 0 & 0 & 0 & 0 & 0 & 0 \\ 0 & F & F_d & F_{dd} & F_{3d} & 0 & 0 & 0 & 0 & 0 \\ 0 & 0 & F & F_d & F_{dd} & F_{3d} & 0 & 0 & 0 & 0 \\ C & C_d & C_{dd} & 0 & 0 & 0 & 0 & 0 & 0 & 0 \\ 0 & C & C_d & C_{dd} & 0 & 0 & 0 & 0 & 0 & 0 \\ 0 & 0 & C & C_d & C_{dd} & 0 & 0 & 0 & 0 & 0 \\ 0 & 0 & 0 & C & C_d & C_{dd} & 0 & 0 & 0 & 0 \\ CA & C_\tau & C_{2\tau} & C_{3\tau} & C_{dd}A_{dd} & 0 & CD & C_dD & C_{dd}D & 0 \\ 0 & CA & C_\tau & C_{2\tau} & C_{3\tau} & C_{dd}A_{dd} & 0 & CD & C_dD & C_{dd}D \end{bmatrix}, \tag{18.27}$$

$$\begin{aligned} \Upsilon_{3\tau} = \big[& FA \quad FA_d + F_dA \quad FA_{dd} + F_dA_d + F_{dd}A \quad F_dA_{dd} + F_{dd}A_d + F_{3d}A \\ & F_{dd}A_{dd} + F_{3d}A_d \quad F_{3d}A_{dd} \quad FD \quad F_dD \quad F_{dd}D \quad F_{3d}D \big]. \end{aligned} \tag{18.28}$$

Remark 3 In Sect. 9.2.1, for the case where condition (9.9) holds, the unknown input $d_1(t)$ can be estimated by designing an observer to estimate the function $\bar{z}(t) = -\bar{A}_{21}\bar{x}_1(t) - \bar{A}_{d21}\bar{x}_1(t-\tau)$ of time-delay system (9.10)–(9.11). This is the same as the observer design problem for the case where $A_{dd} = 0_{n\times n}$, $C_{dd} = 0_{p\times n}$, and $F_{dd} = F_{3d} = 0_{m\times n}$. Thus, from (18.1)–(18.2) and (18.3), we consider the following time-delay system

$$\dot{x}(t) = Ax(t) + A_dx(t-\tau) + Bu(t) + Dd(t), \tag{18.29}$$

$$y(t) = Cx(t) + C_dx(t-\tau), \tag{18.30}$$

and we address the problem of estimating the following vector

$$z_\tau(t) = Fx(t) + F_dx(t-\tau). \tag{18.31}$$

Thus, by letting $A_{dd} = 0_{n\times n}$, $C_{dd} = 0_{p\times n}$, and $F_{dd} = F_{3d} = 0_{m\times n}$, the existence conditions (18.7)–(18.21) are stated as below.

Existence Conditions: The estimate $\hat{z}_\tau(t)$ converges asymptotically to $z_\tau(t)$ for any initial condition $w(\theta)$, $u(t)$, and $d(t)$, provided that conditions (18.7)–(18.9) and the following conditions are met

$$N_dF_d + N_{dd}F + \hat{U}C + \hat{W}C_d + MC_dA_d + M_dC_\tau - F_dA_d = 0, \tag{18.32}$$

$$N_{dd}F_d + \hat{Y}C + \hat{U}C_d + M_dC_dA_d = 0, \tag{18.33}$$

$$\hat{Y}C_d = 0, \tag{18.34}$$

$$MCD - FD = 0, \tag{18.35}$$

$$MC_dD + M_dCD - F_dD = 0, \tag{18.36}$$

$$M_dC_dD = 0, \tag{18.37}$$

$$J = FB - MCB, \tag{18.38}$$

$$J_d = F_dB - MC_dB - M_dCB, \tag{18.39}$$

$$J_{dd} = -M_dC_dB. \tag{18.40}$$

Since we have $J_{3d} = 0_{m\times r}$, $\hat{z}_\tau(t)$ takes the following form

$$\hat{z}_\tau(t) = w(t) + My(t) + M_dy(t-\tau), \tag{18.41}$$

$$\begin{aligned}\dot{w}(t) &= Nw(t) + N_dw(t-\tau) + N_{dd}w(t-2\tau) + Gy(t)\\ &+ G_dy(t-\tau) + G_{dd}y(t-2\tau) + G_{3d}y(t-3\tau) + Ju(t)\\ &+ J_du(t-\tau) + J_{dd}u(t-2\tau).\end{aligned} \tag{18.42}$$

From (18.35)–(18.37), we obtain the following functional observer matching condition

$$\text{rank}\begin{bmatrix} CD & C_dD & 0_{p\times q}\\ 0_{p\times q} & CD & C_dD\\ FD & F_dD & 0_{m\times q}\end{bmatrix} = \text{rank}\begin{bmatrix} CD & C_dD & 0_{p\times q}\\ 0_{p\times q} & CD & C_dD\end{bmatrix}. \tag{18.43}$$

18.3 Numerical Examples

In this section, we first consider three numerical examples (Examples 1-3) to illustrate the design of generalized observers for a class of time-delay systems discussed in Remark 3. Then in Example 4, we consider an example to show the design of a generalized observer for system (18.1)–(18.2).

Numerical Example 1: In this example, we consider time-delay system (18.29)–(18.30) with an external unknown input, where $\tau = 1s$, and matrices A, A_d, B D, C and C_d are given below

$$A = \begin{bmatrix} -10 & -2\\ 0 & -7\end{bmatrix}, A_d = \begin{bmatrix} -3 & 2\\ -3 & -6\end{bmatrix}, B = \begin{bmatrix} 1\\ 2\end{bmatrix}, D = \begin{bmatrix} 1\\ -1\end{bmatrix}, C = \begin{bmatrix} 1 & 0\\ 0 & 0\end{bmatrix} \text{ and } C_d = \begin{bmatrix} 1 & 1\\ 1 & 0\end{bmatrix}.$$

Our aim is to design an observer to estimate $z_\tau(t)$ defined by (18.31), where $F = \begin{bmatrix}0 & 1\end{bmatrix}$ and $F_d = \begin{bmatrix}1 & 1\end{bmatrix}$.

Here, the system contains two outputs, namely, $y_1(t)$ and $y_2(t)$, where

$$y_1(t) = x_1(t) + x_1(t-\tau) + x_2(t-\tau), \tag{18.44}$$

$$y_2(t) = x_1(t-\tau), \tag{18.45}$$

but it is not possible to explicitly express $z_\tau(t)$ in terms of the two outputs since

$$\text{rank}\begin{bmatrix} C & C_d \\ F & F_d \end{bmatrix} = 3. \tag{18.46}$$

Thus, it is necessary to design an observer to estimate $z_\tau(t)$. In the following, we discuss the design of an observer, $\hat{z}_\tau(t)$, based on the existence conditions presented in Remark 3.

Firstly, the functional observer matching condition (18.43) is checked and found to be satisfied. Next, we can solve for the 9 unknown matrices, namely, N, N_d, N_{dd}, $\tilde{T}$, $\hat{U}$, $\hat{W}$, $\hat{Y}$, M and M_d, such that the coupled matrix equations (18.8)–(18.9) and (18.32)–(18.37) are satisfied. For this, these equations can be expressed as follows

$$X\Theta_\tau = \Upsilon_\tau, \tag{18.47}$$

where

$$\Theta_\tau = \begin{bmatrix} F & F_d & 0 & 0 & 0 & 0 & 0 & 0 \\ 0 & F & F_d & 0 & 0 & 0 & 0 & 0 \\ 0 & 0 & F & F_d & 0 & 0 & 0 & 0 \\ C & C_d & 0 & 0 & 0 & 0 & 0 & 0 \\ 0 & C & C_d & 0 & 0 & 0 & 0 & 0 \\ 0 & 0 & C & C_d & 0 & 0 & 0 & 0 \\ 0 & 0 & 0 & C & C_d & 0 & 0 & 0 \\ CA & C_\tau & C_dA_d & 0 & 0 & CD & C_dD & 0 \\ 0 & CA & C_\tau & C_dA_d & 0 & 0 & CD & C_dD \end{bmatrix}, \tag{18.48}$$

$$\Upsilon_\tau = [FA \;\; FA_d + F_dA \;\; F_dA_d \;\; 0 \;\; 0 \;\; FD \;\; F_dD \;\; 0], \tag{18.49}$$

and X is as defined by (15.32).

Then, we check whether the following condition is satisfied or not, where

$$\text{rank}\begin{bmatrix} \Theta_\tau \\ \Upsilon_\tau \end{bmatrix} = \text{rank}\begin{bmatrix} \Theta_\tau \end{bmatrix}. \tag{18.50}$$

In this example, condition (18.50) is found to be satisfied. Furthermore, it is also found that the solution to (18.47) results in a stable error system (18.7). Thus, we can complete the design of observer (18.41)–(18.42). Here, we note that observer (18.41)–(18.42) contains multiple internal delays, i.e., $N_d w(t-\tau)$ and $N_{dd} w(t-2\tau)$.

On the other hand, it is desirable to obtain a simpler observer and thus we proceed to determine if condition (18.50) can be satisfied for the following two cases: (i) $N_d = N_{dd} = 0$, and (ii) $N_{dd} = 0$. Accordingly, in Case (i), we first remove the second and third rows of Θ_τ from Θ_τ and then we test condition (18.50). However, it is found that condition (18.50) is not satisfied. This implies that an observer (18.41)–(18.42) without internal delays, i.e., $N_d = N_{dd} = 0$, does not exist.

In Case (ii), we test condition (18.50) where we remove the third row of Θ_τ. It is found that (18.50) is indeed satisfied and the solution to (18.47) gives the following matrices,
$N = -9$, $N_d = -3$, $N_{dd} = 0$, $\tilde{T} = [-10 \;\; -2.6995]$, $\hat{W} = [-4.3005 \;\; -4.2332]$, $\hat{U} = [0.3109 \;\; 0.1554]$, $\hat{Y} = [0\;0]$, $M = [-1 \;\; -0.0777]$ and $M_d = [0.0777 \;\; 0]$.

With the above obtained matrices, we can again double check that (18.8)–(18.9), and (18.32)–(18.37) are satisfied. We can also confirm that the error system (18.7), i.e., $\dot{e}(t) = -9e(t) - 3e(t-1)$ is stable. Finally, from (18.38)–(18.40), (10.84) and (15.22)-(15.24), we obtain the rest of the observer matrices for observer (18.41)–(18.42), where $J = 3$, $J_d = 6$, $J_{dd} = -0.2332$, $G = [-1 \;\; -2]$, $G_d = [-2 \;\; -4]$, $G_{dd} = [0.0777 \;\; 0.1554]$ and $G_{3d} = [0\;0]$.

Figure 18.1 shows trajectories of $\hat{z}_\tau(t)$ and $z_\tau(t)$. Figure 18.2 shows the trajectories of $\hat{z}_\tau(t)$ and $z_\tau(t)$ with a different type of disturbance $d(t)$. From the figures, it is clear that asymptotic estimation has been achieved, regardless of the types of unknown input, $d(t)$, affecting the system.

Remark 4 Let us revisit Case (i) and design a higher-order observer with no internal delays to estimate $z_\tau(t)$. Recall that it was not possible to design a first-order observer with no internal delays to estimate $z_\tau(t)$. As we have discussed extensively in this book, we can insert an additional row into both F and F_d and proceed to design a second-order observer with no internal delays. For this, we consider an additional function, denoted by $z_{\tau a}(t)$, where

$$z_{\tau a}(t) = [1\;1]x(t) + [2\;0]x(t-\tau). \tag{18.51}$$

Thus, we now have

$$F = \begin{bmatrix} 0 & 1 \\ 1 & 1 \end{bmatrix}, \; F_d = \begin{bmatrix} 1 & 1 \\ 2 & 0 \end{bmatrix}, \tag{18.52}$$

and we proceed to design a second-order observer with no internal delays (i.e., $N_d = N_{dd} = 0_{2\times 2}$). For this, we first note that with F and F_d given above, condition (18.43) is satisfied. Also, we can easily check that

$$\text{rank}\begin{bmatrix} C & C_d \\ F & F_d \end{bmatrix} = 4, \tag{18.53}$$

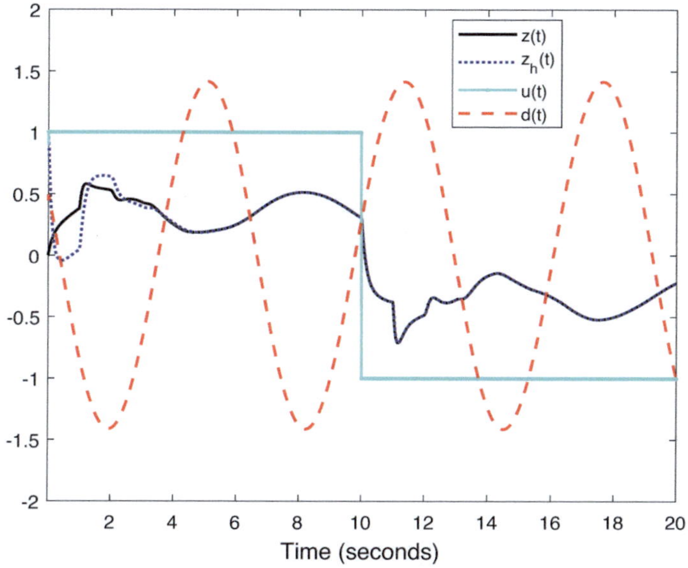

Fig. 18.1 Trajectories of $\hat{z}_\tau(t)$ and $z_\tau(t)$ (Numerical Example 1, first-order observer with an internal delay, in this chapter)

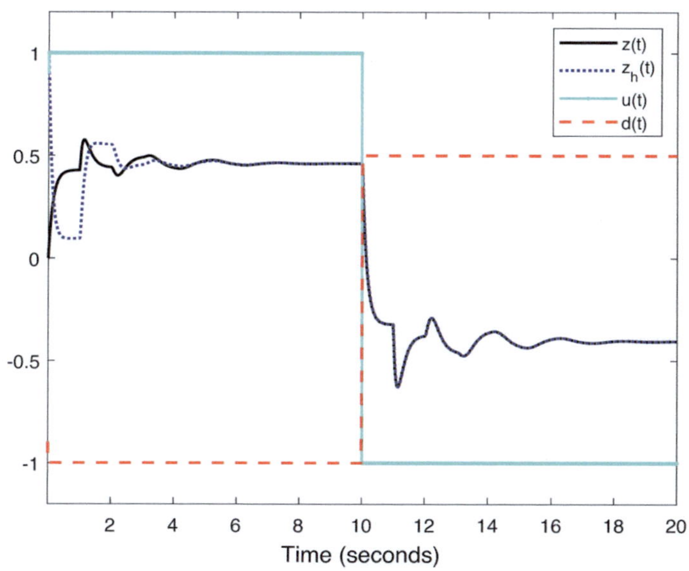

Fig. 18.2 Trajectories of $\hat{z}_\tau(t)$ and $z_\tau(t)$ (Numerical Example 1, first-order observer with an internal delay, in this chapter)

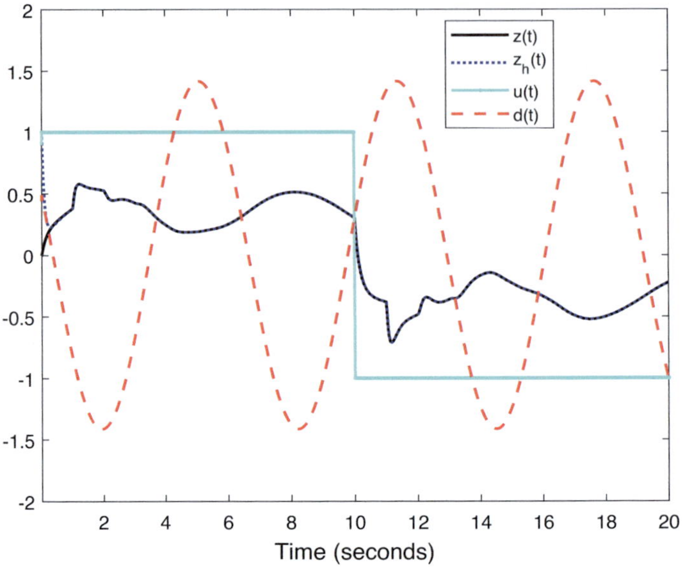

Fig. 18.3 Trajectories of $\hat{z}_\tau(t)$ and $z_\tau(t)$ (Numerical Example 1, second-order observer with no internal delays, in this chapter)

which implies that $z_{\tau a}(t)$ is independent of $z_\tau(t)$ and $y(t)$. This is a necessary requirement to have any chance of satisfying condition (18.50).

Next, we check condition (18.50) with F and F_d given in (18.52) and for the case where $N_d = N_{dd} = 0_{2\times 2}$. It is found that condition (18.50) is satisfied and the solution to (18.47) leads to a stable matrix N. As a result, we obtain the following second-order observer

$$\hat{z}_\tau(t) = w_1(t) + M_1 y(t) + M_{d1} y(t-\tau), \tag{18.54}$$

$$\begin{aligned}\dot{w}(t) &= Nw(t) + Gy(t) + G_d y(t-\tau) + G_{dd} y(t-2\tau) \\ &\quad + Ju(t) + J_d u(t-\tau) + J_{dd} u(t-2\tau),\end{aligned} \tag{18.55}$$

where
$w(t) = \begin{bmatrix} w_1(t) \\ w_2(t) \end{bmatrix}$, $M_1 = \begin{bmatrix} -1 & -0.3575 \end{bmatrix}$, $M_{d1} = \begin{bmatrix} 0.3575 & 0 \end{bmatrix}$, $N = \begin{bmatrix} -10.5 & 1.5 \\ -1.5 & -7.5 \end{bmatrix}$, $\lambda(N) = \{-9, -9\}$, $G = \begin{bmatrix} -1 & 3.6114 \\ -1 & -2.3886 \end{bmatrix}$, $G_d = \begin{bmatrix} -7.6114 & -5.0725 \\ 0.3886 & -1.8497 \end{bmatrix}$, $J = \begin{bmatrix} 3 \\ 3 \end{bmatrix}$, $G_{dd} = \begin{bmatrix} 1.4301 & 0.7150 \\ 2.4663 & 1.2332 \end{bmatrix}$, $J_d = \begin{bmatrix} 6 \\ 0 \end{bmatrix}$ and $J_{dd} = \begin{bmatrix} -1.0725 \\ -1.8497 \end{bmatrix}$.

Figure 18.3 shows the trajectories of $\hat{z}_\tau(t)$ and $z_\tau(t)$. Note that since $\lambda(N) = \{-9, -9\}$, $\hat{z}_\tau(t)$ converges to $z_\tau(t)$ in a much shorter time than the case of a first-order observer (see, Figs. 18.1 and 18.2).

Numerical Example 2: In this example, we discuss the case where $C = 0_{p\times n}$ in (18.29)–(18.30). Let us consider system (18.29)–(18.30) with an unknown input $d(t)$, where $\tau = 1s$, matrices A, A_d, B, D and C_d are given below

$$A = \begin{bmatrix} -4 & 3 \\ 0 & -8 \end{bmatrix}, A_d = \begin{bmatrix} -13 & 2 \\ -3 & -16 \end{bmatrix}, B = \begin{bmatrix} 1 \\ 2 \end{bmatrix}, D = \begin{bmatrix} 1 \\ 0 \end{bmatrix} \text{and } C_d = \begin{bmatrix} 1 & 0 \\ 1 & 1 \end{bmatrix}.$$

Since $C = 0_{2\times 2}$, we have $y(t) = C_d x(t-\tau)$, i.e.,

$$y_1(t) = x_1(t-\tau), \tag{18.56}$$
$$y_2(t) = x_1(t-\tau) + x_2(t-\tau), \tag{18.57}$$

and we obtain

$$x_1(t-\tau) = y_1(t), \tag{18.58}$$
$$x_2(t-\tau) = y_2(t) - y_1(t). \tag{18.59}$$

Thus, from the available output measurements $y_1(t)$ and $y_2(t)$, we can only obtain the delayed state variables $x_1(t-\tau)$ and $x_2(t-\tau)$. On the other hand, if for instance, $x_2(t)$ is needed for the purpose of system monitoring, then from (18.59), we have

$$x_2(t) = y_2(t+\tau) - y_1(t+\tau). \tag{18.60}$$

However, it is not possible to realize (18.60) because we do not have the knowledge of the output measurements $y_1(t+\tau)$ and $y_2(t+\tau)$. Indeed, this practical case is intriguing, and to solve this problem, we can turn to the problem of designing an observer to estimate $x_2(t)$ using the output vector $y(t) = C_d x(t-\tau)$. This is the first time in this book that we deal with this problem, where the time-delay systems contains an external unknown input $d(t)$.

The results developed in Remark 3 can now be readily used to estimate $x_2(t)$. For this, we select F and F_d as follows

$$F = [0\ 1],\ \ F_d = [0\ 0], \tag{18.61}$$

and the function to be estimated is $z_\tau(t) = Fx(t) + F_d x(t-\tau) = x_2(t)$.

Next, with $C = 0_{2\times 2}$, and C_d, F and F_d given above, we test condition (18.43) and find it satisfied. Note that when $C = 0_{p\times n}$, condition (18.43) is more stringent as it is now nescessary that $FD = 0_{m\times q}$, which implies that F must belong to the left null-space of D. In this example, we have $FD = 0$.

We then consider a first-order observer with no internal delays (i.e., $N_d = N_{dd} = 0$). For this, condition (18.50) is satisfied, and we complete the design task and obtain the following stable first-order observer

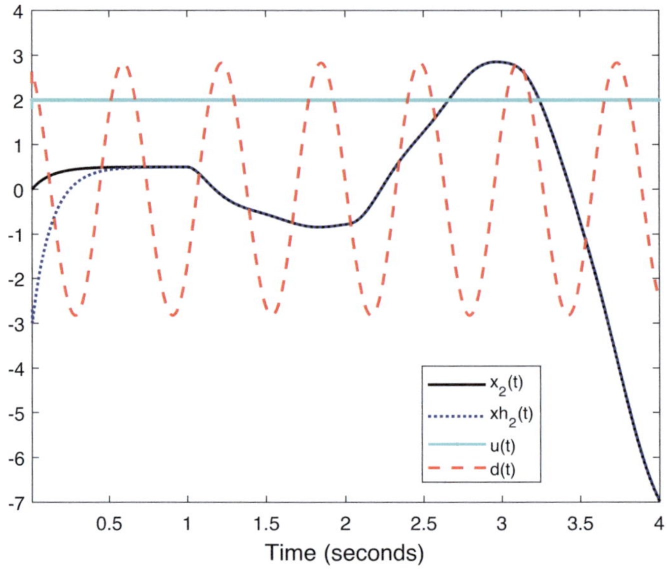

Fig. 18.4 Trajectories of $\hat{x}_2(t)$ and $x_2(t)$ (Numerical Example 2, first-order observer with no internal delays, in this chapter)

$$\hat{x}_2(t) = w(t) + My(t) + M_d y(t-\tau), \tag{18.62}$$

$$\dot{w}(t) = Nw(t) + Gy(t) + G_d y(t-\tau) + G_{dd} y(t-2\tau) + Ju(t) + J_d u(t-\tau) + J_{dd} u(t-2\tau), \tag{18.63}$$

where $M = [-0.5065\ 0.5065]$, $M_d = [0.2117\ \ -0.2117]$, $N = -8$, $G = [13\ \ -16]$, $G_d = [-6.5849\ 8.1045]$, $G_{dd} = [2.7526\ \ -3.3878]$, $J = 2$, $J_d = -1.0131$ and $J_{dd} = 0.4235$.

Figure 18.4 shows the trajectories of $x_2(t)$, and $\hat{x}_2(t)$ and it is clear that we can estimate the instantaneous state variable $x_2(t)$ by using an observer.

Here, the constraint $C = 0_{p\times n}$ implies that we require rank $\begin{bmatrix} F \\ FA \end{bmatrix} = \text{rank}(F)$. For the case where $m \le n$, the satisfaction of this condition resulted in $N = FAF^+$ which implies that the eigenvalues of N are fixed and belong to the subset of $\lambda(A)$.

Numerical Example 3: In this example, again, we discuss the case where $C = 0_{p\times n}$ in (18.29)–(18.30). Let us consider system (18.29)–(18.30) with an unknown input $d(t)$, where $\tau = 1s$, and matrices A, A_d, B, D and C_d are given below

$$A = \begin{bmatrix} -10 & -3 \\ 0 & -7 \end{bmatrix},\ A_d = \begin{bmatrix} -5 & 2 \\ -3 & -15 \end{bmatrix},\ B = \begin{bmatrix} 1 \\ 2 \end{bmatrix},\ D = \begin{bmatrix} 1 \\ -1 \end{bmatrix} \text{ and } C_d = \begin{bmatrix} 1 & 2 \end{bmatrix}.$$

Our aim is to design a generalized functional observer to estimate following function

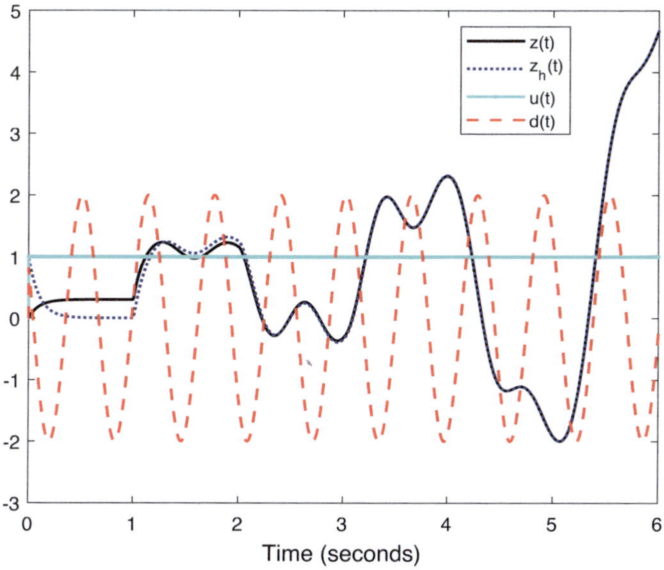

Fig. 18.5 Trajectories of $\hat{z}_\tau(t)$ and $z_\tau(t)$ (Numerical Example 3, first-order observer with an internal delay, in this chapter)

$$z_\tau(t) = Fx(t) + F_d x(t-\tau) = \begin{bmatrix} 1 & 1 \end{bmatrix} x(t) + \begin{bmatrix} 3 & 4 \end{bmatrix} x(t-\tau). \tag{18.64}$$

We first test condition (18.43) and find it satisfied. Then, we can design a first-order observer with an internal delay, $N_d w(t-1)$, where

$$\hat{z}_\tau(t) = w(t) + y(t), \tag{18.65}$$

$$\dot{w}(t) = -10w(t) - 3w(t-1) - 5y(t) - 10y(t-1) + 3u(t) + 6u(t-1). \tag{18.66}$$

Figure 18.5 shows the trajectories of $z_\tau(t)$ and $\hat{z}_\tau(t)$. It is clear that $\hat{z}_\tau(t)$ converges asymptotically to $z_\tau(t)$.

Finally, we reaffirm that estimating $z_\tau(t)$ directly results in a simpler observer than any other available options. This point has been demonstrated on several occasions in this book.

Note that in this example, we cannot estimate $x(t)$ directly (i.e., $F = I_2$, $F_d = 0_{2\times 2}$), since condition (18.43) is not satisfied. Thus, it is not possible for us to estimate $x(t)$ first and obtain $\hat{z}_\tau(t) = F\hat{x}(t) + F_d\hat{x}(t-\tau)$.

Another available option is estimate $z_\tau(t)$ by designing two observers, one for $Fx(t)$ and the other for $F_d x(t-\tau)$, i.e.,

$$z_\tau(t) = z_1(t) + z_2(t), \tag{18.67}$$

where $z_1(t) = \begin{bmatrix} 1 & 1 \end{bmatrix} x(t)$ and $z_2(t) = \begin{bmatrix} 3 & 4 \end{bmatrix} x(t-\tau)$. Here, we obtain two observers, where

$$\hat{z}_1(t) = w_1(t), \tag{18.68}$$

$$\dot{w}_1(t) = -10w_1(t) - 3w_1(t-1) - 5y(t) + 3u(t), \tag{18.69}$$

and

$$\hat{z}_2(t) = w_2(t) + y(t), \tag{18.70}$$

$$\dot{w}_2(t) = -10w_2(t) - 3w_2(t-1) - 10y(t-1) + 6u(t-1). \tag{18.71}$$

Thus, $\hat{z}_\tau(t)$ is obtained as follows

$$\hat{z}_\tau(t) = \hat{z}_1(t) + \hat{z}_2(t) = w_1(t) + w_2(t) + y(t), \tag{18.72}$$

where $w_1(t)$ and $w_2(t)$ are defined by (18.69) and (18.71), respectively. Observe that realization (18.65) requires fewer computation blocks and is simpler than realization (18.72), which requires two observers.

Numerical Example 4: In this example, we consider time-delay system (18.1)–(18.2) with an external unknown input, where $\tau = 1s$, and matrices A, A_d, A_{dd}, B, D, C, C_d and C_{dd} are as given below

$$A = \begin{bmatrix} -10 & -7 \\ 0.5 & -20 \end{bmatrix}, A_d = \begin{bmatrix} -2.5 & -8 \\ 1 & -2.5 \end{bmatrix}, A_{dd} = \begin{bmatrix} 0 & -1 \\ 0 & 2 \end{bmatrix}, B = \begin{bmatrix} 1 \\ 2 \end{bmatrix}, D = \begin{bmatrix} 1 \\ 0 \end{bmatrix},$$
$$C = \begin{bmatrix} 1 & 0 \\ 0 & 0 \end{bmatrix}, C_d = \begin{bmatrix} 0 & -1 \\ 0.5 & 1 \end{bmatrix} \text{ and} C_{dd} = \begin{bmatrix} 0 & 2 \\ 0 & -0.5 \end{bmatrix}.$$

Our aim is to design a generalized observer to estimate $z_{3\tau}(t)$ defined by (18.3), where $F = \begin{bmatrix} 1 & 1 \end{bmatrix}$, $F_d = \begin{bmatrix} 2.5 & 19 \end{bmatrix}$, $F_{dd} = \begin{bmatrix} -1 & 0.5 \end{bmatrix}$ and $F_{3d} = \begin{bmatrix} 0 & -2 \end{bmatrix}$. Here, $\hat{z}_{3\tau}(t)$ is required to estimate of an unknown input as discussed in Numerical Example 3 in Chap. 9.

Again, note that even though there are two output measurements in this example, it is still not possible to explicitly obtain $x_1(t)$ and $x_2(t)$. This is clear since

$$y_1(t) = x_1(t) - x_2(t-\tau) + 2x_2(t-2\tau), \tag{18.73}$$

$$y_2(t) = 0.5x_1(t-\tau) + x_2(t-\tau) - 0.5x_2(t-2\tau), \tag{18.74}$$

and thus it is not possible to algebraically obtain $x_1(t)$ and $x_2(t)$ in terms of measurements $y_1(t)$ and $y_2(t)$. Hence, an observer is required to estimate $z_{3\tau}(t)$.

Let us now design a generalized observer of form (18.4)–(18.5). For this, we first test condition (18.24) and find it satisfied. Then, as discussed in Remarks 1 and 2, we solve for the 9 unknown matrices defined by (15.32), such that conditions (18.7)–(18.17) are satisfied. It is found that a stable first-order observer with internal delays N_d, $N_{dd} \neq 0$ exists, and as a result, we obtain the following generalized observer to estimate $z_{3\tau}(t)$

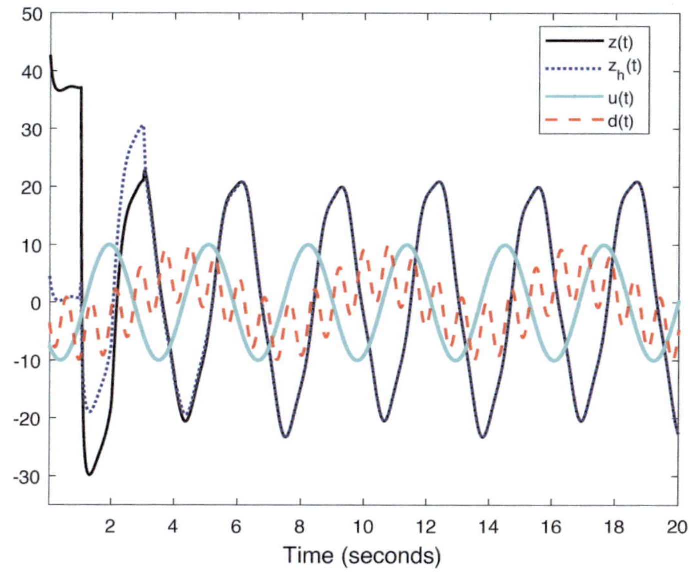

Fig. 18.6 Trajectories of $\hat{z}_{3\tau}(t)$ and $z_{3\tau}(t)$ (Numerical Example 4, first-order observer with internal delays, in this chapter)

$$\hat{z}_{3\tau}(t) = w(t) + \begin{bmatrix}1 & 1.2945\end{bmatrix} y(t) + \begin{bmatrix}1.8528 & -2\end{bmatrix} y(t-1), \tag{18.75}$$

$$\begin{aligned}\dot{w}(t) = &-20w(t) - 4w(t-1) + 2w(t-2) + \begin{bmatrix}0.5 & 2\end{bmatrix} y(t) + \begin{bmatrix}9.3528 & 37.4110\end{bmatrix} y(t-1)\\ &+ \begin{bmatrix}1.5 & 6\end{bmatrix} y(t-2) + \begin{bmatrix}-3.3528 & -13.4110\end{bmatrix} y(t-3) + 2u(t) + 37.4110u(t-1)\\ &+ 6u(t-2) - 13.4110u(t-3).\end{aligned} \tag{18.76}$$

Figure 18.6 shows the trajectories of $z_{3\tau}(t)$ and $\hat{z}_{3\tau}(t)$, and it is clear that $\hat{z}_{3\tau}(t)$ converges asymptotically to $z_{3\tau}(t)$.

Again, just like in Numerical Example 3, we reaffirm the advantage of estimating $z_{3\tau}(t)$ directly. Let us realize $\hat{z}_{3\tau}(t)$ by first estimating the entire state vector $x(t)$, and then $\hat{z}_{3\tau}(t)$ is obtained as follows

$$\hat{z}_{3\tau}(t) = F\hat{x}(t) + F_d\hat{x}(t-\tau) + F_{dd}\hat{x}(t-2\tau) + F_{3d}\hat{x}(t-3\tau). \tag{18.77}$$

Thus, if this option is to be adopted, we first need to design an observer to estimate $x(t)$, i.e., we estimate $z(t) = Fx(t)$ where $F = I_2$ and the rest $F_d = F_{dd} = F_{3d} = 0_{2\times 2}$.

With the above given matrices, we test condition (18.24) and find it satisfied. We then test condition (18.50) for the case where $N_{dd} = 0_{2\times 2}$. The condition is also satisfied, and as a result, we can design a second-order observer with an internal delay, i.e., $N_d \neq 0_{2\times 2}$. The rest of the observer parameters can be obtained. The key message here is that this option results in a more complex observer than the one given by (18.75)–(18.76). Also, similar to Numerical Example 3, estimating $\hat{z}_{3\tau}(t)$

directly led to directly leads to lower computational requirement and is simpler than using the scheme (18.77).

On the other hand, if our main objective is to estimate the entire state vector, $x(t)$, then in addition to a second-order observer with an internal delay as discussed above, we can also design a third-order observer with no internal delay, i.e., $N_d = N_{dd} = 0_{3\times 3}$. This is possible thanks to our generalized observer design framework. This point has been consistently emphasized throughout in this book. In this example, this achieved by estimating the following functional

$$z_a(t) = F_a x(t) + F_{ad} x(t - \tau),$$

where we select $F_a = \begin{bmatrix} 1 & 0 \\ 0 & 1 \\ 0 & 0 \end{bmatrix}$ and $F_{ad} = \begin{bmatrix} 0 & 0 \\ 0 & 0 \\ 1 & 2 \end{bmatrix}$. Once and observer is obtained (this can be done and is left to the reader to complete), then $\hat{x}(t)$ is obtained as follows

$$\hat{x}(t) = \begin{bmatrix} 1 & 0 & 0 \\ 0 & 1 & 0 \end{bmatrix} \hat{z}_a(t).$$

18.4 Concluding Remark

In this chapter, we have considered the problem of designing a generalized observer to estimate a generalized functional (18.3) of a class of time-delay systems (18.1)–(18.2) with an unknown input vector and multiple delays in both the state and output vectors. Thanks to our unified framework, we can readily obtain observer existence conditions, and via Numerical Examples 1–4, we have demonstrated in detail the design of various types of observers and their salient features.

References

1. Darouach, M., Zasadzinski, M., Xu, S.J.: Full-order observers for linear systems with unknown inputs. IEEE Trans. Automat. Contr. **39**(3), 606–609 (1994)
2. Hou, M., Müller, P.C.: Design of observers for linear systems with unknown inputs. IEEE Trans. Automat. Contr. **37**(6), 871–875 (1992)
3. Tran, H.M., Trinh, H.: Distributed functional observer based fault detection for interconnected time-delay systems. IEEE Systems J. **13**(1), 940–951 (2019)
4. Trinh, H., Fernando, T.: Functional Observers for Dynamical Systems. Springer, Berlin, Heidelberg (2012)
5. Trinh, H., Tran, T.D., Fernando, T.: Disturbance decoupled observers for systems with unknown inputs. IEEE Trans. Automat. Contr. **53**(10), 2397–2402 (2008)

Chapter 19
Generalized Observers Mixed Non-Commensurate Delays

19.1 System Description and Problem Statement

In this chapter, we consider the following linear system with mixed non-commensurate time delays in both state and output vectors

$$\begin{aligned}\dot{x}(t) &= Ax(t) + A_d x(t-\tau) + A_{dd} x(t-2\tau) + A_h x(t-h) \\ &\quad + A_{dh} x(t-\tau-h) + Bu(t), \qquad (19.1)\\ y(t) &= Cx(t) + C_d x(t-\tau) + C_{dd} x(t-2\tau) + C_h x(t-h) \\ &\quad + C_{dh} x(t-\tau-h), \qquad (19.2)\end{aligned}$$

where $x(t) \in \mathbb{R}^n$, $y(t) \in \mathbb{R}^p$ and $u(t) \in \mathbb{R}^r$ are, respectively, the state vector, output vector and control input vector. Matrices $A, A_d, A_{dd}, A_h, A_{dh} \in \mathbb{R}^{n\times n}$, $B \in \mathbb{R}^{n\times r}$ and $C, C_d, C_{dd}, C_h, C_{dh} \in \mathbb{R}^{p\times n}$ are known constant matrices. Here, $0 < \tau < h$, where τ and h are known constant non-commensurate delays.

Our aim is to design a generalized observer to estimate the following vector

$$\begin{aligned}z_{3\tau h}(t) &= Fx(t) + F_d x(t-\tau) + F_{dd} x(t-2\tau) + F_{3d} x(t-3\tau) \\ &\quad + F_h x(t-h) + F_{dh} x(t-\tau-h) + F_{2dh} x(t-2\tau-h), \qquad (19.3)\end{aligned}$$

where $z_{3\tau h}(t) \in \mathbb{R}^m$, and F, F_d, F_{dd}, F_{3d}, F_h, F_{dh} and F_{2dh} are known constant matrices of appropriate dimensions. We note here that the problem of designing an observer to asymptotically estimate vector $z_{3\tau h}(t)$ for system (19.1)–(19.2) has not yet been considered in the literature. In this book, the need to address this problem stems from the requirement to estimate the unknown input, $d(t)$, as defined by (8.25).

H. Trinh et al., *Unknown Inputs Estimation in Linear Time-Delay Systems Using Generalized Functional Observers*, Studies in Systems, Decision and Control 648,
https://doi.org/10.1007/978-3-032-11870-7_19

19.2 Proposed Observer with a General Structure

Thanks to our general framework for solving the estimation problem as discussed so far in this book, we can proceed with the design of the following observer for time-delay system (19.1)–(19.2),

$$\begin{aligned}\hat{z}_{3\tau h}(t) &= w(t) + My(t) + M_d y(t-\tau) + M_{dd} y(t-2\tau) + M_h y(t-h)\\ &\quad + M_{dh} y(t-\tau-h) + M_{2dh} y(t-2\tau-h), \qquad (19.4)\\ \dot{w}(t) &= Nw(t) + N_d w(t-\tau) + N_h w(t-h) + Gy(t) + G_d y(t-\tau) + G_{dd} y(t-2\tau)\\ &\quad + G_{3d} y(t-3\tau) + G_h y(t-h) + G_{hh} y(t-2h) + G_{dh} y(t-\tau-h)\\ &\quad + G_{2dh} y(t-2\tau-h) + G_{3dh} y(t-3\tau-h) + G_{d2h} y(t-\tau-2h)\\ &\quad + G_{2d2h} y(t-2\tau-2h) + G_{3d2h} y(t-3\tau-2h) + Ju(t) + J_d u(t-\tau) + J_h u(t-h)\\ &\quad + J_{dd} u(t-2\tau) + J_{dh} u(t-\tau-h) + J_{hh} u(t-2h) + J_{2dh} u(t-2\tau-h)\\ &\quad + J_{d2h} u(t-\tau-2h) + J_{2d2h} u(t-2\tau-2h) + J_{3d} u(t-3\tau) + J_{3dh} u(t-3\tau-h)\\ &\quad + J_{3d2h} u(t-3\tau-2h) + J_{4d} u(t-4\tau) + J_{4dh} u(t-4\tau-h), \qquad (19.5)\end{aligned}$$

where $w(\theta) = \rho(\theta),\ \forall\theta \in [0, h]$, $w(t) \in \mathbb{R}^m$, and $\hat{z}_{3\tau h}(t)$ is the estimate of $z_{3\tau h}(t)$. A total of 35 unknown matrices, namely, M, M_d, M_{dd}, M_h, M_{dh}, M_{2dh}, N, N_d, N_h, $G, G_d, G_{dd}, G_{3d}, G_h, G_{hh}, G_{dh}, G_{2dh}, G_{3dh}, G_{d2h}, G_{2d2h}, G_{3d2h}, J, J_d, J_h, J_{dd}, J_{hh}$, J_{dh}, J_{2dh}, J_{d2h}, J_{2d2h}, J_{3d}, J_{3dh}, J_{3d2h}, J_{4d} and J_{4dh} need to be determined such that $\hat{z}_{3\tau h}(t)$ converges asymptotically to $z_{3\tau h}(t)$. In the following, we present existence conditions of observer (19.4)–(19.5).

19.3 Observer Existence Conditions

The estimate $\hat{z}_{3\tau h}(t)$ converges asymptotically to $z_{3\tau h}(t)$ for any initial condition $w(\theta)$ and any input $u(t)$, provided that the following conditions hold:

$$\dot{e}(t) = Ne(t) + N_d e(t-\tau) + N_h e(t-h) \text{ is asymptotically stable,} \qquad (19.6)$$

$$\mathscr{L}_{(17.7)} = 0,\ \mathscr{L}_{(17.8)} = 0,\ \mathscr{L}_{(17.9)} = 0,\ \mathscr{L}_{(17.11)} = 0, \qquad (19.7)$$

$$J_{dd} - F_{dd}B + MC_{dd}B + M_d C_d B + M_{dd} CB = 0, \qquad (19.8)$$

$$J_{dh} - F_{dh}B + MC_{dh}B + M_d C_h B + M_h C_d B + M_{dh} CB = 0, \qquad (19.9)$$

$$J_{2dh} - F_{2dh}B + M_dC_{dh}B + M_{dd}C_hB + M_hC_{dd}B + M_{dh}C_dB + M_{2dh}CB = 0, \tag{19.10}$$

$$J_{d2h} + M_hC_{dh}B + M_{dh}C_hB = 0, \tag{19.11}$$

$$J_{2d2h} + M_{dh}C_{dh}B + M_{2dh}C_hB = 0, \tag{19.12}$$

$$J_{3d} - F_{3d}B + M_dC_{dd}B + M_{dd}C_dB = 0, \tag{19.13}$$

$$J_{3dh} + M_{dd}C_{dh}B + M_{dh}C_{dd}B + M_{2dh}C_dB = 0, \tag{19.14}$$

$$J_{3d2h} + M_{2dh}C_{dh}B = 0, \tag{19.15}$$

$$J_{4d} + M_{dd}C_{dd}B = 0, \tag{19.16}$$

$$J_{4dh} + M_{2dh}C_{dd}B = 0, \tag{19.17}$$

$$\mathscr{L}_{(17.15)} = 0, \mathscr{L}_{(17.16)} = 0, \mathscr{L}_{(17.17)} = 0, \tag{19.18}$$

$$NF_{dd} + N_dF_d + \tilde{U}_{dd}C + \hat{W}C_d + \tilde{T}C_{dd} + MC_{2\tau} + M_dC_\tau + M_{dd}CA - FA_{dd} \\ - F_dA_d - F_{dd}A = 0, \tag{19.19}$$

$$NF_{dh} + N_dF_h + N_hF_d + \tilde{U}_{dh}C + \hat{W}_hC_d + \hat{W}C_h + \tilde{T}C_{dh} + MC_{\tau h} + M_d\tilde{C}_h \\ + M_hC_\tau + M_{dh}CA - FA_{dh} - F_dA_h - F_hA_d - F_{dh}A = 0, \tag{19.20}$$

$$N_hF_h + \tilde{U}_{hh}C + \hat{W}_hC_h + MC_hA_h + M_h\tilde{C}_h - F_hA_h = 0, \tag{19.21}$$

$$N_hF_{dh} + \tilde{U}_{d2h}C + \tilde{U}_{hh}C_d + \tilde{U}_{dh}C_h + \hat{W}_hC_{dh} + MC_{\tau 2h} + M_dC_hA_h + M_hC_{\tau h} \\ + M_{dh}\tilde{C}_h - F_hA_{dh} - F_{dh}A_h = 0, \tag{19.22}$$

$$NF_{2dh} + N_dF_{dh} + N_hF_{dd} + \hat{U}_{2dh}C + \tilde{U}_{dh}C_d + \hat{W}_hC_{dd} + \tilde{U}_{dd}C_h + \hat{W}C_{dh} + MC_{2\tau h} + M_dC_{\tau h} + M_{dd}\tilde{C}_h + M_hC_{2\tau} + M_{dh}C_\tau + M_{2dh}CA - F_dA_{dh} - F_{dd}A_h - F_hA_{dd} - F_{dh}A_d - F_{2dh}A = 0, \tag{19.23}$$

$$N_hF_{2dh} + \tilde{U}_{2d2h}C + \tilde{U}_{d2h}C_d + \tilde{U}_{hh}C_{dd} + \hat{U}_{2dh}C_h + \tilde{U}_{dh}C_{dh} + MC_{dh}A_{dh} + M_dC_{\tau 2h} + M_{dd}C_hA_h + M_hC_{2\tau h} + M_{dh}C_{\tau h} + M_{2dh}\tilde{C}_h - F_{dh}A_{dh} - F_{2dh}A_h = 0, \tag{19.24}$$

$$NF_{3d} + N_dF_{dd} + \tilde{U}_{3d}C + \tilde{U}_{dd}C_d + \hat{W}C_{dd} + MC_{3\tau} + M_dC_{2\tau} + M_{dd}C_\tau - F_dA_{dd} - F_{dd}A_d - F_{3d}A = 0, \tag{19.25}$$

$$N_dF_{2dh} + N_hF_{3d} + \tilde{U}_{3dh}C + \hat{U}_{2dh}C_d + \tilde{U}_{dh}C_{dd} + \tilde{U}_{3d}C_h + \tilde{U}_{dd}C_{dh} + MC_{3\tau h} + M_dC_{2\tau h} + M_{dd}C_{\tau h} + M_hC_{3\tau} + M_{dh}C_{2\tau} + M_{2dh}C_\tau - F_{dd}A_{dh} - F_{3d}A_h - F_{dh}A_{dd} - F_{2dh}A_d = 0, \tag{19.26}$$

$$G_{3d2h}C + \tilde{U}_{2d2h}C_d + \tilde{U}_{d2h}C_{dd} + \tilde{U}_{3dh}C_h + \hat{U}_{2dh}C_{dh} + M_dC_{dh}A_{dh} + M_{dd}C_{\tau 2h} + M_hC_{3\tau h} + M_{dh}C_{2\tau h} + M_{2dh}C_{\tau h} - F_{2dh}A_{dh} = 0, \tag{19.27}$$

$$\tilde{U}_{hh}C_h + M_hC_hA_h = 0, \tag{19.28}$$

$$\tilde{U}_{d2h}C_h + \tilde{U}_{hh}C_{dh} + M_hC_{\tau 2h} + M_{dh}C_hA_h = 0, \tag{19.29}$$

$$\tilde{U}_{2d2h}C_h + \tilde{U}_{d2h}C_{dh} + M_hC_{dh}A_{dh} + M_{dh}C_{\tau 2h} + M_{2dh}C_hA_h = 0, \tag{19.30}$$

$$G_{3d2h}C_h + \tilde{U}_{2d2h}C_{dh} + M_{dh}C_{dh}A_{dh} + M_{2dh}C_{\tau 2h} = 0, \tag{19.31}$$

$$N_dF_{3d} + \tilde{U}_{3d}C_d + \tilde{U}_{dd}C_{dd} + MC_{dd}A_{dd} + M_dC_{3\tau} + M_{dd}C_{2\tau} - F_{dd}A_{dd} - F_{3d}A_d = 0, \tag{19.32}$$

$$\tilde{U}_{3dh}C_d + \hat{U}_{2dh}C_{dd} + \tilde{U}_{3d}C_{dh} + M_dC_{3\tau h} + M_{dd}C_{2\tau h} + M_hC_{dd}A_{dd} + M_{dh}C_{3\tau} + M_{2dh}C_{2\tau} - F_{3d}A_{dh} - F_{2dh}A_{dd} = 0, \tag{19.33}$$

$$G_{3d2h}C_d + \tilde{U}_{2d2h}C_{dd} + \tilde{U}_{3dh}C_{dh} + M_{dd}C_{dh}A_{dh} + M_{dh}C_{3\tau h} + M_{2dh}C_{2\tau h} = 0, \tag{19.34}$$

$$G_{3d2h}C_{dh} + M_{2dh}C_{dh}A_{dh} = 0, \tag{19.35}$$

$$\tilde{U}_{3d}C_{dd} + M_dC_{dd}A_{dd} + M_{dd}C_{3\tau} - F_{3d}A_{dd} = 0, \tag{19.36}$$

$$\tilde{U}_{3dh}C_{dd} + M_{dd}C_{3\tau h} + M_{dh}C_{dd}A_{dd} + M_{2dh}C_{3\tau} = 0, \tag{19.37}$$

$$G_{3d2h}C_{dd} + M_{2dh}C_{3\tau h} = 0, \tag{19.38}$$

$$M_{dd}C_{dd}A_{dd} = 0, \tag{19.39}$$

$$M_{2dh}C_{dd}A_{dd} = 0. \tag{19.40}$$

In the above conditions, $\tilde{T}$, $\hat{W}$, $\hat{W}_h$, $\tilde{U}_{hh}$, $\tilde{U}_{dh}$, $\tilde{U}_{d2h}$, C_τ, $C_{2\tau}$ and $C_{3\tau}$ are defined by (17.28), (17.29), (17.30), (17.32), (17.33), (17.35), (16.18), (18.22) and (18.23), espectively, and

$$\tilde{C}_h = CA_h + C_hA, \tag{19.41}$$

$$C_{3\tau h} = C_{dd}A_{dh} + C_{dh}A_{dd}, \tag{19.42}$$

$$C_{\tau h} = CA_{dh} + C_dA_h + C_hA_d + C_{dh}A, \tag{19.43}$$

$$C_{2\tau h} = C_dA_{dh} + C_{dd}A_h + C_hA_{dd} + C_{dh}A_d, \tag{19.44}$$

$$C_{\tau 2h} = C_hA_{dh} + C_{dh}A_h, \tag{19.45}$$

$$\tilde{U}_{dd} = G_{dd} - N_dM_d - NM_{dd}, \tag{19.46}$$

$$\tilde{U}_{3d} = G_{3d} - N_dM_{dd}, \tag{19.47}$$

$$\tilde{U}_{3dh} = G_{3dh} - N_d M_{2dh}, \tag{19.48}$$

$$\tilde{U}_{2d2h} = G_{2d2h} - N_h M_{2dh}, \tag{19.49}$$

$$\hat{U}_{2dh} = G_{2dh} - N_h M_{dd} - N_d M_{dh} - N M_{2dh}. \tag{19.50}$$

Proof: The above existence conditions are obtained by first considering an error vector $e(t) = (\hat{z}_{3\tau h}(t) - z_{3\tau h}(t))$, and from (19.3)–(19.4), we have

$$\begin{aligned} e(t) &= \hat{z}_{3\tau h}(t) - z_{3\tau h}(t) \\ &= w(t) + My(t) + M_d y(t-\tau) + M_{dd} y(t-2\tau) + M_h y(t-h) + M_{dh} y(t-\tau-h) \\ &+ M_{2dh} y(t-2\tau-h) - Fx(t) - F_d x(t-\tau) - F_{dd} x(t-2\tau) - F_{3d} x(t-3\tau) \\ &- F_h x(t-h) - F_{dh} x(t-\tau-h) - F_{2dh} x(t-2\tau-h). \end{aligned} \tag{19.51}$$

Secondly, we take the derivative of both sides of (19.51), and after quite a drawn-out process, we obtain the following error dynamic equation,

$$\begin{aligned} \dot{e}(t) &= Ne(t) + N_d e(t-\tau) + N_h e(t-h) + \mathscr{L}_{(17.7)} u(t) + \mathscr{L}_{(17.8)} u(t-\tau) \\ &+ \mathscr{L}_{(17.9)} u(t-h) + \mathscr{L}_{(17.11)} u(t-2h) + \mathscr{L}_{(19.8)} u(t-2\tau) + \mathscr{L}_{(19.9)} u(t-\tau-h) \\ &+ \mathscr{L}_{(19.10)} u(t-2\tau-h) + \mathscr{L}_{(19.11)} u(t-\tau-2h) + \mathscr{L}_{(19.12)} u(t-2\tau-2h) \\ &+ \mathscr{L}_{(19.13)} u(t-3\tau) + \mathscr{L}_{(19.14)} u(t-3\tau-h) + \mathscr{L}_{(19.15)} u(t-3\tau-2h) \\ &+ \mathscr{L}_{(19.16)} u(t-4\tau) + \mathscr{L}_{(19.17)} u(t-4\tau-h) + \mathscr{L}_{(17.15)} x(t) + \mathscr{L}_{(17.16)} x(t-\tau) \\ &+ \mathscr{L}_{(17.17)} x(t-h) + \mathscr{L}_{(19.19)} x(t-2\tau) + \mathscr{L}_{(19.20)} x(t-\tau-h) \\ &+ \mathscr{L}_{(19.21)} x(t-2h) + \mathscr{L}_{(19.22)} x(t-\tau-2h) + \mathscr{L}_{(19.23)} x(t-2\tau-h) \\ &+ \mathscr{L}_{(19.24)} x(t-2\tau-2h) + \mathscr{L}_{(19.25)} x(t-3\tau) + \mathscr{L}_{(19.26)} x(t-3\tau-h) \\ &+ \mathscr{L}_{(19.27)} x(t-3\tau-2h) + \mathscr{L}_{(19.28)} x(t-3h) + \mathscr{L}_{(19.29)} x(t-\tau-3h) \\ &+ \mathscr{L}_{(19.30)} x(t-2\tau-3h) + \mathscr{L}_{(19.31)} x(t-3\tau-3h) + \mathscr{L}_{(19.32)} x(t-4\tau) \\ &+ \mathscr{L}_{(19.33)} x(t-4\tau-h) + \mathscr{L}_{(19.34)} x(t-4\tau-2h) + \mathscr{L}_{(19.35)} x(t-4\tau-3h) \\ &+ \mathscr{L}_{(19.36)} x(t-5\tau) + \mathscr{L}_{(19.37)} x(t-5\tau-h) + \mathscr{L}_{(19.38)} x(t-5\tau-2h) \\ &+ \mathscr{L}_{(19.39)} x(t-6\tau) + \mathscr{L}_{(19.40)} x(t-6\tau-h), \end{aligned} \tag{19.52}$$

where $\mathscr{L}_{(.)}$ is used to denote all the terms on the left-hand side of equation (.).

Thirdly, from (19.52), conditions to ensure the error vector $e(t)$ converges asymptotically to zero are derived and given by (19.6)–(19.40). As such, the proof has been completed.

Remark 1 : Based on observer existence conditions (19.6)–(19.40), we can proceed to find the 35 unknown matrices of observer (19.4)–(19.5). Firstly, from (19.7)–(19.17), we can directly obtain 14 matrices J, J_d, J_h, J_{dd}, J_{hh}, J_{dh}, J_{2dh}, J_{d2h}, J_{2d2h}, J_{3d}, J_{3dh}, J_{3d2h}, J_{4d} and J_{4dh}. The remaining 21 unknown matrices of observer (19.4)–(19.5), namely, M, M_d, M_{dd}, M_h, M_{dh}, M_{2dh}, N, N_d, N_h, G, G_d, G_{dd}, G_{3d}, G_h, G_{hh}, G_{dh}, G_{2dh}, G_{3dh}, G_{d2h}, G_{2d2h} and G_{3d2h} are determined to satisfy conditions (19.6) and (19.18)–(19.40). Our approach to solve for these 21 unknown matrices is based on the same approach we have employed throughout in this book (see, for example, Chapter 17). First, we first derive a necessary and sufficient condition for the solvability of the 25 matrix equations (19.18)–(19.40). Then, upon satisfying their solvability conditions, we consider the stability/stabilization of the time-delay error system (19.6). Here, equations (19.18)–(19.40) contain 21 unknown matrices, and we augment them as follows

$$\begin{aligned} X_{3\tau h} = [N\ N_d\ N_h\ \tilde{T}\ \hat{W}\ \hat{W}_h\ \tilde{U}_{dd}\ \tilde{U}_{dh}\ \tilde{U}_{hh}\ \hat{U}_{2dh}\ \tilde{U}_{d2h}\ \tilde{U}_{2d2h} \\ \tilde{U}_{3d}\ \tilde{U}_{3dh}\ G_{3d2h}\ M\ M_d\ M_h\ M_{dh}\ M_{dd}\ M_{2dh}], \end{aligned} \tag{19.53}$$

where $X_{3\tau h} \in \mathbb{R}^{m\times(3m+18p)}$.

Thus, equations (19.18)–(19.40) can be expressed in the following form

$$X_{3\tau h}\Theta_{3\tau h} = \Upsilon_{3\tau h}, \tag{19.54}$$

where $\Upsilon_{3\tau h} \in \mathbb{R}^{m\times 25n}$ and $\Theta_{dh} \in \mathbb{R}^{(3m+18p)\times 25n}$, and we would like to leave it to the reader to determine $\Theta_{3\tau h}$ and $\Upsilon_{3\tau h}$.

Hence, a necessary and sufficient condition for the solvability of (19.54) is derived and upon its satisfaction, (19.54) is solved for $X_{3\tau h}$. Once asymptotic stability of time-delay error system (19.6) is established, the rest of the observer parameters can be obtained.

Remark 2 A close examination of (19.39) and (19.40) reveals that for the case where $C_{dd}A_{dd} = 0_{p\times n}$, then the 25 matrix equations (19.18)–(19.40) are reduced to 23 matrix equations (19.18)–(19.38). On the other hand if $C_{dd}A_{dd} \neq 0_{p\times n}$ and is a full rank matrix, then condition described by (19.18)–(19.40) is met if and only if $M_{dd} = M_{2dh} = 0_{m\times p}$. In both cases, we employ the approach to solve for the unknown observer matrices as outlined in Remark 1.

19.4 Concluding Remark

In this chapter, we have looked at the problem of designing generalized observers for a class of systems with mixed non-commensurate time delays in both state and output vectors. We have presented a general observer structure (19.4)–(19.5) together with their existence conditions (19.6)–(19.40). It should be noted here that the necessary and sufficient condition for the solvability of (19.54), i.e., rank $\begin{bmatrix} \Theta_{3\tau h} \\ \Upsilon_{3\tau h} \end{bmatrix} =$ rank$\begin{bmatrix} \Theta_{3\tau h} \end{bmatrix}$, involves high dimensional matrices $\Theta_{3\tau h}$ and $\Upsilon_{3\tau h}$. When the condition is not satisfied, we should increase the order of the designed observer by estimating additional functionals in a similar way as we have discussed extensively in this book. However, due to the complexity in dealing with high dimensional matrices, at this moment, we do not yet know how to solve this problem. As well as we stop short of finding numerical examples to illustrate the design procedure, and from employing such examples in the development of unknown input observers, as discussed in Case 2 in Sect. 8.2.2 of Chap. 8. Thus, the problem of designing generalized observers to estimate $z_{3\tau h}(t)$ for a class of systems with mixed non-commensurate time delays in both state and output vectors (i.e., systems (19.1)–19.2)) remains an open research problem for future research.

Chapter 20
Stabilization of Time-Delay Systems A Linear Matrix Inequalities Approach

For the purpose of having a self-contained book, this short chapter includes a discussion on the β-exponential stability of the following error time-delay systems

$$\dot{e}(t) = (N_1 + ZN_2)e(t) + (N_{d1} + ZN_{d2})e(t - \tau), \tag{20.1}$$

$$\dot{e}(t) = (N_1 + ZN_2)e(t) + (N_{d1} + ZN_{d2})e(t - \tau) + (N_{dd1} + ZN_{dd2})e(t - 2\tau), \tag{20.2}$$

$$\dot{e}(t) = (N_1 + ZN_2)e(t) + (N_{d1} + ZN_{d2})e(t - \tau) + (N_{h1} + ZN_{h2})e(t - h). \tag{20.3}$$

For the above systems, N_1, N_{d1}, N_{dd1} and N_{h1} are known constant matrices of dimension $n \times n$. Matrices N_2, N_{d2}, N_{dd2} and N_{h2} are known constant of dimension $m \times n$. The stabilization problem to be considered is stated as follows. For given time delays, τ and h, and any arbitrary initial error function $e(\theta)$, we need to find a matrix $Z \in \mathbb{R}^{n\times m}$ such that the time-delay systems (20.1), (20.2) and (20.3) are β-exponentially stable [1, 2], i.e., the error vector $e(t)$ converges to zero with a convergence rate of β. Note that a solution to this problem is required in the design of observers with internal delays for time-delay systems (see, for example, (13.44), (12.32) and (17.68)).

In [2], the authors considered the β-exponential stability of the following system with multiple delays

$$\dot{e}(t) = \sum_{i=0}^{k} N_i e(t - \tau_i), \tag{20.4}$$

where $N_i \in \mathbb{R}^{n\times n}$, $i = 0, 1, \ldots, k$ and $0 = \tau_0 < \tau_1 < \cdots < \tau_k$ are known constant delays. They derived an effective LMI-type condition that involves minimal computation for the β-stability of system (20.4) (see, Theorem 3 in [2]). Thus, we can adapt their stability condition to solve the stabilization problem of systems (20.1),

H. Trinh et al., *Unknown Inputs Estimation in Linear Time-Delay Systems Using Generalized Functional Observers*, Studies in Systems, Decision and Control 648,
https://doi.org/10.1007/978-3-032-11870-7_20

(20.2) and (20.3). Accordingly, we derive the following β-exponential stabilizability conditions.

β-exponential stabilization of system (20.1): For a given τ and a positive constant β, if there exist two real $n \times n$ positive definite matrices P, Q and a matrix $G \in \mathbb{R}^{n\times m}$, such that the following LMI holds

$$\begin{bmatrix} \mathscr{M}_1 & PN_{d1} + GN_{d2} \\ N_{d1}^T P + N_{d2}^T G^T & -e^{-2\beta\tau} Q \end{bmatrix} < 0, \tag{20.5}$$

where

$$\mathscr{M}_1 = PN_1 + N_1^T P + GN_2 + N_2^T G^T + 2\beta P + Q. \tag{20.6}$$

Then, the error time-delay system (20.1) is β-exponentially stabilizable. Furthermore, matrix Z is given by $Z = P^{-1}G$.

For a given τ, we can employ the one-dimensional search method to find the largest β. The feasibility of LMI (20.5) can be readily checked by using MATLAB LMI Control Toolbox.

Numerical Example 1: Let us consider Numerical Example 5 (Chap. 13), Option (i), where instead of letting $Z = 0$ as in (13.43), we now consider the stabilization problem of finding a Z, so that $e(t)$ converges to zeros with a convergence rate of β. For this example, matrices N_1, N_2, N_{d1} and N_{d2} are given as follows

$$N_1 = \begin{bmatrix} -6.6296 & 4.9259 \\ -0.6914 & -5.7284 \end{bmatrix}, N_{d1} = \begin{bmatrix} 0.9259 & -0.8148 \\ 4.2716 & -3.3457 \end{bmatrix}, N_2 = \begin{bmatrix} 0.1235 & 0.3086 \end{bmatrix}$$

and $N_{d2} = \begin{bmatrix} 0.3086 & 0.0617 \end{bmatrix}$.

Note that in the case where $\tau = 1s$ and $Z = 0$, the stability condition presented in Theorem 1 of [2] is feasible for $\beta_{max} = 1.1392$.

On the other hand, as discussed in Example 5 (Chap. 13), since $N_2 \neq 0$ and $N_{d2} \neq 0$, we can search for a matrix Z, such that the error vector $e(t)$ converges to zero within a shorter time than for the case where $Z = 0$. Accordingly, the stabilization problem (20.5) is feasible for $\beta_{max} := \beta_z = 1.3481$ and $Z = \begin{bmatrix} -0.4672 \\ -3.2017 \end{bmatrix}$. In this case, since $\beta_z > \beta_{max}$, it is expected that the error vector $e(t)$ converges to zero faster than for the case where $Z = 0$. Indeed, Figs. 20.1 and 20.2 show the trajectories of $e_1(t)$ and $e_2(t)$ with the initial condition $e(\theta) = \begin{bmatrix} 15 \\ 10 \end{bmatrix}$, $\forall\theta \in [-1, 0]$ for both cases where $Z = 0$ and $Z \neq 0$.

β-exponential stability of system (20.2): As for system (20.2) with $N_2 = N_{d2} = N_{dd2} = 0$, the authors [2] derived the following β-exponential stability condition.

[2]: *For a given τ and a positive constant β, if there exist three real $n \times n$ positive definite matrices P, Q_1 and Q_2 such that the following LMI holds*

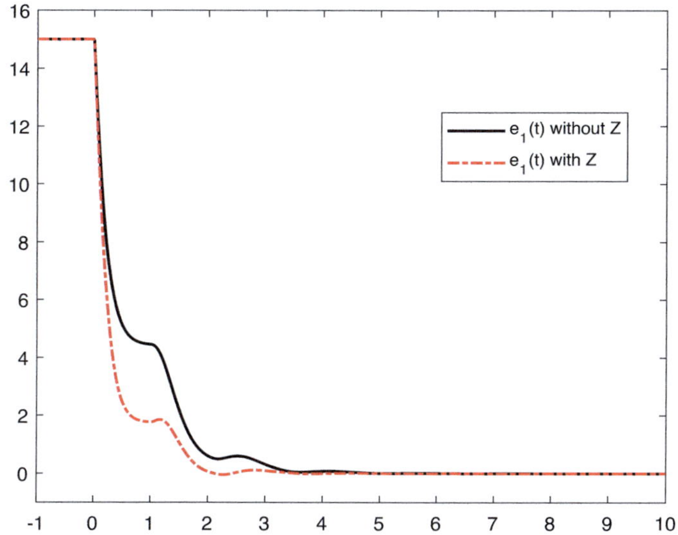

Fig. 20.1 Trajectories of $e_1(t)$ (Numerical Example 1, in this chapter)

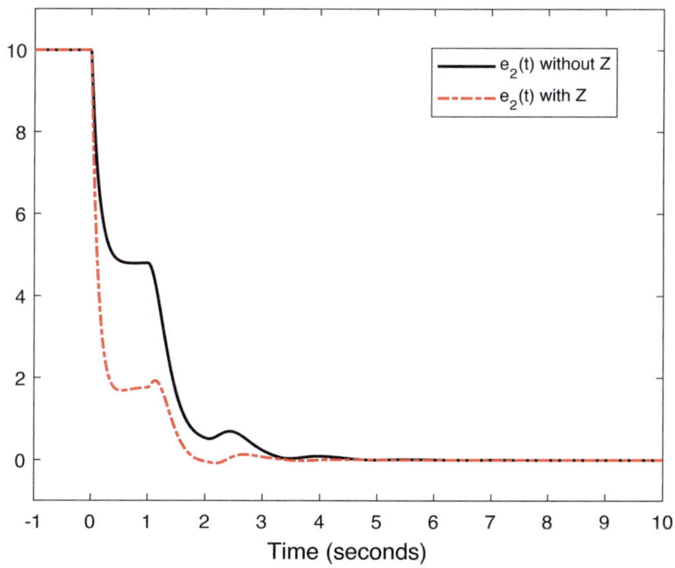

Fig. 20.2 Trajectories of $e_2(t)$ (Numerical Example 1, in this chapter)

$$\begin{bmatrix} \mathcal{M}_2 & PN_{d1} & PN_{dd1} \\ N_{d1}^T P & -e^{-2\beta\tau} Q_1 & 0 \\ N_{dd1}^T P & 0 & -e^{-4\beta\tau} Q_2 \end{bmatrix} < 0, \tag{20.7}$$

where

$$\mathcal{M}_2 = N_1^T P + PN_1 + 2\beta P + Q_1 + Q_2, \tag{20.8}$$

then the error delay system (20.2) *with* $N_2 = N_{d2} = N_{dd2} = 0$ *is* β*-exponentially stable.*

[2]: *For given* τ, h *and a positive constant* β*, if there exist three real* $n \times n$ *positive definite matrices* P*,* Q_1 *and* Q_2 *such that the following LMI holds*

$$\begin{bmatrix} \mathcal{M}_2 & PN_{d1} & PN_{h1} \\ N_{d1}^T P & -e^{-2\beta\tau} Q_1 & 0 \\ N_{h1}^T P & 0 & -e^{-2\beta h} Q_2 \end{bmatrix} < 0, \tag{20.9}$$

then the error delay system (20.3) *with* $N_2 = N_{d2} = N_{h2} = 0$ *is* β*-exponentially stable.*

Thus, we can now adapt conditions (20.7) and (20.9) to solve the β-exponential stabilization of system (20.2) and system (20.3), respectively. This is stated as follows.

β-exponential stabilization of system (20.2): For a given τ and a positive constant β, if there exist three real $n \times n$ positive definite matrices P, Q_1, Q_2 and a matrix $G \in \mathbb{R}^{n\times m}$ such that the following LMI holds

$$\begin{bmatrix} \bar{\mathcal{M}_2} & PN_{d1} + GN_{d2} & PN_{dd1} + GN_{dd2} \\ N_{d1}^T P + N_{d2}^T G^T & -e^{-2\beta\tau} Q_1 & 0 \\ N_{dd1}^T P + N_{dd2}^T G^T & 0 & -e^{-4\beta\tau} Q_2 \end{bmatrix} < 0, \tag{20.10}$$

where

$$\bar{\mathcal{M}_2} = N_1^T P + PN_1 + GN_2 + N_2^T G^T + 2\beta P + Q_1 + Q_2, \tag{20.11}$$

then the error time-delay system (20.2) is β-exponentially stabilizable. Furthermore, matrix Z is given by $Z = P^{-1}G$.

β-exponential stabilization of system (20.3): For given τ, h and a positive constant β, if there exist three real $n \times n$ positive definite matrices P, Q_1, Q_2 and a matrix $G \in \mathbb{R}^{n\times m}$ such that the following LMI holds

$$\begin{bmatrix} \bar{\mathcal{M}_2} & PN_{d1} + GN_{d2} & PN_{h1} + GN_{h2} \\ N_{d1}^T P + N_{d2}^T G^T & -e^{-2\beta\tau} Q_1 & 0 \\ N_{h1}^T P + N_{h2}^T G^T & 0 & -e^{-2\beta h} Q_2 \end{bmatrix} < 0. \tag{20.12}$$

Then, error time-delay system (20.3) is β-exponentially stabilizable. Furthermore, matrix Z is given by $Z = P^{-1}G$.

Numerical Example 2: Let us consider system (20.2), with $\tau = 1s$ and matrices N_1, N_{d1} and N_{dd1} are adopted from Example 3 in [2], i.e., $N_1 = \begin{bmatrix} -7 & -1 \\ 0.5 & -5.5 \end{bmatrix}$, $N_{d1} = \begin{bmatrix} 10.1 & 14.2 \\ -6.6 & -10.2 \end{bmatrix}$ and $N_{dd1} = \begin{bmatrix} -5.2 & -6.2 \\ 3.6 & 5.1 \end{bmatrix}$.

In this case, LMI (20.7) is feasible for $\beta = 0.1033$. Thus, the system is β-exponentially stable.

Let us next consider the stabilization of (20.2), where N_2, N_{d2} and N_{dd2} are presented as follows

$$N_2 = \begin{bmatrix} 1 & 1 \end{bmatrix}, N_{d2} = \begin{bmatrix} 0 & 1 \end{bmatrix} \text{ and } N_{dd2} = \begin{bmatrix} 1 & 0 \end{bmatrix}.$$

It is found that LMI (20.10) is feasible for $\beta = 0.8088$, and thus matrix Z is obtained, where $Z = \begin{bmatrix} 0.3771 \\ 0.3672 \end{bmatrix}$.

References

1. Hien, L.V., Trinh, H.: Exponential stability of time-delay systems via new weighted integral inequalities. Appl. Math. Comput. **275**, 335–344 (2016)
2. Mondié, S., Kharitonov, V.L.: Exponential estimates for retarded time-delay systems: an lmi approach. IEEE Trans. Autom. Control **50**(2), 268–273 (2005)